Modeling and Simulating Cardiac Electrical Activity

About the Series

The Biophysical Society and IOP Publishing have forged a new publishing partnership in biophysics, bringing the world-leading expertise and domain knowledge of the Biophysical Society into the rapidly developing IOP ebooks program.

The program publishes textbooks, monographs, reviews, and handbooks covering all areas of biophysics research, applications, education, methods, computational tools, and techniques. Subjects of the collection will include: bioenergetics; bioengineering; biological fluorescence; biopolymers *in vivo*; cryo-electron microscopy; exocytosis and endocytosis; intrinsically disordered proteins; mechanobiology; membrane biophysics; membrane structure and assembly; molecular biophysics; motility and cytoskeleton; nanoscale biophysics; and permeation and transport.

Modeling and Simulating Cardiac Electrical Activity

Edited by
Trine Krogh-Madsen and David J Christini
Weill Cornell Medicine, New York, USA

IOP Publishing, Bristol, UK

Multimedia content is available for this book from http://iopscience.iop.org/book/978-0-7503-2064-1.

ISBN 978-0-7503-2064-1 (ebook)
ISBN 978-0-7503-2062-7 (print)
ISBN 978-0-7503-2065-8 (myPrint)
ISBN 978-0-7503-2063-4 (mobi)

DOI 10.1088/978-0-7503-2064-1

Version: 20201201

IOP ebooks

British Library Cataloguing-in-Publication Data: A catalogue record for this book is available from the British Library.

Published by IOP Publishing, wholly owned by The Institute of Physics, London

IOP Publishing, Temple Circus, Temple Way, Bristol, BS1 6HG, UK

US Office: IOP Publishing, Inc., 190 North Independence Mall West, Suite 601, Philadelphia, PA 19106, USA

Contents

Preface

Mathematical models of cardiomyocyte electrophysiology have been developed over decades and used as research tools. Model simulations have been carried out for a multitude of conditions using different types of models, but an overall focus has been on simulations for understanding the fundamental mechanisms underlying normal and diseased cardiac activity. Advancing from mainly basic research tools, cardiac models are currently used increasingly to make quantitative predictions of antiarrhythmic effects of pharmacological agents or genetic variations. Recently, models have entered the clinical realm and been used to help guide treatment decisions.

When we began working in the field of cardiac modeling in the 1990s, the first step in a modeling study was to implement the model to be simulated, typically from equations published in an appendix and prone to typos. Such re-implementation was redundant, therefore wasting time and effort, and introduced the possibility of implementation mistakes. Now, model source code is readily accessible from the authors, their publications, and/or from model repositories such as CellML (https://www.cellml.org). In addition, several software packages for cardiac modeling are available, including Chaste (http://www.cs.ox.ac.uk/chaste), Web Lab (https://chaste.cs.ox.ac.uk/WebLab), Abubu (https://abubujs.org), Continuity (https://continuity.ucsd.edu), cbcbeat (https://cbcbeat.readthedocs.io), and openCARP (https://opencarp.org). Still, considerable insights and sometimes critical understanding can be gained from a comprehension of the underlying equations even if you don't have to implement the model equations.

One of the goals of this book is to provide graduate students and researchers new to the field with a thorough introduction to modeling cardiac electrical activity. The book also aims to provide a resource for both newcomers and more experienced researchers on the more specialized topics that are tightly coupled to the electrophysiology of the heart. The focus is on methodology rather than particular modeling applications as the latter is more readily available in the primary literature.

Many experts have contributed to this book and we thank all of them for their contributions and dedication to this work.

Trine Krogh-Madsen

David J Christini

Editor biographies

Trine Krogh-Madsen

Trine Krogh-Madsen received an MSc degree in applied physics from The Technical University of Denmark and a PhD degree in physiology from McGill University. She is an Associate Research Professor in the Department of Medicine and at the Institute for Computational Biomedicine at Weill Cornell Medicine, New York, where she also directs an MS program in computational biology. Her research is in the area of computational cardiac electrophysiology, using interdisciplinary approaches for cardiac model development, simulation, and computation.

David J Christini

David J Christini received a BS degree in electrical engineering from the Pennsylvania State University and MS and PhD degrees in biomedical engineering from Boston University. After postdoctoral training in cardiac electrophysiology at Weill Cornell Medical College, he then served there as a member of the faculty for nearly twenty years. He is currently a Professor of Physiology & Pharmacology at SUNY Downstate Health Sciences University, Brooklyn NY, where he also serves as Senior Vice President for Research for the campus. He uses computational and experimental methods to study cellular- to organ-level cardiac electrophysiological dynamics, with an emphasis on understanding the mechanisms underlying arrhythmia initiation and in developing new arrhythmia therapies.

Contributors

Elizabeth M Cherry
School of Computational Science and Engineering,
Georgia Institute of Technology,
Atlanta, Georgia, USA

David J Christini
Department of Medicine, Division of Cardiology,
Weill Cornell Medicine,
New York, New York, USA

Caroline Mendonca Costa
School of Biomedical Engineering & Imaging Sciences,
King's College London, London, UK

Edmund J Crampin
School of Mathematics and Statistics, and Department of Biomedical Engineering,
University of Melbourne,
Melbourne, Australia

Andrew G Edwards
Department of Pharmacology,
UC Davis,
Davis, California, USA

Flavio H Fenton
School of Physics,
Georgia Institute of Technology,
Atlanta, Georgia, USA

Shourya Ghosh
Department of Biomedical Engineering,
University of Melbourne,
Melbourne, Australia

Eleanora Grandi
Department of Pharmacology,
UC Davis,
Davis, California, USA

Daniel Gratz
The Frick Center for Heart Failure and Arrhythmia,
Dorothy M. Davis Heart and Lung Research Institute,
The Ohio State University Wexner Medical Center,
Department of Biomedical Engineering,
The Ohio State University,
Columbus, Ohio, USA

Amara Greer-Short
The Frick Center for Heart Failure and Arrhythmia,
Dorothy M. Davis Heart and Lung Research Institute,
The Ohio State University Wexner Medical Center,
Department of Biomedical Engineering,
The Ohio State University,
Columbus, Ohio, USA

Thomas J Hund
The Frick Center for Heart Failure and Arrhythmia,
Dorothy M. Davis Heart and Lung Research Institute,
Department of Internal Medicine, College of Medicine,
The Ohio State University Wexner Medical Center,
Department of Biomedical Engineering,
The Ohio State University,
Columbus, Ohio, USA

Claire Yanyan Ji
School of Physics,
Georgia Institute of Technology,
Atlanta, Georgia, USA

Trine Krogh-Madsen
Department of Medicine, Division of Cardiology,
Institute for Computational Biomedicine,
Weill Cornell Medicine,
New York, New York, USA

Kathryn Mangold
Department of Biomedical Engineering,
Washington University in St. Louis,
St. Louis, Missouri, USA

Steven Niederer
School of Biomedical Engineering & Imaging Sciences,
King's College London,
London, UK

Michael Pan
School of Mathematics and Statistics, and Department of Biomedical Engineering,
University of Melbourne,
Melbourne, Australia

Vijay Rajagopal
Department of Biomedical Engineering,
University of Melbourne,
Melbourne, Australia

Orod Razeghi
School of Biomedical Engineering & Imaging Sciences,
King's College London,
London, UK

Caroline Roney
School of Biomedical Engineering & Imaging Sciences,
King's College London,
London, UK

Jonathan R Silva
Department of Biomedical Engineering,
Washington University in St. Louis,
St. Louis, Missouri, USA

Marina Strocchi
School of Biomedical Engineering & Imaging Sciences,
King's College London,
London, UK

Kenneth Tran
Auckland Bioengineering Institute,
University of Auckland,
Auckland, New Zealand

Edward J Vigmond
IHU Liryc, Electrophysiology and Heart Modeling Institute,
Université de Bordeaux,
Bordeaux, France

Alexander Winkle
Department of Biomedical Engineering,
The Ohio State University,
Columbus, Ohio, USA

IOP Publishing

Modeling and Simulating Cardiac Electrical Activity

Trine Krogh-Madsen and David J Christini

Chapter 1

Quantitative description of cardiac action potentials

Trine Krogh-Madsen and David J Christini

Cardiac action potentials are generated and modulated by the concurrent activity of a myriad of proteins that transport ions in different ways. Despite this complexity, modern mathematical models of cardiac electrical activity successfully capture many key features and may also simulate action potential abnormalities underlying cardiac arrhythmias and disease. In this chapter, we introduce the biophysical mechanisms underlying the action potential and describe how to quantify and model ionic currents. We outline how quantitative descriptions of individual ionic fluxes come together to form a cardiac cell model and we discuss recent advances in modern development aimed at improving model specificity and representing population heterogeneity.

1.1 Cardiac action potentials

For each heart beat, a wave of electrical activity travels through the heart, triggering mechanical contraction of the cardiac muscle cells and the pumping of blood. The wave of electrical activity is a propagating action potential. An action potential is a transient depolarization in the electrical potential across the membrane of a cardiac myocyte. Cardiac action potentials differ quantitatively from action potentials in other cells types (e.g. neurons), but the biophysical principles underlying cardiac action potentials are the same as those underlying action potentials in other cells. We will discuss these principles below.

1.1.1 Ionic gradients and action potential generation

Ionic gradients across the cell membrane

Generation of cardiac action potentials relies on the presence of gradients between the intra- and the extracellular environment in several ionic species, the most important of which are K^+, Na^+, and Ca^{2+}. The concentration of K^+ ions inside a

doi:10.1088/978-0-7503-2064-1ch1

cardiac myocyte (around 140 mM) is a lot larger than the physiological concentration outside the cell (around 5.4 mM). The situation is reversed for Na^+ ions. Here, the inside concentration is around 10 mM, while the outside concentration is around 140 mM.

For Ca^{2+} ions the numbers are different. The physiological external Ca^{2+} concentration is around 2 mM and is much higher than the intracellular concentration. Calcium is a signaling molecule involved in a large number of processes inside the cell and the low concentration is necessary for allowing the changes required for its signaling functions without moving large numbers of Ca^{2+} ions around. For the cardiac myocyte, the key signaling process is the increase in intracellular calcium that occurs concurrently with the action potential and allows for mechanical contraction of the cell. This increase in calcium concentration is called the calcium transient and involves the concentration changing from about 0.1 μM to as much as 1 μM.

The gradients in Na^+ and K^+ ions concentrations are generated and maintained by molecular pumps embedded in the cell membrane—the Na^+/K^+-ATPase or the Na^+/K^+-pump. This protein uses energy (from hydrolysis of ATP) to transport K^+ and Na^+ ions against their chemical gradients, i.e. K^+ into the cell and Na^+ out of the cell.

The concentration difference in Ca^{2+} is generated primarily by a protein called the Na^+/Ca^{2+}-exchanger. The Na^+/Ca^{2+} exchanger does not use chemical energy to pump ions, but instead couples an energetically favorable transport of Na^+ into the cell with an energetically unfavorable transport of Ca^{2+} out of the cell.

Transmembrane potential

The action potential is a time course of the electrical potential across the cell membrane (i.e. the transmembrane potential, or sometimes just the membrane potential). This membrane potential describes the difference in net ionic charge between the inside and the outside of the cell. Hence, if the net ionic charge on the outside of a cell is exactly the same as the net ionic charge on the inside of the cell, then the membrane potential is zero. The standard convention is to define the potential as the charge difference inside the cell with respect to the outside of the cell. With this convention, a positive transmembrane potential describes a net surplus of positive charges inside the cell while a negative transmembrane potential describes a net surplus of positive charges outside of the cell.

Ionic channels

In addition to the transporter proteins (Na^+/K^+-pump and Na^+/Ca^{2+} exchanger), cardiac cell membranes host a number of proteins called ion channels that allow fluxes of ions between the inside and the outside of the cell. Ion channels are typically highly selective for a particular ionic species. Cardiac myocytes express K^+ channels which allow K^+ ions to flux through, Na^+ channels for Na^+ flux, and Ca^{2+} channels that primarily allow the passage of Ca^{2+}. Channels open and close, typically in time- and voltage-dependent manners, and it is these open/close dynamics that give rise to the cardiac action potential.

Resting membrane potential

Consider an electrically balanced cell with equal net charge on the inside and the outside of its membrane. Say that only K^+ channels are open. Under normal physiological conditions there will then be a net flux of K^+ ions out of the cell through the open K^+ channels, due to the chemical gradient in K^+ ions. This efflux of positively charged ions will render the transmembrane of the cell increasingly negative, thus enhancing the electrical driving force for movement of positively charged ions into the cell. At a particular membrane potential, the electrical driving force into the cell will be of the same magnitude as the chemical driving force out of the cell. When the cell is at this potential, it is in electrochemical balance, with influx equal to efflux, i.e. a net flux of zero. This potential is termed the Nernst potential (E_K) and can be calculated from the following equation:

$$E_K = \frac{RT}{zF} \ln \frac{[K^+]_o}{[K^+]_i}, \tag{1.1}$$

where R is the gas constant, T is the temperature, F is Faraday's constant and z is valence (=1 for K^+). At physiological temperature ($T = 293K$), the Nernst potential for K^+ is

$$E_K = 26.7 \text{ mV} \ln \frac{5.4 \text{ mM}}{140 \text{ mM}} = -87 \text{ mV}. \tag{1.2}$$

If the membrane potential is negative to E_K, the influx of K^+ ions due to the electrical gradient surpasses the efflux, reversing the net flux from outward to inward. For this reason, the Nernst potential is also referred to as the reversal potential.

This means that a cell with normal physiological ion concentrations and with only K^+ channels open will come to rest (electrochemical equilibrium) at a potential of around −87 mV. Most cardiac cells actually do come to rest at a potential (termed the resting membrane potential) in the range of −85 to −90 mV and they do so because at rest, their cell membranes are permeable predominantly to K^+ ions.

Action potential

Now, consider a cell that is at its resting membrane potential and suppose that positive charges enter it. This influx of positive charge could occur via a current injection through an electrode in an *in vitro* experiment or it could be current from a neighboring cell in intact tissue. Injection of such current ('stimulus current') makes the membrane potential less negative, i.e. depolarizes the cell membrane. If sufficiently large, the stimulus triggers a positive feedback loop with depolarization causing voltage-dependent Na^+ channels to open, allowing an influx of positive charge into the cell, further depolarizing the membrane. The depolarization brings the membrane potential towards the Nernst potential for Na^+ ions:

$$E_{Na} = \frac{RT}{zF} \ln \frac{[Na^+]_o}{[Na^+]_i} = 26.7 \text{ mV} \ln \frac{10 \text{ mM}}{140 \text{ mM}} = 70 \text{ mV}. \tag{1.3}$$

This process is the upstroke of the action potential.

Ca^{2+} channels also open with depolarization but both Na^+ and Ca^{2+} channels close with prolonged depolarization. Indeed, Na^+ channels close before the membrane potential reaches E_{Na}, so that the peak potential is around 30–40 mV. When Na^+ and Ca^{2+} channels have closed, K^+ channels are again the major ionic pathway and the membrane potential returns (i.e. repolarizes) to its resting membrane potential. This, in essence, is the time course of the cardiac action potential. In the next section we will discuss the ionic currents in more detail.

But first, let's address the following question: doesn't the flux of ions across the membrane change the ionic concentrations? For example, during the action potential upstroke, doesn't the influx of Na^+ ions increase $[Na^+]_i$? To address this, one can do a rough estimate of the amount of sodium entering the cell during the action potential. As we'll see in the next section, in a typical cardiac myocyte, the sodium current is around 300 pA pF^{-1} in amplitude and is on for a fraction of a millisecond. (Normalizing current to capacitance is a standard way of normalizing for cell size in electrophysiology.) This means that the charge, ΔQ, carried by the sodium current to generate the upstroke is

$$\Delta Q = 300\ \text{pA pF}^{-1} \cdot \frac{1}{3}\ \text{ms} = 100\ \text{pA pF}^{-1}\ \text{ms}. \tag{1.4}$$

Given a typical cell capacitance of 100 pF, we get

$$\Delta Q = 100\ \text{pA pF}^{-1}\ \text{ms} \cdot 100\ \text{pF} = 10^4\ \text{pA ms} = 10\ \text{pC}. \tag{1.5}$$

In moles, this amount of charge corresponds to:

$$\Delta n = \frac{\Delta Q}{F} = \frac{10\ \text{pC}}{10^5\ \text{C mol}^{-1}} = 10^{-16}\ \text{mol}, \tag{1.6}$$

with F again being Faraday's constant.

We can compare this amount to the amount of Na^+ ions in the cell, n_{Na}, computed from the cell's volume and Na^+ concentration, assuming the cell is cylindric with length 100 μm and radius 10 μm

$$\begin{aligned} n_{Na} &= C \cdot V = 10\ \text{mM} \cdot \pi \cdot 10^2\ \mu\text{m}^2 \cdot 100\ \mu\text{m} \\ &= 10 \times 10^{-3}\ \text{mol/L} \cdot 10^3\ \text{L/m}^3 \cdot 3 \times 10^4\ \mu\text{m}^3 \\ &= 3 \times 10^{-13}\ \text{mol} \end{aligned} \tag{1.7}$$

The amount of Na^+ that passes through the sodium channels during an action potential is thus only a tiny fraction of the amount of Na^+ present in the cell and the bulk Na^+ concentration does not change appreciably during an action potential.

Importantly, as stated above, for Ca^{2+} the intracellular concentration does change during an action potential. Two key features of calcium contribute to the change in intracellular concentration during the calcium transient. Most importantly, $[Ca^{2+}]_i$ is several orders of magnitude smaller than $[Na^+]_i$ (sub μM vs 10 mM). Secondly, the influx of Ca^{2+} ions into the cell causes the release of an even larger Ca^{2+} influx from the sarcoplasmic reticulum, an intracellular calcium storage compartment. Indeed, this calcium-induced calcium release is a critical component of the excitation–contraction coupling that ensures mechanical contraction of a cardiac cell in response to an action potential. Finally, it is important to emphasize that although the change in intracellular Na^+ is small on the time scale of a single action potential (and likewise for K^+), these small changes accumulate over time and the cell requires maintained work by the Na^+/K^+-pump in order to maintain its concentration gradients.

1.1.2 Ventricular action potential

Cardiac action potentials vary in morphology across different regions of the heart. We will focus first on ventricular action potentials as the ventricles are the main pumping chambers of the heart and ventricular myocytes make up the bulk of the heart.

The action potential can be described in terms of different phases (figure 1.1). Phase 0 is the action potential upstroke. It is generated by rapid opening of Na^+ channels, activating an inward Na^+ current. The Na^+ current also inactivates (due to channels closing) very quickly, such that the Na^+ current is very brief, on the order of a millisecond. You can think of the Na^+ channel as a sprinter—it turns on rapidly but remains on for only a very short time.

Many ion channels come in distinct versions and for the Ca^{2+} channels there are two major flavors. The dominant one in ventricular myocytes is the L-type Ca^{2+} channel; the resulting current is termed $I_{Ca,L}$. This channel also activates during the action potential upstroke but not nearly as fast as the Na^+ channels. Ca^{2+} channels also inactivate much more slowly than Na^+ channels. In keeping with the running analogy, Ca^{2+} channels are more like marathoners and the current remains on for hundreds of milliseconds (figure 1.1).

One of the interesting features of cardiac electrophysiology is that cardiac myocytes express many different types of K^+ channels. One of these gives rise to a current called the transient outward current (I_{to}). Inactivation of Na^+ channels and activation of I_{to} changes depolarization to repolarization. In some cardiomyocytes where I_{to} is quite large, a rapid initial phase of repolarization is observed—this is phase 1 of the action potential (not present in the action potential in figure 1.1; see figure 4.5 for an example).

Some K^+ channels activate more slowly with depolarization. Primary examples are the rapid and the slow delayed rectifier K^+ currents (I_{Kr} and I_{Ks}, respectively). These K^+ currents help balance the Ca^{2+} current such that ionic flux in and out of the cell is nearly equal, meaning that the membrane potential is nearly constant. This prolonged phase of maintained depolarization is phase 2 of the action potential (figure 1.1). With Ca^{2+} channel inactivation and K^+ channel activation, the cell membrane eventually repolarizes (phase 3 of the action potential; figure 1.1). Note that I_{Ks} is often, as in the model used to generate figure 1.1, quite small under baseline conditions. However, it may increase and play an important role in repolarization in other conditions, including during fast heart rates.

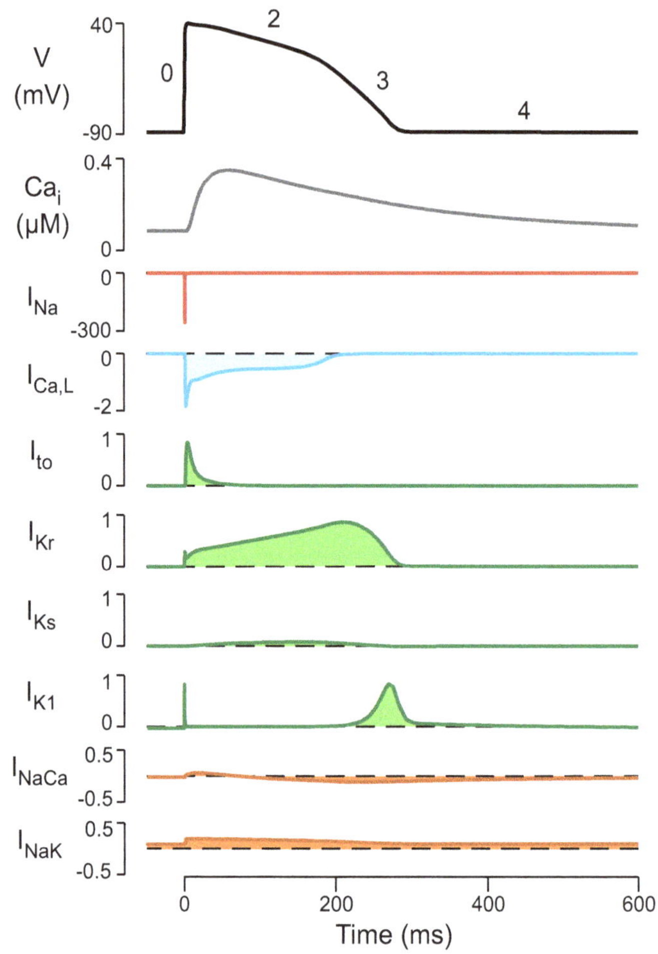

Figure 1.1. Action potential, calcium transient, and major ionic currents during a simulated human ventricular action potential. Simulations were performed using the ORd model [21], with endocardial cell parameters, and a pacing rate of 1 s. Currents are in A F^{-1}. Note different scales for I_{Na}, $I_{Ca,L}$, and the rest of the currents. Positive currents carry a net flux of positive charge out of the cell and repolarize the cell membrane, whereas negative current corresponds to positive charge moving into the cell, depolarizing the cell membrane.

Activation of yet another type of K^+ channel, the inward rectifier (I_{K1}), at fairly hyperpolarized potentials, contributes to late phase 3 repolarization. The I_{K1} channel remains open when the cell reaches its resting membrane potential, providing the main conductance for maintaining the resting membrane potential close to E_K (phase 4).

The different K^+ currents are thus active during different phases of the action potential and carry out different functional roles, controlling initial repolarization (phase 1; I_{to}), plateau and regular repolarization (phases 2 and 3; I_{Kr}, I_{Ks}), or maintaining stable resting membrane potential (phase 4, I_{K1}). The channels generating I_{Kr} and I_{Ks}, which are both mostly active during phases 2 and 3, belong to the same family of channels, while others come from different channel families, with distinct protein structures and therefore markedly different current dynamics.

The pumps and exchangers working to maintain ionic concentration gradients also carry a current. The Na^+/K^+-pump extrudes more Na^+ ions than it pumps in K^+ ions (it has a stoichiometry of 3:2) and therefore results in a net repolarizing current (figure 1.1).

During most of the cardiac cycle, the Na^+/Ca^{2+}-exchanger, having a 3:1 stoichiometry, moves less positive charge out the cell (Ca^{2+}) than it takes in (Na^+), thus generating a net depolarizing current. However, due to its dependence on Na^+ and voltage, the exchanger may operate in its reverse mode during the initial part of the action potential and generate a repolarizing current (figure 1.1).

Control of the ventricular action potential

So, a number of different ionic currents determine the shape of the action potential. If these currents are perturbed, e.g. by pharmacological agents, features of the action potential may change. For example, block of a repolarizing current such as I_{Kr} will tend to delay repolarization and thereby increase the duration of the action potential (figure 1.2A). Inversely, block of the depolarizing I_{CaL} current will shorten the action potential duration (figure 1.2B).

In situ, ionic currents also vary with physiological regulation of the activity of the heart. The sympathetic branch of the autonomic nervous system is the dominant pathway for regulating the ionic activity in ventricular cardiac cells. Activation of this system, e.g. during exercise, triggers several intracellular modulatory pathways, the main effect of which is to increase I_{CaL} and I_{Ks}. The boosted I_{CaL} plays an important functional role in that it leads to increased intracellular Ca^{2+} levels and therefore more forceful pumping by the heart. However, if occurring by itself, increased I_{CaL} would result in prolongation of the action potential, when in fact it is critical to shorten the action potential duration during fast heart rates to allow sufficient time for both contraction and relaxation of the myocytes. The increase in I_{Ks} with sympathetic activation works to counter the increased I_{CaL} in terms of

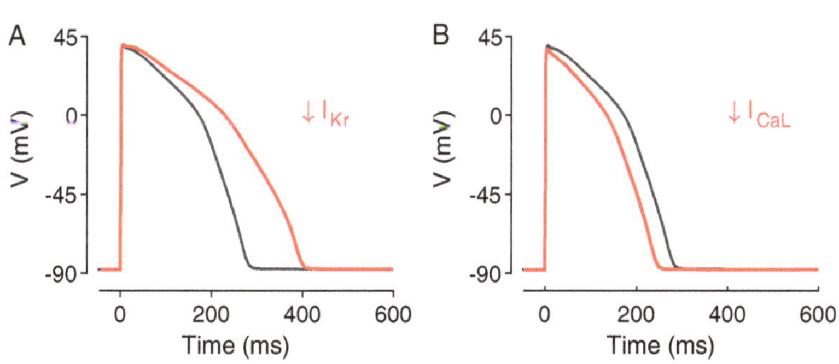

Figure 1.2. (A) Simulated partial block of I_{Kr} by reducing its conductance by 50%. (B) Simulated partial block of I_{CaL} by reducing its conductance by 50%. Simulations used the ORd model [21]; black traces show baseline action potentials.

repolarization and thus plays an important role in contributing to shortening of the action potential duration.

1.1.3 Sinoatrial node action potential

Your heart is autonomous in the sense that it triggers its own electrical excitation, i.e. it does not require nervous input to generate each beat. A completely denervated heart beats. This automaticity originates in the sinoatrial node, a small region in the upper right atrium of the heart, and it comes about because sinoatrial node cells generate action potentials spontaneously. The morphology of sinoatrial node action potentials is different from those in the ventricles. In particular, sinoatrial node cells do not rest at around E_K but depolarize slowly until reaching a voltage threshold for action potential take-off (figure 1.3). Electrophysiological differences among cardiac cells occur because they express different channels and a different balance of channels, as will be discussed below.

The slow phase of depolarization characteristic of spontaneous activity makes up phase 4 of the sinoatrial node action potential. A key contributor to this slow depolarization is an inward current called the pacemaker current or the funny current (I_f). The current is generated by ionic channels called pacemaker channels or hyperpolarization-activated, cyclic nucleotide-gated (HCN) channels. These channels conduct Na^+ and K^+ ions almost equally well and therefore have a reversal potential that is close to the simple average of the two (i.e. around −30 mV). The channels generate an inward current for potentials hyperpolarized below −30 mV and therefore provide a depolarizing current during phase 4 of the action potential (figure 1.3). The channel is activated by hyperpolarization, which is atypical, hence the name 'funny'.

The upstroke (phase 0) of the sinoatrial node action potential is generated by I_{CaL} whereas I_{Na} is minimal (figure 1.3). Consequently, the upstroke in sinoatrial node cells is much slower than in ventricular myocytes where it is driven by a rapid and very large I_{Na}.

As with ventricular myocytes, sinoatrial node cells repolarize when Ca^{2+} channels inactivate and voltage-gated K channels (mainly I_{Kr}) activate (figure 1.3). Additional K^+ currents contribute to repolarization in various phases of the sinoatrial node action potential, including I_{to}, the ultra-rapid delayed rectifier I_{Kur}, and I_{Ks} (figure 1.3). Sinoatrial node cells express little or no I_{K1}, which contributes to these cells not coming to rest at around E_K.

In addition to I_f, I_{NaCa} and the T-type Ca^{2+} current (I_{CaT}) provide inward current during phase 4 of the sinoatrial node action potential and therefore contribute to pacemaking. During phase 4, these inward currents are only slightly larger than the outward currents I_{NaK} and I_{Kr}. Indeed, the total current required to depolarize the cell from the minimal voltage reached during the action potential cycle (the so-called maximal diastolic potential, around −60 mV) to the take-off potential (beginning of the action potential upstroke, around −40 mV) over the roughly 600 ms duration of this phase is only 20 mV/600 ms = 0.03 A F^{-1}. Thus, during phase 4 of the sinoatrial node action potential, the small total current which sets the rate of depolarization is

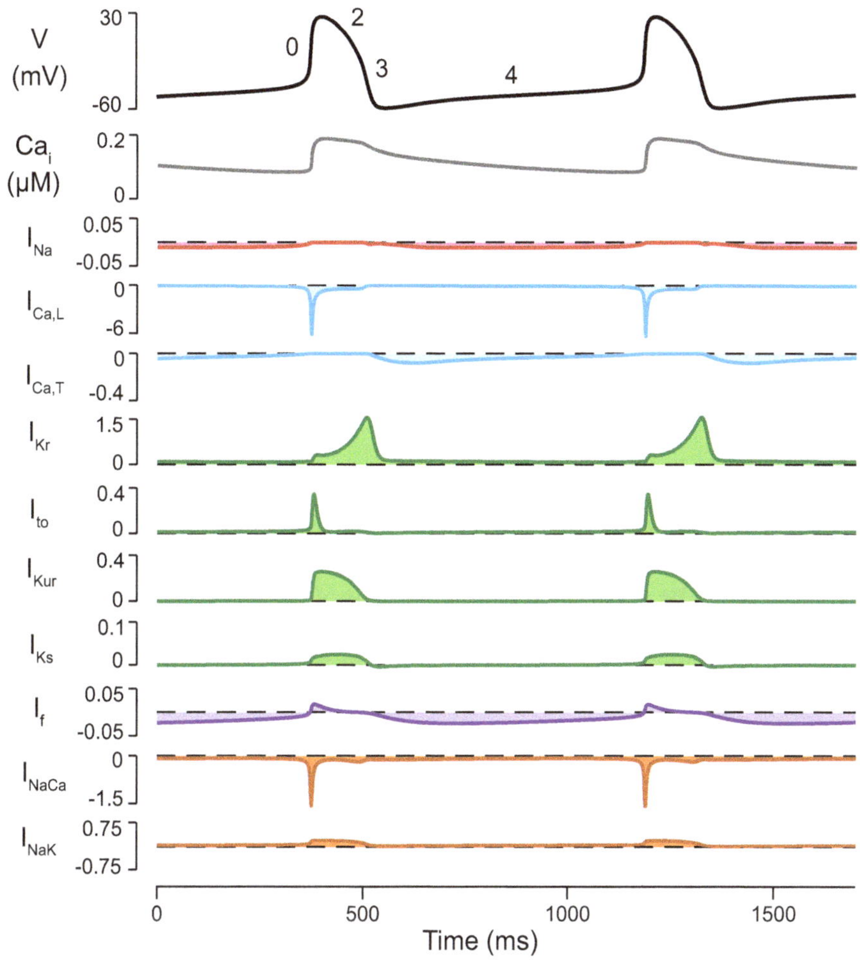

Figure 1.3. Action potential, calcium transient, and major ionic currents during simulated human sinoatrial node action potentials. Sinoatrial node cells do not come to rest, but generate spontaneous action potentials when slow phase 4 depolarization brings the membrane potential to threshold for opening of Ca^{2+} channels. Simulations were performed using the model by Fabbri *et al* [6]. Currents are in A F^{-1}. Note different scales for the currents.

generated as the algebraic sum of several much larger inward and outward currents nearly balancing out. This can make determining exactly how individual currents contribute to pacemaking a non-trivial task [6, 15].

Regulation of heart rate

The rate of spontaneous action potential generation in the sinoatrial node sets the heart rate of the intact heart. As you know, heart rate can vary a lot: it falls from its normal value of 60–70 beats/min by 10–20 beats/min during sleep and it can increase several-fold during heavy exercise.

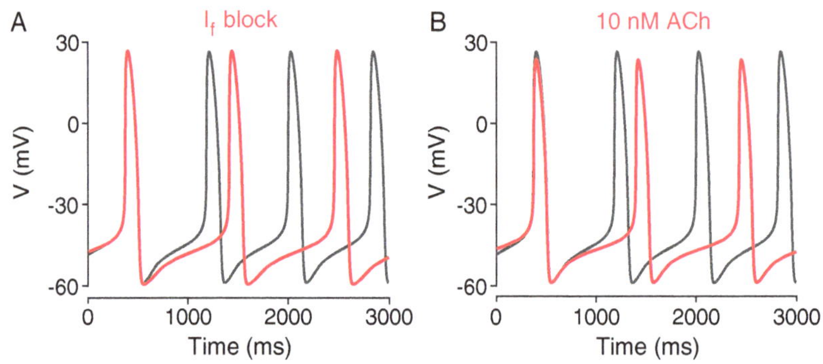

Figure 1.4. (A): Simulated action potentials with complete block of I_f. (B) Simulated effects of 10 nM ACh, the primary effects of which are to decrease I_f and activate $I_{K,ACh}$. Simulations used the Fabbri *et al* model [6]; black traces show baseline action potentials.

Heart rate is controlled primarily by the parasympathetic branch of the autonomic nervous system, with increase in parasympathetic activity slowing down heart rate and decrease in parasympathetic activity accelerating it. How this control works can be understood from the effects of parasympathetic activity on the sinoatrial node action potential. Activation of the parasympathetic division causes release of acetylcholine (ACh) which in turn leads to decreased intracellular levels of cyclic AMP (cAMP), a second messenger involved in numerous biological processes. One effect of reduced cAMP is to decrease I_f which you may recall is generated by the hyperpolarization-activated, *cyclic nucleotide-gated* channel. Release of ACh also activates a repolarizing current, the ACh-activated K^+ current $I_{K,ACh}$. The reduced I_f and increased $I_{K,ACh}$ both work to lower the rate of phase 4 depolarization and hence slow the action potential rate (figure 1.4). Clinically, a decrease in heart rate can also be obtained by administration of a blocker of the I_f channel, ivabradine.

1.2 Modeling cardiac action potentials

So, how can we model these electrophysiological events? The standard approach is to model the cell membrane as an electrical circuit (figure 1.5). The circuit contains multiple resistance pathways that represent the different ion channel pathways, i.e. Na^+ channels, L-type Ca^{2+} channels, and the several types of K^+ channels, etc. We often use the reciprocal measure of conductance to quantify the current through these pathways. The conductance depends on the state of the channels (how many are open) and can vary with time and voltage. Recall from above that the current through a conductance pathway is zero when the voltage is at the reversal potential (i.e. the Nernst potential in case of a completely specific ion channel) and a net driving force for ionic flux exists when the membrane potential is away from its reversal potential. In the circuit diagram this is represented by the batteries. The circuit has a capacitance because the cell membrane (an electrically insulating lipid structure) separates two conductors (the intracellular from the extracellular fluid).

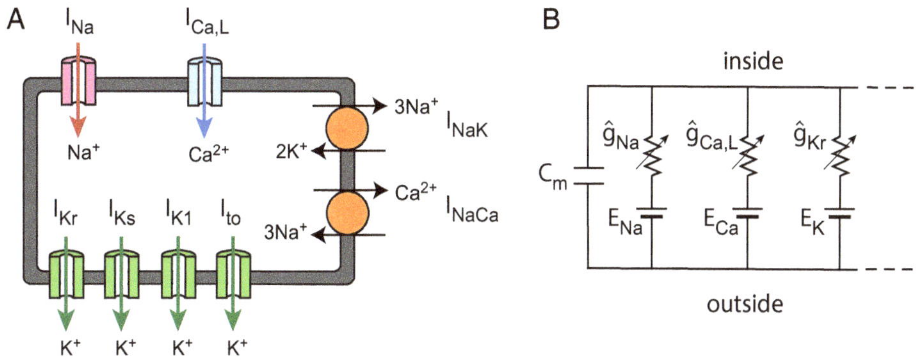

Figure 1.5. (A) Schematic of ventricular cardiomyocyte with transmembrane currents. (B) Equivalent electrical circuit. Each ionic current in the circuit depends on its conductance (the inverse of the resistance) and the deviation from its reversal potential: $I_x = \hat{g}_x(V, t)(V - E_x)$. Only a subset of the currents is shown in this circuit.

Based on this simple circuit model, one can derive a dynamic model of the membrane potential, i.e. a mathematical model that describes how the voltage changes with time.

The amount of charge that a capacitor can hold is proportional to the potential across it:

$$Q = C_m V, \tag{1.8}$$

where C_m is the membrane capacitance and Q the stored charge.

The current across the capacitor, I_C, is therefore

$$I_C = \frac{dQ}{dt} = C_m \frac{dV}{dt}. \tag{1.9}$$

Because of charge conservation, we have

$$I = I_C + I_{\text{ion}} = 0, \tag{1.10}$$

where I_{ion} is the sum of the different conductance pathways, in this example,

$$I_{\text{ion}} = I_{\text{Na}} + I_{\text{Ca,L}} + I_{\text{Kr}} + \cdots. \tag{1.11}$$

Combining equations (1.9) and (1.10) we get

$$\frac{dV}{dt} = -\frac{I_{\text{ion}}}{C_m}. \tag{1.12}$$

This differential equation is at the heart of all ionic models of cardiac electrophysiology and allows us to integrate the membrane potential from the sum of the individual ionic currents.

We can understand equation (1.12) intuitively. It simply states that the transmembrane potential, which is a measure of charge difference across the membrane, changes in proportion to the net current, i.e. in proportion to the net flux of charged particles. Note also that the sign of I_{ion} is opposite that of dV/dt. This means that an inward flow of positive charge, which we defined as a negative current (e.g. I_{Na} or $I_{Ca,L}$), will work towards making dV/dt positive, i.e. increase the membrane potential or depolarize the cell. Similarly, an outward flow of positive charge (i.e. a positive current such as the K^+ currents), will work towards making dV/dt negative, i.e. decrease the membrane potential or repolarize the cell. This agrees with our qualitative description of action potential generation in the previous section.

In the simplest current formulations, each ionic current in equation (1.11) is given by Ohm's law:

$$I_x = \hat{g}_x(V, t)(V - E_x), \tag{1.13}$$

where $\hat{g}_x$ varies with time and voltage as channels open and close. As we will see below, much of the complexity of cardiac ionic models involves describing the time- and voltage dependence of all the ionic currents.

1.2.1 Voltage clamp recordings

The quantitative description that we have of bioelectricity in cardiac and other cells come from voltage clamp experiments. Very basically, voltage clamping works by the experimentalist forcing the membrane potential to remain at a fixed (clamped) level by injecting current into the cell.

When we inject a current, I_{inj}, equation (1.12) becomes

$$\frac{dV}{dt} = -\frac{I_{ion} + I_{inj}}{C_m}. \tag{1.14}$$

During a voltage clamp experiment, $dV/dt = 0$, and so

$$I_{ion} = -I_{inj}. \tag{1.15}$$

That is, the net current though the ionic channels is simply the opposite of the (known) amount of current injected into the cell.

So what do experimental electrophysiologists find when they perform voltage clamp experiments on cardiac cells? An example of a particular voltage step, a 750 ms depolarizing step to +40 mV from a holding potential of −30 mV, is shown in figure 1.6. The resulting current shows several distinct features. The first is that the current spikes when the membrane potential is changed. This is due to capacitative current: the amount of charge that the capacitor can hold changes with voltage (equation (1.8)). The capacitative current can be used to determine the cell capacitance. For purposes of investigating ionic currents, however, the capacitative current is typically removed, which is possible because it happens on a faster time scale than the ionic currents.

It is clear that the recorded ionic current in figure 1.6 is voltage dependent: its magnitude is different at +40 mV and at −30 mV. The current also shows dynamics, i.e. changes over time. It increases slowly during the step to +40 mV and declines when the potential is stepped back down to −30 mV. In addition, there are abrupt changes in the size of the ionic current when the membrane potential is changed.

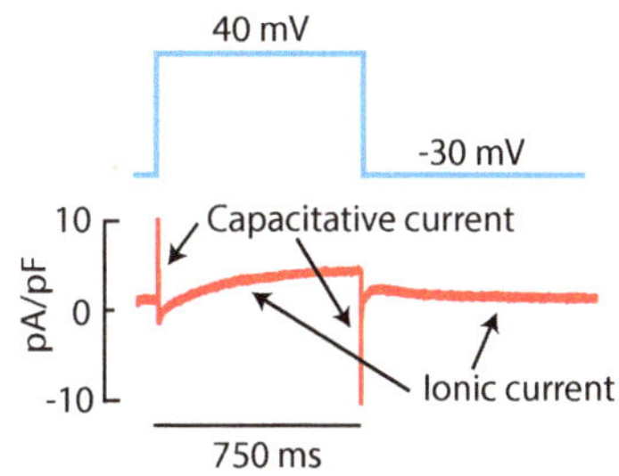

Figure 1.6. Recording of total current from an isolated guinea pig ventricular cell under a 750 ms voltage clamp step from −30 mV to +40 mV. The current trace is composed of capacitative current spikes and ionic current. The capacitative current spikes occur on a microsecond time scale immediately following the abrupt changes in voltage (their size have been reduced for clarity). The ionic current exhibits both time- and voltage dependence.

You may have also noticed that the changes in the current are not symmetrical: the response to the depolarizing step is clearly larger than the response to repolarization.

In short, even the response to a simple voltage step can be complex. This is partly because the recorded current in this setup is the total membrane current, i.e. the sum of all the individual ionic currents (equation (1.11)), each of which may respond differently to the voltage step. In practice, experimentalists strive to isolate individual currents, e.g. $I_{\mathrm{Ca,L}}$ or I_{Kr}, in order to obtain quantitative current measurements. Such isolation is most often accomplished by adding a drug that preferentially blocks the current of interest. Performing the same experiment in the absence and in the presence of the drug will yield the drug-sensitive current, i.e. the particular current blocked by the drug. A well-known challenge in this regard is that many drugs are not completely specific to one particular current. An alternative approach is to express the channel of interest in an expression system that lacks the ionic currents endogenous to cardiac myocytes. However, these cells also lack the regulatory systems that give rise to the normal functioning of many ionic channels.

1.2.2 Gates and channels

Where does voltage- and time dependence of ion channels come from? We'll present a simplified view here—the topic of channel structure is presented in chapter 2. The change in the conductance that occurs over time or across voltages for a particular current is due to changes in the fraction of open channels. Channels have *voltage sensors*, small positively charged structures within each ion channel. Movement of these sensors in response to a change in voltage causes secondary movements of larger structures, *gates*, that control whether the channel is open or closed.

As an example, take a K^+ channel with one voltage-dependent gate. The current is then given by

$$I_{\mathrm{K}} = g_{\mathrm{K}} n (V - E_{\mathrm{K}}), \tag{1.16}$$

where n is a so-called gating variable that represents the fraction of gates that are open (i.e. with a range of values between 0 and 1). This gating variable will depend on time and voltage. In contrast, g_K is a constant, representing the maximal conductance which can be attained when all gates are open.

1.2.3 Quantitative description of voltage clamp data

Let us consider how to model the gating dynamics of a particular gate, x. We'll use the variable x to denote the fraction of the gates that are open. Then $1 - x$ denotes the fraction of gates that are closed. Gates open with an opening rate α_x and close with a closing rate β_x. Although both of these rates are typically voltage dependent—we write them as $\alpha_x(V)$ and $\beta_x(V)$—they are often referred to as rate constants.

We can represent this description as a reaction scheme for a two-state model

$$x \underset{\alpha_x(V)}{\overset{\beta_x(V)}{\rightleftharpoons}} 1 - x. \tag{1.17}$$

To generate a dynamic model of gating, we can write down a differential equation describing the rate of change in x:

$$\frac{dx}{dt} = \alpha_x(V)(1 - x) - \beta_x(V)x. \tag{1.18}$$

This differential equation describes the change in x as a balance between two terms: $\alpha_x(V)(1 - x)$ (i.e. the opening rate times the fraction of closed gates) and $\beta_x(V)x$ (i.e. the closing rate times the fraction of open gates). You can understand this differential equation intuitively as *change in open gates = closed gates that open – open gates that close.*

You can also see from equation (1.18) that if the system is in a state where most gates are closed and the opening rate is larger that the closing rate, the first term will dominate and make *dx*/*dt* positive, meaning that x will increase, i.e. gates will open. Likewise, if the system is in a state where most gates are open and the closing rate is larger than the opening rate, the $\beta_x(V)x$ term will dominate, *dx*/*dt* will be negative, and gates will close.

Even though this model is very simple, the current dynamics can be somewhat complex because the opening and closing rates are voltage dependent. However, if you set V constant, i.e. do a mathematical voltage clamp experiment, it becomes easy to solve equation (1.18). To do so, it is convenient to first introduce the *steady-state gating variable* (why this is convenient will be come clear after equation (1.22))

$$x_\infty(V) = \frac{\alpha_x(V)}{\alpha_x(V) + \beta_x(V)}, \tag{1.19}$$

and the *time constant*

$$\tau_x(V) = \frac{1}{\alpha_x(V) + \beta_x(V)}, \tag{1.20}$$

Then we can rewrite equation (1.18) as

$$\begin{aligned}\frac{dx}{dt} &= \alpha_x(1 - x) - \beta_x x \\ &= \frac{x_\infty - x}{\tau_x},\end{aligned} \tag{1.21}$$

where we have dropped the functional dependence of the gating parameters on V to emphasize that the following solution is for constant values of V.

We can then separate the variables involving x from those involving t and integrate to obtain the solution:

$$\begin{aligned}\frac{dx}{dt} &= \frac{x_\infty - x}{\tau_x} \\ \frac{dx}{x_\infty - x} &= \frac{dt}{\tau_x} \\ \int_{x_0}^{x} \frac{dx'}{x_\infty - x'} &= \int_0^t \frac{dt'}{\tau_x} \\ [-\ln(x_\infty - x')]_{x_0}^{x} &= \frac{t}{\tau_x} \\ \ln\frac{x_\infty - x}{x_\infty - x_0} &= -\frac{t}{\tau_x} \\ \frac{x_\infty - x}{x_\infty - x_0} &= e^{-t/\tau_x} \\ x = x_\infty - (x_\infty - x_0)&e^{-t/\tau_x},\end{aligned} \tag{1.22}$$

where x_0 is the initial value of x. Note that when $t \to \infty$, we see that $x \to x_\infty$, the steady-state gating variable—this makes sense. Note furthermore that this steady state is approached exponentially with time constant τ_x. Examples of these dynamics are shown in figure 1.7.

We can use this gating model to describe quantitatively the current during a voltage clamp step. The recordings in figure 1.8 show I_{Ks} in response to five different

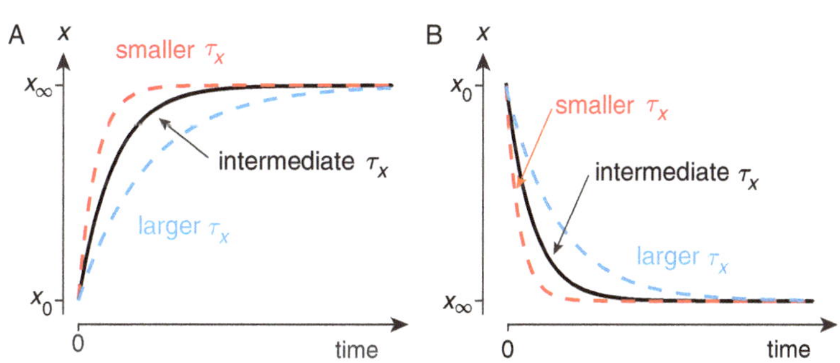

Figure 1.7. Dynamics of a single gating variable x in response to a voltage step. Starting from an initial condition x_0, x approaches its steady-state value x_∞ with a characteristic time constant τ_x. A: $x_0 < x_\infty$; B: $x_0 > x_\infty$.

depolarizing voltage steps. The current activates over several seconds upon depolarization to potentials positive to −20 mV with voltage-dependent increases in the maximal current reached during the depolarizing step (figure 1.8(A) and (B)). Part of this current increase is due to an increase in the driving force ($V_{test} - E_{Ks}$, with $E_{Ks} \approx -75$ mV in these experiments). By dividing the maximal current by the driving force, one obtains the time- and voltage-dependent conductance $\hat{g}_{Ks}(V, t)$ (equation (1.13)). This conductance also is increased for depolarized potentials (figure 1.8(C)), illustrating the activation or opening of the I_{Ks} channel with depolarization.

When clamping back to the repolarized potential (V_{rest} in figure 1.8(A)), the current is diminished instantaneously by the reduced driving force and then deactivates to turn the current off completely (arrow in figure 1.8(A)). This deactivation phase ('tail current') resembles the exponential decay of figure 1.7(B).

If you look closely at the onset of the current in the experimental recordings of figure 1.8(A), you may notice that the onset is slightly different from that of the gating variable (figure 1.7(A)). Indeed, experimental recordings of I_{Ks} data typically exhibit a sigmoid (s-shaped) onset. This is most clearly observed in the current

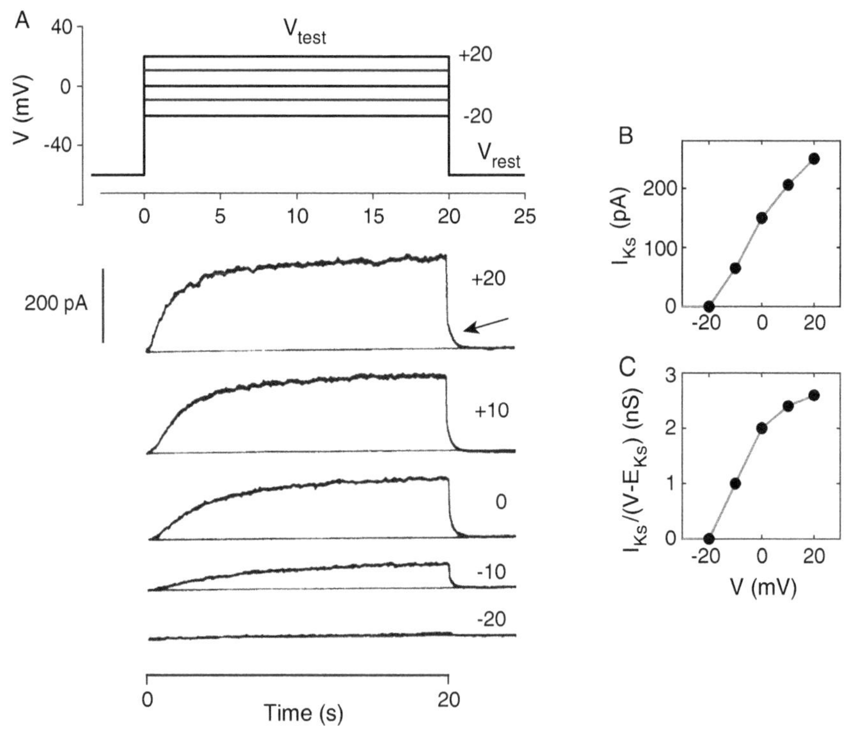

Figure 1.8. Voltage clamp recordings of I_{Ks} in embryonic chick ventricular myocytes. (A) Voltage was stepped from a resting potential (V_{rest}) at −60 mV to a testing potential (V_{test}) of −20, −10, 0, 10, or 20 mV in successive runs resulting in a family of current responses. Arrow indicates tail current. (B) Steady-state current measured at the end of the depolarizing voltage step is an increasing function of V_{test}. (C) Voltage dependence of the I_{Ks} conductance calculated by dividing the steady-state current by the driving force. Current recordings are reproduced from [2] with permission. Copyright 1988 by the American Physical Society.

response to the step to +10 mV (figure 1.8(A)). This sigmoid onset can be replicated in the model of I_{Ks} by raising the activation variable n to a power (figure 1.9(A)). It turns out that raising n to the second power captures experimental data well and so n^2 is typically used to model I_{Ks}. This model can be interpreted as a channel having two identical gates, both of which must be open for the channel to be open, such that n^2 gives the fraction of open channels. Two gates having to open in order to open the channel results in a delay in the channel opening time compared to the one-gate channel. The delay is not observed in the tail current, the decay of which is controlled by the first of the two gates closing. Do note that this model is a simplification as the I_{Ks} channel is known to have four identical voltage-sensing domains—chapter 2 discusses modeling approaches based more closely on channel structure.

To quantify the time- and voltage-dependence of n, one can fit the model current

$$I_{Ks} = g_{Ks} n^2 (V - E_{Ks}), \tag{1.23}$$

where

$$n = n_\infty - (n_\infty - n_0) e^{-t/\tau_n}, \tag{1.24}$$

to the experimentally recorded traces. In this way, one can determine the steady-state value for n (n_∞) and its time constant (τ_n), or alternatively, the rate constants (α_n and β_n), at each voltage step value. Given a set of voltage steps to a range of

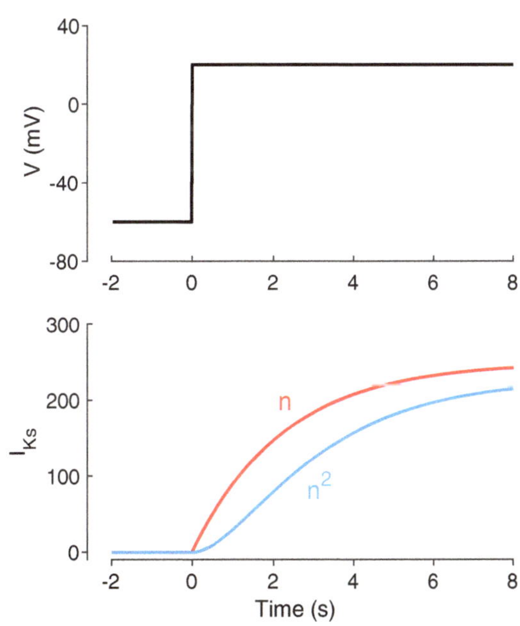

Figure 1.9. Current activation with one-gate and two-gate models. The one-gate model $I_{Ks} = g_{Ks} n(V - E_{Ks})$ (red) has a faster onset than the two-gate model $I_{Ks} = g_{Ks} n^2 (V - E_{Ks})$ (blue).

potentials, one can then generate the voltage-dependent curves for each these variables. These curves are shown in figure 1.10.

Note that the opening rate constant α_n is an increasing function of voltage, while the closing rate constant β_n is a decreasing function (figure 1.10(A)). This corresponds with our understanding that the n gate opens with depolarization, which is also seen in the steady-state curve (figure 1.10(C)).

The time constant τ_n forms a bell-shaped curve with a peak value close to −5 mV (figure 1.10(B)), meaning that the activation of I_{Ks} is slowest at around −5 mV. You can see this in the I_{Ks} traces in figure 1.8(A) (compare, e.g. the trace at 0 mV to that at 20 mV). The time constant is relatively small at repolarized potentials, which can also be seen in figure 1.8(A): when clamping back to −60 mV the deactivation of the tail current occurs in less than 1 s.

Based on these gating properties, one can predict the dynamics of I_{Ks} during an action potential. Take a ventricular action potential that rests at −85 mV. At this potential, I_{Ks} is not activated. Upon the action potential upstroke, I_{Ks} will start to activate and will do so towards a steady-state value and with a time constant each given by their values at the plateau potential of around +20 mV. From figure 1.9, this would mean that channels would begin to open with a time constant of about 2 s. Given the much shorter duration of the action potential, I_{Ks} is far from fully activated during the normal action potential (figure 1.1).

The formalism and the approach of obtaining the steady-state gating variables and their time constants (or, alternatively, the rate constants), was developed by Hodgkin and Huxley in the mid 20th century. Remarkably, it forms the basis of much of the cardiac modeling that we do today.

1.2.4 Example current models

Sodium current

Many currents have dynamics that are more complex than that of I_{Ks}. For example, I_{Na} has both activation and inactivation processes. This can be modeled by introducing multiple gating variables. Typically, I_{Na} is modeled as

$$I_{Na} = g_{Na} m^3 hj(V - E_{Na}), \quad (1.25)$$

where m is an activation variable given by

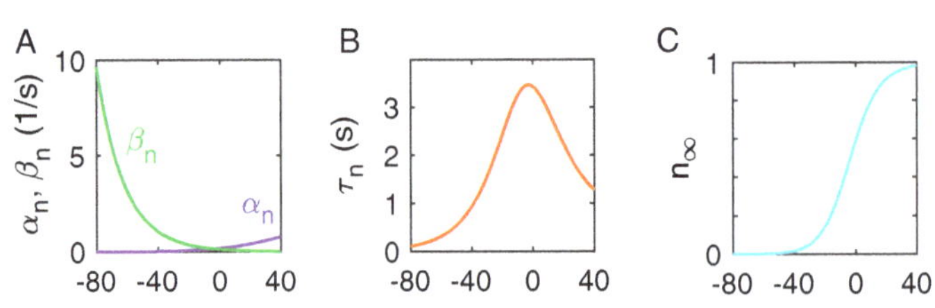

Figure 1.10. Gating properties of I_{Ks}. (A) Rate constants; (B) time constant; and (C) steady-state activation.

$$\frac{dm}{dt} = \frac{m_\infty - m}{\tau_m}, \tag{1.26}$$

and h and j are two different inactivation gating variables,

$$\frac{dh}{dt} = \frac{h_\infty - h}{\tau_h}, \qquad \frac{dj}{dt} = \frac{j_\infty - j}{\tau_j}. \tag{1.27}$$

As was the case for I_{Ks}, the product of the gating variables (here, m^3hj) represents the fraction of open channels.

The voltage dependence of the gating variables is shown in figure 1.11. The steady-state value of the activation variable, m_∞, increases with depolarization, while the two inactivation variables are given by the same decreasing function of V (figure 1.11(A)). Activation is very fast, with τ_m less than 1 ms across the physiologically relevant voltage range (figure 1.11(B) and (C)). Inactivation occurs on two time scales: an intermediate time scale for τ_h and a slow time scale for τ_j (figure 1.11(B)).

Given these dependencies, we can understand the dynamics of I_{Na} during an action potential. The activation variable m increases very quickly with depolarization such that full activation ($m \approx 1$) is reached within 1 ms (figure 1.11(E)).

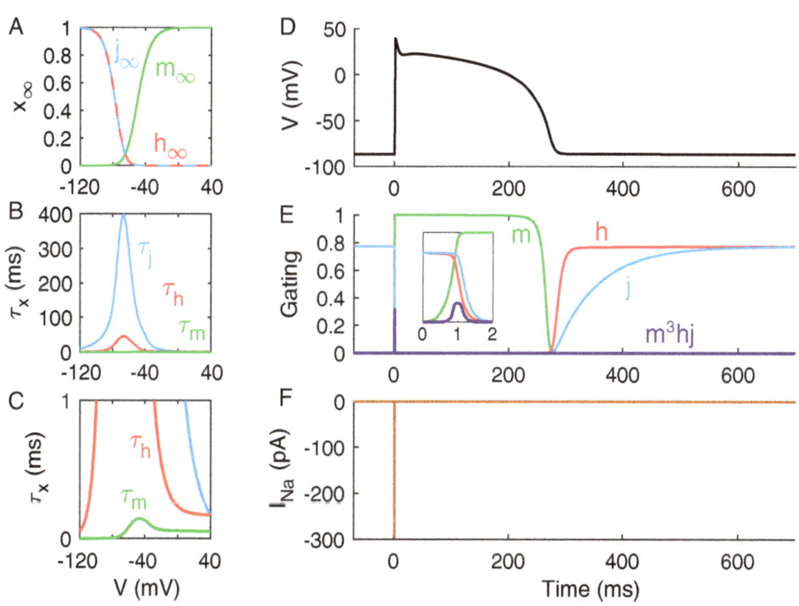

Figure 1.11. Sodium current gating and dynamics. (A) Steady-state gating variables for activation (m_∞) and inactivation (h_∞ and j_∞). (B) Time constants of activation (τ_m) and inactivation (τ_h and τ_j). (C) Zoom-in on fast time constants. (D) Simulated action potential. (E) Gating variables and channel opening (m^3hj) during simulated action potential. Inset shows first 2 ms after action potential initiation. (F) simulated $I_{Na} = g_{Na}m^3hj(V - E_{Na})$ showing brief spike during the action potential upstroke. Model equations from ten Tusscher *et al* [30].

The inactivation variables then decrease due to depolarization and do so quickly as τ_h and τ_j are small at depolarized potentials. Thus, the fraction of channels being open (m^3hj) is non-zero for a brief duration, peaking at around 1 ms (figure 1.11(E)), resulting in the characteristic brief current spike (figure 1.11F).

With repolarization, the m-gate decreases ('deactivates') prior to return of the inactivation variables to non-zero levels ('recovery from inactivation'). This means that the product of the gating variables, and therefore the current, remains ≈0 during repolarization (figure 1.11(E) and (F)).

Note that the j-gate recovers quite slowly from inactivation after the action potential as τ_j is much larger than τ_h at voltages close to the resting membrane potential. This means that if one were to attempt to trigger another action potential immediately after repolarization of the preceding one (e.g. at around 280 ms in figure 1.11), j would not have recovered from inactivation, making $I_{Na} \approx 0$. With no I_{Na}, no action potential would normally be triggered. With a later attempt at triggering a second action potential (e.g. at around 350 ms), j would still not have recovered completely from inactivation, causing less I_{Na} to be induced, which in turn would make the action potential upstroke slower and the amplitude smaller relative to the first action potential. In this way, the j gate contributes to determining the refractoriness of the cell.

Inward rectifier K^+ current

The inward rectifier, I_{K1}, is often modeled using an equation of the form

$$I_{K1} = g_{K1}R_{K1}(V - E_K), \tag{1.28}$$

where R_{K1} is a voltage-dependent rectification variable that varies between 0 and 1. Biologically, rather than a gate that opens and closes to allow/block ionic flux through the channel, the process of rectification is caused by spermine and Mg^{2+} ions blocking the I_{K1} channel at depolarized potentials. However, regardless of the mechanism controlling the channel conducting/non-conducting process, it can be modeled phenomologically by a decreasing function of voltage and the rectification variable can be thought of as a time-independent inactivation variable that turns off at depolarized potentials (figure 1.12(A)).

Multiplying this decreasing sigmoid function by the linearly increasing driving force (figure 1.12(B)) results in a current that increases roughly linearly for hyperpolarized potentials where the current is inward (negative) and shuts off at depolarized potentials (figure 1.12(C)), i.e. mimicking a classic rectifier allowing current in only one direction. In this way, the voltage-dependent conductance (i.e. the slope of the current–voltage curve in figure 1.12(C)) is large around the resting membrane potential, allowing the current to contribute to maintaining the resting membrane without providing an overwhelming repolarizing current during the action potential.

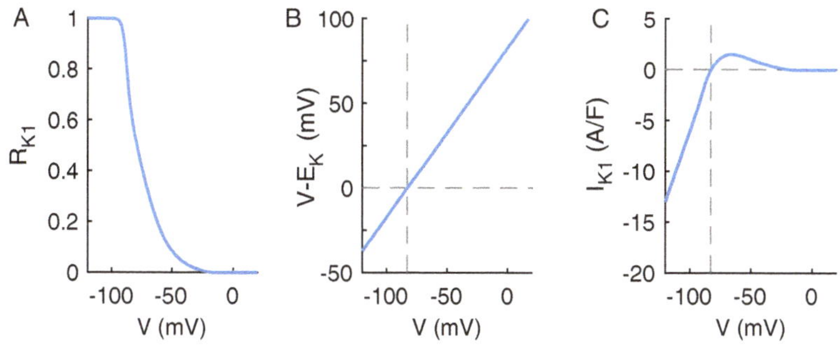

Figure 1.12. Inward rectifier K^+ properties. (A) Rectification variable. (B) Driving force. (C) The current is proportional to the product of the rectification variable and the driving force. Model equations from the Grandi *et al* human ventricular model [7].

L-type Ca^{2+} *current*

The L-type Ca^{2+} current is of critical importance to the myocyte as its influx of Ca^{2+} ions triggers the calcium transient and mechanical contraction. The current is typically modeled using three gating variables:

$$I_{\mathrm{Ca,L}} = dff_{Ca}\bar{I}_s, \tag{1.29}$$

where $\bar{I}_s$ is the maximal current, d is a time- and voltage-dependent activation gating variable, f is a time- and voltage-dependent inactivation variable, and f_{Ca} is a voltage-independent, time- and Ca^{2+}-dependent inactivation variable modeling the experimentally observed process of L-type Ca^{2+} channel inactivation by Ca^{2+} ions.

Rather than computing the maximal current $\bar{I}_s$ as the product of a maximal conductance and a linear driving force, it is better quantified by the Goldman–Hodgkin–Katz current equation:

$$\bar{I}_s = P_{\mathrm{Ca}}z^2\frac{VF^2}{RT}\frac{\gamma_i[Ca^{2+}]_i e^{zVF/RT} - \gamma_o[Ca^{2+}]_o}{e^{zVF/RT} - 1}, \tag{1.30}$$

where P_{Ca} is the constant permeability and γ_i and γ_o ionic activity coefficients of Ca^{2+} ions in the intra- and extracellular spaces, respectively, [31].

Even though the L-type Ca^{2+} channel is fairly specific to Ca^{2+} ions, L-type Ca^{2+} channels carry a significant flux of Na^+ and K^+ ions because these are present in much higher concentrations than Ca^{2+} ions. To account for these fluxes, the total current through the L-type Ca^{2+} channel typically has additional terms for Na^+ and K^+, governed by the same gating as in equation (1.29) above.

Modeling of $I_{\mathrm{Ca,L}}$ is described in detail in chapter 3.

1.2.5 Beyond the Hodgkin–Huxley formalism

Modern models can include many more currents than those introduced in section 1.2.4, including non-gated currents like the Na^+/K^+-pump current and Na^+/Ca^{2+} exchanger current. As an example, figure 1.13 shows the currents included in the widely used ORd model [21]. While this chapter provides a framework for

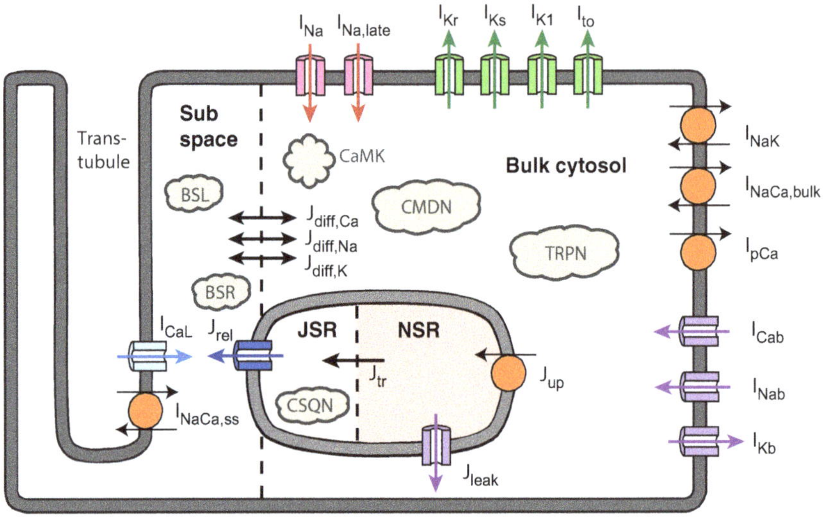

Figure 1.13. Schematic of ventricular cardiomyocyte model. The schematic is for the ORd model used to generate figure 1.1 [21]. The model consists of 14 transmembrane currents, the major ones of which were shown in figure 1.1. The background currents (I_{Cab}, I_{Nab}, and I_{Kb}), the sarcolemmal Ca^{2+} pump current (I_{pCa}), and the late Na^+ current ($I_{Na,late}$) were not included in that figure. Note that, physiologically, $I_{Na,late}$ is carried by the fast Na^+ channel but features of its gating are different and it is therefore shown here as a separate channel. The inside of the myocyte is divided into four separate compartments. The largest of these is the bulk cytosol. The sub(sarcolemmal) space is a smaller compartment representing the cytosolic space surrounding the transverse tubules (for simplicity one transverse tubule is shown—in myocytes they form an array spaced about 2 μm apart along the length of the cell). All the L-type Ca^{2+}-channels are modeled as located in the transverse tubules, while the Na^+/Ca^{2+} exchanger is modeled as located in both the transverse tubules (80%) and on the non-tubule sarcolemmal membrane (20%). The Ca^{2+}-storing sarcoplasmic reticulum is separated into a junctional space (JSR) in close proximity to the transverse tubules and a bulk network space (NSR). Ca^{2+} ions diffuse between both the sub space and the bulk cytosolic compartments (with flux $J_{diff,Ca}$) and between the two SR compartments (flux J_{tr}). Na^+ and K^+ ions also diffuse between the sub space and the bulk cytosolic compartments ($J_{diff,Na}$ and $J_{diff,K}$). Ca^{2+} ions exit the JSR through calcium release channels (ryanodine receptors; J_{rel}). Some Ca^{2+} may leak from the NSR to the cytosol (J_{leak}) and Ca^{2+} ions are pumped back into the NSR (J_{up}) by a protein called the sarco/endoplasmic reticulum Ca^{2+}-ATPase to end the calcium transient. Ca^{2+} ions are heavily buffered by binding to various proteins, including calmodulin (CMDN), troponin (TRPN), anionic sarcoplasmic reticulum binding sites (BSR), anionic sarcolemmal binding sites (BSL), and calsequestrin (CSQN). Binding of Ca^{2+} to the Ca^{2+}/calmodulin-dependent protein kinase II (CaMK) modifies a number of ionic currents and fluxes.

understanding a quantitative description of voltage-gated currents, further details and more currents are introduced in subsequent chapters (2, 3 and 4).

Some channels generate currents that are more accurately modeled by Markov models rather than by Hodgkin–Huxley-type equations. For example, the Hodgkin–Huxley formalism assumes independence between gates in a channel, which can be an oversimplification. Chapter 2 gives an introduction to Markovian models and details how these models relate to both functional (currents) and structural channel data.

Dynamics of ionic concentrations

In addition to the membrane ionic currents, modern models also include dynamics of intracellular Ca^{2+} fluxes. This is typically done by dividing the intracellular space into multiple compartments, such as the sub-membrane space, the bulk cytosolic space, the network (bulk) sarcoplasmic reticulum (NSR), and the junctional sarcoplasmic reticulum (JRS) (figure 1.13). Each compartment has its own Ca^{2+} concentration which changes with time as a consequence of Ca^{2+}-carrying trans-membrane currents and intracellular Ca^{2+} fluxes. In addition, Ca^{2+} ions are heavily buffered in cardiac myocytes. Consequently, dynamic changes in the amount of buffering also influence Ca^{2+} concentrations. In general, concentrations of the individual compartments are therefore computed by mass balance differential equations following this principle:

$$\frac{dC}{dt} = \sum \text{flux in} - \sum \text{flux out} + \Delta C_{\text{buf}}, \tag{1.31}$$

where ΔC_{buf} gives the change in concentration per time unit within the compartment due to buffering.

In the simplest cases, Ca^{2+} flux between compartments is passive and can be computed consistently with Fick's first law of diffusion. This is the case, e.g. for the translocation of Ca^{2+} from the NSR to the JSR:

$$J_{\text{tr}} = k_{\text{tr}}([Ca^{2+}]_{\text{NSR}} - [Ca^{2+}]_{\text{JSR}}), \tag{1.32}$$

where the parameter k_{tr} sets the time scale of the diffusional flux. Other fluxes are more complex, e.g. the release of Ca^{2+} from the JSR, which is a Ca^{2+}-gated process. This can be modeled phenomenologically (e.g. in the ORd model) or by specifically including the release channels:

$$J_{\text{rel}} = k_{\text{rel}} P_o([Ca^{2+}]_{\text{JSR}} - [Ca^{2+}]_{ss}), \tag{1.33}$$

where P_o gives the fraction of open Ca^{2+} release channels, which varies dynamically with the local Ca^{2+} concentration. Fluxes that rely on energy consumption, including pump-driven Ca^{2+} transport, are not well-described by linear equations. Chapters 3 and 6 cover modeling approaches to such transport. Chapter 3 also discusses modeling of intracellular Ca^{2+} dynamics in general.

Most modern models also include dynamics for the concentrations of intracellular Na^+ and K^+. Although these concentrations do not vary significantly during the course of an action potential, they can change appreciably over the course of minutes, for example in response to a change in the pacing rate. During fast pacing, the intracellular Na^+ concentration rises as Na^+ influx overcomes Na^+ extrusion. This increase in intracellular Na^+ can have several secondary effects on the action potential and the calcium transient, primarily via modulation of the Na^+/K^+-pump and the Na^+/Ca^{2+}-exchanger [12, 23, 33].

The dynamics of intracellular Na^+ and K^+ can be modeled by mass balance differential equations. When doing so, care must be taken to maintain overall charge conservation. This topic is discussed in chapter 6.

1.2.6 Cardiac model development

Limitations of the traditional approach

The process of model development outlined in the previous sections, in which models of individual currents are developed based on voltage clamp experiments, has been foundational to the field. The approach worked extremely well for Hodgkin and Huxley in their pioneering work. After determining the maximal conductances as well as all the parameters that define the functional forms of the rate constants in the model, they computed action potentials using the model and found a close similarity between their computed and previously recorded action potentials [9].

However, this approach of determining all model parameters based on voltage clamp data and then assembling current formulations into a composite cell model without further parameter modification, rarely works for modern cardiomyocyte models. Several factors contribute to this deficiency. For one, there are intrinsic shortcomings to the traditional voltage-clamp-based approach such as an inability to completely separate activation and inactivation processes for currents containing both (e.g. the Na^+ current [16]). Also, as mentioned in section 1.2.1, characterization of individual currents often relies on isolation by channel-blocking drugs which lack full specificity.

Secondly, because a large number of currents are included in modern models and due to the specialized skills necessary for acquiring the different ionic currents, the individual current formulations that go into a model are often based on data obtained from a number of laboratories [20]. As different laboratories may use different protocols and experimental conditions, this introduces inconsistency across the data on which a model is based. In addition, models often contain a source of divergence because data may not be available for all the relevant ionic currents for the particular species and cell-type of choice (e.g. a human ventricular myocyte model). Modelers then make do with what they deem is the best substitute (e.g. data from guinea pig ventricular myocytes). Current data used for model development can also be lacking in the sense that the data were not acquired for the specific purpose of informing a thorough quantitative description [21, 30]. For example, rate constants may be determined at a few voltages only, whereas a model requires values over the full physiological voltage range. Modelers then have to extrapolate from limited data.

Thirdly, action potentials of individual cardiomyocytes vary inherently, both among individuals but also among cells from within the same heart, and even from within the same region of a heart figure 1.14(A) [8]. Mathematical cardiomyocyte models typically do not include cell-to-cell variability, but instead use average data to formulate a 'representative' cell. This approach can be limiting for some applications as we don't know how representative the average is [11, 24, 28].

Given these limitations, it may not be surprising that parameter values often need adjustment when assembling the individual current equations into a composite cell model. Traditionally, this calibration process has been done manually, laboriously tuning parameters until the model outcome is deemed satisfactory in some subjective

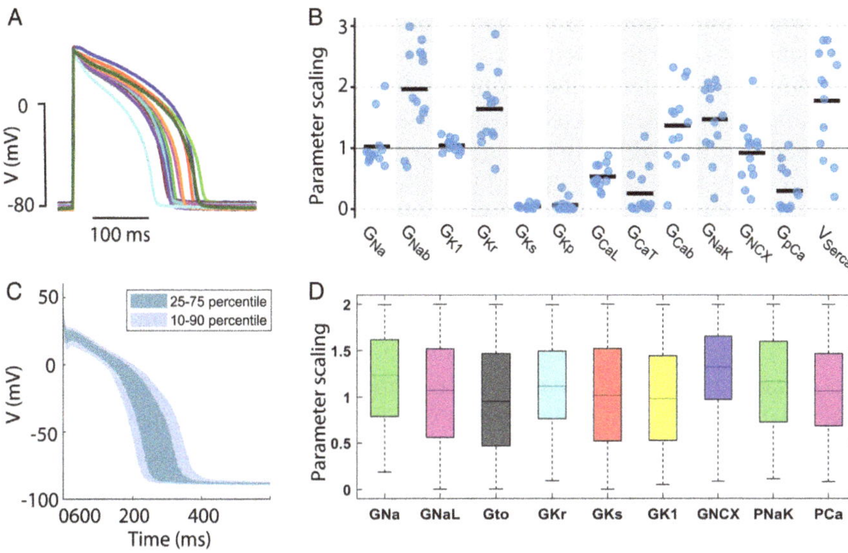

Figure 1.14. Cell-specific and population-based action potential models. (A) Cell-to-cell variability in action potentials recorded from 14 myocytes isolated from the base of guinea pig left ventricles. Variations are pronounced in the action potential duration and in the range of voltages occurring during phase 2 of the action potential. Reproduced from [14] with permission. (B) Conductance scalings for 13 conductance parameters determined specifically to 12 guinea pig left ventricular myocytes showing significant cell-to-cell variations in conductances. Reproduced from [4] with permission. (C,D) Population of human ventricular cell models generated by sampling nine conductances between 0% and 200% of their baseline values and filtering out models producing output inconsistent with normal action potentials. Reproduced from [31] with permission from eLife. The population of models covers a range of action potential durations and phase 2 voltages, shown in the percentile-based summary plot (C). The distribution of conductances among models in the population is shown as a box plot (D), for which the central mark is the median of the population, box limits are the 25th and 75th percentiles, and whiskers extend to the most extreme data points.

sense. Using simple experimental data (e.g. just an action potential) to calibrate a model is inadequate, because simple outputs can be generated by different combinations of parameter values [5, 8, 26, 34]. Take, for example, a model with a set of parameter values that can reproduce an action potential. It may then be possible to find, say, a model version with increased g_{CaL} and g_{Kr} values that also reproduces that action potential. However, as I_{CaL} and I_{Kr} respond differently to changes in pacing rates, both model versions would not be expected to reproduce action potential alteration occurring as a consequence of pacing rate variations.

Improving models through optimization and complex calibration data

To improve parameter estimation, modern approaches leverage improvements in computational power by fitting models to complex data with optimization methods [11, 31, 32]. There are many possible optimization methods, but genetic algorithms are perhaps the most used strategy for cardiac ionic models [1, 5, 8, 10, 29, 31].

Genetic algorithms test many different parameter combinations in a manner guided by reducing the difference between experimental target data and model output.

Optimization is generally computationally expensive. The computational cost of an optimization problem increases immensely with the number of parameters to be determined, as the addition of each parameter adds another dimension to the parameter space to be searched, and because more output data need to be simulated to constrain a larger number of parameters. To keep computational times at manageable levels, one approach is to focus on maximal conductances only. One rationale for doing so is that ion-channel kinetics appear relatively constant among (healthy) subjects while conductances vary as a result of expression level variations.

Using rich data as the target for optimization is an important step in determining model parameters. Such data may include more varied action potential data, e.g. action potentials recordings obtained at various pacing intervals or with application of different ionic channel blockers, and may also include measurements of other variables, e.g. current measured during complex voltage clamp protocols or estimates of the calcium transient [4, 5, 8, 10, 13, 14, 29, 31].

When the target data set can be acquired from individual cells, this approach can generate cell-specific models, thereby overcoming the problems of cell-to-cell variability and data inconsistency. This requires the protocol generating the target data to be sufficiently short to be run in single cells before significant rundown of ionic currents or other damage occurs. As an example, figure 1.14(B) shows the conductance parameters determined for 12 cell-specific models, each generated by optimizing a baseline model to cell-specific data. Such cell-specific models may allow us to generate accurate quantitative predictions (e.g. the response to a therapeutic agent) in a given cell with its particular mix of currents. Cell-specific models can also be valuable tools for studying population heterogeneity, including intra-heart variations and inter-subject variability [19, 25].

To the extent that target data is available from a given patient, this approach may be used to generate a patient-specific model [3, 17]. Such patient-specific models may be coupled with a personalized heart anatomy, obtained from clinical imaging, as significant inter-patient structural variations are prevalent in many cardiac pathologies. Chapter 8 discusses methodologies and challenges for building computational models of patient-specific electrical activity.

Population heterogeneity may also be studied by generating a population of model versions with randomized parameter values and then pruning the ones whose output falls outside of experimental or clinical ranges (figure 1.14(C) and (D)) [18, 19]. Given such a population of models, one may then simulate a variety of perturbations, e.g. various drug applications, and obtain a quantitative prediction of the frequency of action potential abnormalities occurring in response to each perturbation. Populations of models can also aid in characterizing subpopulations, e.g. in sensitivity to drugs [22, 27].

Quantifying heterogeneity in electrophysiological properties is currently gaining significant attention. One of the goals of studying heterogeneity is to enable mechanistic insights into why some patients are more prone than others to experience cardiac arrhythmias with particular interventions. Model-based quantitative predictions into proneness or resistance to arrhythmias may ultimately help inform pharmacological treatment decisions for individual patients.

References

[1] Cairns D I, Fenton F H and Cherry E M 2017 Efficient parameterization of cardiac action potential models using a genetic algorithm *Chaos* **27** 093922

[2] Clapham D E and Logothetis D E 1988 Delayed rectifier K^+ current in embryonic chick heart ventricle *Am. J. Physiol.* **254** H192–7

[3] Corrado C, Whitaker J, Chubb H, Williams S, Wright M, Gill J, O'Neill M D and Niederer S A 2020 Personalized models of human atrial electrophysiology derived from endocardial electrograms *IEEE Trans. Biomed. Eng.* **64** 735–42

[4] Devenyi R A, Ortega F A, Groenendaal W, Krogh-Madsen T, Christini D J and Sobie E A 2017 Differential roles of two delayed rectifier potassium currents in regulation of ventricular action potential duration and arrhythmia susceptibility *J. Physiol.* **595** 2301–17

[5] Dokos S and Lovell N H 2004 Parameter estimation in cardiac ionic models *Prog. Biophys. Mol. Biol.* **85** 407–31

[6] Fabbri A, Fantini M, Wilders R and Severi S 2017 Computational analysis of the human sinus node action potential: model development and effects of mutations *J. Physiol.* **595** 2365–96

[7] Grandi E, Pasqualini F S and Bers D 2010 A novel computational model of the human ventricular action potential and Ca transient *J. Mol. Cell. Cardiol.* **48** 112–21

[8] Groenendaal W, Ortega F A, Kherlopian A R, Zygmunt A C, Krogh-Madsen T and Christini D J 2015 Cell-specific cardiac electrophysiology models *PLoS Comput. Biol.* **11** e1004242

[9] Hodgkin A L and Huxley A F 1952 A quantitative description of membrane current and its application to conduction and excitation in nerve *J. Physiol.* **117** 500–44

[10] Kaur J, Nygren A and Vigmond E J 2014 Fitting membrane resistance along with action potential shape in cardiac myocytes improves convergence: application of a multi-objective parallel genetic algorithm *PLoS One* **9** e107984

[11] Kernik D C, Morotti S, Wu H, Garg P, Duff H J, Kurokawa J, Jalife J, Wu J C, Grandi E and Clancy C E 2019 A computational model of induced pluripotent stem-cell derived cardiomyocytes incorporating experimental variability from multiple data sources *J. Physiol.* **597** 4533–64

[12] Krogh-Madsen T and Christini D J 2017 Slow $[Na^+]_i$ dynamics impacts arrhythmogenesis and spiral wave reentry in cardiac myocyte ionic model *Chaos* **27** 093907–12

[13] Krogh-Madsen T, Jacobson A F, Ortega F A and Christini D J 2017 Global optimization of ventricular myocyte model to multi-variable objective improves predictions of drug-induced Torsades de Pointes *Front. Physiol.* **8** 1059

[14] Krogh-Madsen T, Sobie E A and Christini D J 2016 Improving cardiomyocyte model fidelity and utility via dynamic electrophysiology protocols and optimization algorithms *J. Physiol.* **594** 2525–36

[15] Kurata Y, Matsuda H, Hisatome I and Shibamoto T 2010 Roles of hyperpolarization-activated current If in sinoatrial node pacemaking: insights from bifurcation analysis of mathematical models *Am. J. Physiol. Heart Circ. Physiol* **298** H1748–60

[16] Lee J, Smaill B and Smith N 2006 Hodgkin–Huxley type ion channel characterization: An improved method of voltage clamp experiment parameter estimation *J. Theor. Biol.* **242** 123–34

[17] Lombardo D M, Fenton F H, Narayan S M and Rappel W-J 2016 Comparison of detailed and simplified models of human atrial myocytes to recapitulate patient specific properties *PLoS Comput. Biol.* **12** e1005060
[18] Muszkiewicz A *et al* 2016 Variability in cardiac electrophysiology: using experimentally-calibrated populations of models to move beyond the single virtual physiological human paradigm *Prog. Biophys. Mol. Biol.* **120** 115–27
[19] Ni H, Morotti S and Grandi E 2018 A heart for diversity: simulating variability in cardiac arrhythmia research *Front. Physiol.* **9** 958
[20] Niederer S A, Fink M, Noble D and Smith N P 2009 A meta-analysis of cardiac electrophysiology computational models *Exp. Physiol.* **94** 486–95
[21] O'Hara T, Virág L, Varró A and Rudy Y Simulation of the undiseased human cardiac ventricular action potential: model formulation and experimental validation *PLoS Comput. Biol.* **7** e1002061
[22] Passini E, Britton O J, Lu H R, Rohrbacher J, Hermans A N, Gallacher D J, Greig R J H, Bueno-Orovio A and Rodriguez B 2017 Human in silico drug trials demonstrate higher accuracy than animal models in predicting clinical pro-arrhythmic cardiotoxicity *Front. Physiol.* **8** 668
[23] Pueyo E, Husti Z, Hornyik T, Baczkó I, Laguna P, Varró A and Rodríguez B 2010 Mechanisms of ventricular rate adaptation as a predictor of arrhythmic risk *Am. J. Physiol. Heart Circ. Physiol.* **298** H1577–87
[24] Rees C M, Yang J-H, Santolini M, Lusis A J, Weiss J N and Karma A 2018 The Ca^{2+} transient as a feedback sensor controlling cardiomyocyte ionic conductances in mouse populations *eLife* **7** 4417
[25] Sarkar A X, Christini D J and Sobie E A 2012 Exploiting mathematical models to illuminate electrophysiological variability between individuals *J. Physiol.* **590** 2555–67
[26] Sarkar A X and Sobie E A 2010 Regression analysis for constraining free parameters in electrophysiological models of cardiac cells *PLoS Comput. Biol.* **6** e1000914
[27] Sarkar A X and Sobie E A 2011 Quantification of repolarization reserve to understand interpatient variability in the response to proarrhythmic drugs: A computational analysis *Heart Rhythm* **8** 1749–55
[28] Schulz D J, Goaillard J-M and Marder E 2006 Variable channel expression in identified single and electrically coupled neurons in different animals *Nat. Neurosci.* **9** 356–62
[29] Syed Z, Vigmond E, Nattel S and Leon L J 2005 Atrial cell action potential parameter fitting using genetic algorithms *Med. Biol. Eng. Comput.* **43** 561–71
[30] ten Tusscher K H W J, Noble D, Noble P J and Panfilov A V 2004 A model for human ventricular tissue *Am. J. Physiol. Heart Circ. Physiol.* **286** H1573–89
[31] Tomek J *et al* 2019 Development, calibration, and validation of a novel human ventricular myocyte model in health, disease, and drug block *eLife* **8** e48890
[32] Whittaker D G, Clerx M, Lei C L, Christini D J and Mirams G R 2020 Calibration of ionic and cellular cardiac electrophysiology models *WIREs Syst. Biol. Med.* **12** e1482
[33] Xie Y, Liao Z, Grandi E, Shiferaw Y and Bers D M 2015 Slow $[Na]_i$ changes and positive feedback between membrane potential and $[Ca]_i$ underlie intermittent early afterdepolarizations and arrhythmias *Circ Arrhythm. Electrophysiol.* **8** 1472–80
[34] Zaniboni M, Riva I, Cacciani F and Groppi M 2010 How different two almost identical action potentials can be: a model study on cardiac repolarization *Math. Biosci.* **228** 56–70

IOP Publishing

Modeling and Simulating Cardiac Electrical Activity

Trine Krogh-Madsen and David J Christini

Chapter 2

Modeling the molecular details of ion channels

Kathryn Mangold and Jonathan R Silva

Significant progress has been made to allow for models with microscopic and macroscopic channel features that bridge the channel and tissue levels for unprecedentedly accurate arrhythmogenic predictions. Historically, Hodgkin and Huxley's model of the neuronal action potential laid the foundation for cardiac ion channel modeling. Along with the maturation of channel experimentation, channel modeling transitioned to state-dependent (Markov) models to recapitulate more complex dynamics, which can account for functional and rudimentary structural data. With advances in computational power and experimental techniques, molecular dynamics simulations and solved channel structures provide an extraordinary opportunity for inclusion of even more molecular detail.

2.1 Introduction

The heart is an electrical organ, whose activity is reflected on the body surface by potentials that are measured by the electrocardiogram. The myocytes of the heart generate the potentials that are detected on the body surface by allowing ions to move across the cell membrane and alter the membrane potential. These ions transit through ion channels and transporter proteins that selectively allow sodium, potassium, and calcium ions to cross the membrane as a function of time and membrane potential. Precise timing of channel opening gives rise to the myocyte level electrical wave form known as an action potential. Nanoscale changes due to genetic mutations or other environmental factors can disrupt channel function and predispose patients to deadly arrhythmia. However, it is difficult to predict exactly how subtle changes to the ion channel's structure will affect the overall action potential and the ability of the tissue to initiate and sustain an arrhythmia. Detailed models of ion channels have the potential to represent this complicated scenario spanning multi-scales in time and space to direct experimentation and improve our understanding of how the nano and whole organ scales are connected [1].

doi:10.1088/978-0-7503-2064-1ch2

In this chapter, we begin with insight into the molecular structure of ion channels gained from the Hodgkin–Huxley model. Additional channel behavior represented by Markov type kinetic models follows. We then focus on the molecular level with an introduction to the theory behind detailed molecular simulation and the use of molecular dynamics to create detailed ion channel models. Finally, we conclude with paths forward that combine the strengths of both high-level and low-level approaches.

2.2 Lessons from Hodgkin and Huxley

Ion channel modeling began with the seminal work of Hodgkin and Huxley (HH) in 1952 with the squid giant axon [2]. While voltage clamping the axon they attributed positive charges entering the cell to sodium and the positive ions leaving as potassium. HH imagined the cell membrane acting as a capacitor between the extracellular and intracellular spaces with parallel resistors in the membrane in series with a biological battery. They assigned one resistor to each type of current: sodium, potassium and leak. The following equation describes the relationship between the membrane voltage, V_m (mV), the membrane capacitance, C_m (μF cm^{-2}), and the ionic current, I_{ion} (μA cm^{-2}), flowing through the imagined resistors:

$$\frac{dV_m}{dt} = -\frac{1}{C_m} * I_{\text{ion}}. \tag{2.1}$$

Remarkably, HH made this circuitry model without knowing about the existence of ion channels until years later. We now know that the ion channels are the imagined resistors in parallel across the membrane [3].

Using Ohm's law, HH wrote expressions for the currents flowing through the ion channels:

$$I_X = g_x \cdot (V_m - E_X), \tag{2.2}$$

where I_X is the specific ionic current in μA cm^{-2}, g_x (mS cm^{-2}) is the conductance of the specific ion across the membrane and E_x (mV) is the reversal potential.

HH used continuous random variables (gating variables) to describe the voltage and time dependence of channel conductance, g_X [4]. Thus, the product of these gating variables represented the time-dependent ion channel open probability as a function of membrane potential. These gating variables could either be activating or inactivating: activation variables start at zero and maximize at one while inactivation variables start at one and decrease towards zero. In the case of the potassium conductance curve, the profile was best fit with four independent activation gates. HH found that a simple monoexponential decay ($e^{\frac{-t}{\tau}}$) could describe the fall of conductance during inactivation, but to fit the activation profile more terms were needed to describe the sigmoidicity: $(1 - e^{-\frac{t}{\tau}})^4$ [1]. Thus, the conductance of potassium channel, g_K, is described by the following expression where $\bar{g}_K$ (mS μF^{-1}) is the maximum conductance of the potassium channel:

$$g_K = \bar{g}_K * n^4. \tag{2.3}$$

Not only does this expression represent the potassium conductance kinetic curves well in the nerve, but it remarkably corresponds to the tetrameric structure of cardiac voltage-gated potassium channels (figure 2.1(A)). Voltage-gated potassium channels are comprised of four identical subunits each with six transmembrane spanning segments, S1–S6. The first four segments, S1–S4, form the voltage sensing domain (VSD) while S5–S6 form the pore. The S4 segment is special: it contains positive charges that endow the channel with voltage sensitivity. Mutagenesis studies in the years following HH confirm that all four S4 segments respond to changes in membrane voltage [5]. Their kinetic contributions may not be strictly equal as hypothesized by HH [5], but the assumption continues to serve as a model of the potassium channel today [6, 7].

HH noticed that the conductance profile for sodium was more complicated [1]. They decided on a product of three activation variables, m^3, and one inactivation gate, h, to best reproduce the conductance curve with $\bar{g}_{\text{Na}}$ (mS cm^{-2}) as the maximum conductance of the sodium channel:

$$g_{\text{Na}} = \bar{g}_{\text{Na}} * m^3 * h. \tag{2.4}$$

As in the case of potassium, experimentalists have shown that this assumption still holds some structural and functional relevance today. Mammalian voltage-gated sodium channels can be formed by monomers with four homologous domains (DI-DIV) each with the six transmembrane segments S1–S6. The S1–S6 segments

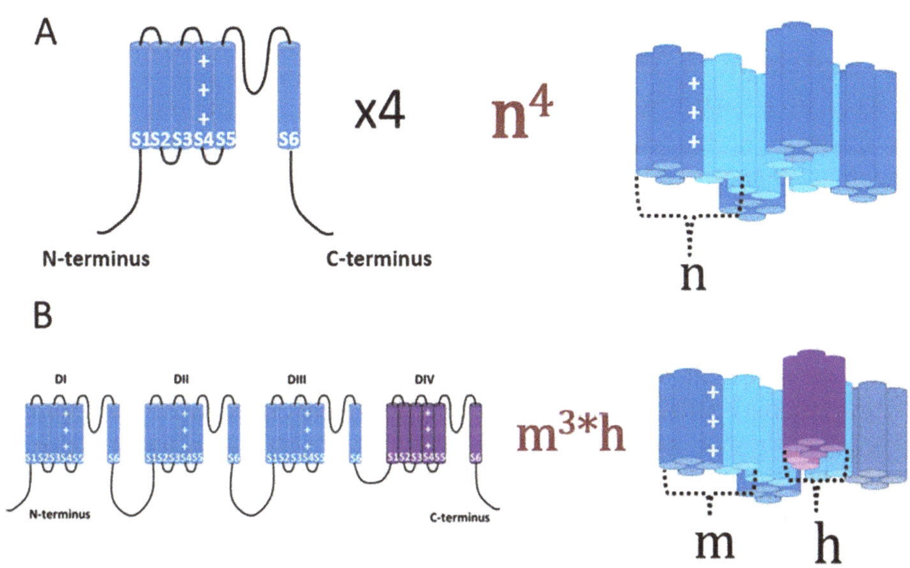

Figure 2.1. Voltage-gated potassium and sodium channels and their relationship to Hodgkin and Huxley's (HH) probabilistic gates. (A) The tetrameric structure of voltage-gated potassium channels and its relationship to the n gate in HH. (B) The four domains of the voltage-gated sodium channel and the m and h gates predicted by HH. Mutagenesis studies with DI-DIII suggest an activation-like role. Studies show that DIV is responsible for the onset of fast inactivation which mirrors the role of the h gate.

arrange to form the VSD and pore as in voltage-gated potassium channels (figure 2.1(B)).

When observing the DIV VSD with experimental methods that track VSD movement in voltage-gated sodium channels, experimentalists found that the DIV VSD moves five times slower than the other three VSDs [8], and it initiates a second pore conformation that precedes inactivation [9, 10]. Further, mutating positive S4 residues to glutamines resulted in slower onset of inactivation [11]. Taking these results together, HH's hypothesis corresponds well with the asynchronous gating model [12] as depicted in figure 2.1(B). The DI, DII and DIII VSDs primarily control the onset of activation, and thus mirror the three activation variables, m^3, while DIV controls the onset of inactivation and corresponds to the inactivation variable, h.

By creating a system of voltage-dependent nonlinear ODEs for the other gating variables and combining with equation (2.1), HH successfully deterministically recapitulated the neuronal action potential [13]. Neher and Sakmann's invention of patch-clamp [14] recording later supported the idea that the macroscopic action potential results from numerous, single ion channel openings and closings. Replacing the continuous random gating variables with a binary random variable that stochastically transitions from open to closed as a function of time and voltage allows for single-channel simulations using the HH equations [13, 15]. We also now know that calcium channels and multiple types of potassium channels also play major roles in the cardiac action potential; nevertheless, this model still serves as the basis for understanding of modeling cardiac electrophysiology.

2.3 Movement to Markov models

In the years since HH's work with giant axon of the squid, in addition to discovering the existence of ion channels, experimentalists have made great strides with numerous mutagenesis studies in understanding ion channel dynamics, which have called into question certain assumptions in the model. Most notably, activation and inactivation are now known as dependent processes rather than as independent. Armstrong and Bezanilla showed that inactivation only proceeds after channel activation, and the voltage dependence of inactivation results from pairing to channel activation [16, 17]. Models must represent that inactivation primarily occurs when the channel is open. HH assumed independence in their model by formulating the conductance as a product of gating variables, m and h^3. Horn and Lange *et al* showed that ion channels could be represented in a Markovian (state-dependent) fashion that accounts for activation and inactivation coupling [18]. This knowledge has pushed the ion channel modeling field to use Markov models of channel kinetic gating instead.

Mathematically, Markov models are graphs with a collection of size N states, or nodes, connected through edges where rate parameters dictate kinetic transitions between the states. Markov models are generalizable: they can handle dependent processes but can also represent the independent HH formulism for stochastic simulations, for example [3, 15]. The voltage-dependent rate constants are commonly parametrized by Eyring transition state theory [19] with a single transitional barrier in the form:

$$\alpha = a_0 * \exp\left(\frac{V_m}{b}\right), \tag{2.5}$$

where a_0 and b are constants to be optimized.

Markov models are abstract representations of the kinetic transitions in the channel; they do not necessarily relate to the channel's structure [20]. The Markov designation dictates that only the current state of the model determines the next state; in other words, the model is 'memoryless.' One may write out a system of linear ordinary differential equations (ODEs) from the connections in the model and assemble a transition rate, 'Q,' matrix enumerating how the probability distribution across the N states evolves throughout the simulation. Starting from the probability distribution $X(t)$ which is a $1 \times N$ state probability vector, the next state of the system is given by a homogenous set of differential equations:

$$\frac{d}{dt}X(t) = Q * X(t), \tag{2.6}$$

where $X(0)$ is given by the initial distribution, X_0, and Q is the transition matrix [21]. Figure 2.2 shows an example of creating the system of ODEs for a sample five-state model and the corresponding transition matrix. The solution $X(t)$ is given by the following where Q is the matrix exponential of the transition matrix [22]:

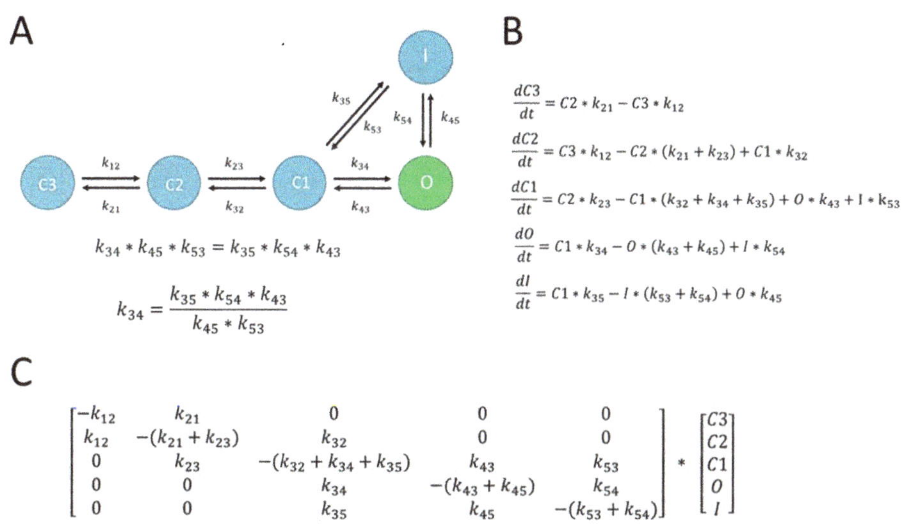

Figure 2.2. Markov model rate derivation. (A) A sample five state ion channel model with labeled forward and backward rates. Note that to impose microscopic reversibility, the loop between states C1 (closed), I (inactivated), and O (open) has one dependent rate, for example, k_{34}. For more information about ensuring microscopic reversibility in Markov models see Colquhoun *et al* [64]. (B) Corresponding system of coupled linear ordinary differential equations from the model structure in A. (C) Transition matrix of the model used in equation (2.8) to solve for the open probability as a function of time.

$$X(t) = e^{Qt}X_0. \tag{2.7}$$

The system of may also be calculated numerically using ordinary differential equation solvers. Once the time evolution of the state probability vector is calculated, the ionic current follows from the following relation:

$$I_x = \bar{g}_{SC,x} * n * O * (V_m - E_x), \tag{2.8}$$

where $\bar{g}_{SC,x}$ is the single-channel conductance, n is number of channels per unit area of membrane, and O is the open probability of the channel [3]. The variables V_m and E_x correspond to the membrane potential and reversal potential of the ion, respectively, and as before.

Many Markov models of ion channel gating kinetics have been created since the 1980s when they rose to prominence. As experimentalists learn more about cardiac sodium channel functions channel functions, for example, modelers add additional states to account for this complexity [23]. The implicit Markov model for the sodium channel under the HH formalism is depicted in figure 2.3(A). Multiple closed states (C1–C3) represent the m^3 activation formalism. Inactivated states (I) are reachable from all C3 to O states to represent the independence of the onset of inactivation ($m^3 * h$). Experimental advances following HH included identifying multiple, cooperative VSD movements prerequisite for activation along with the

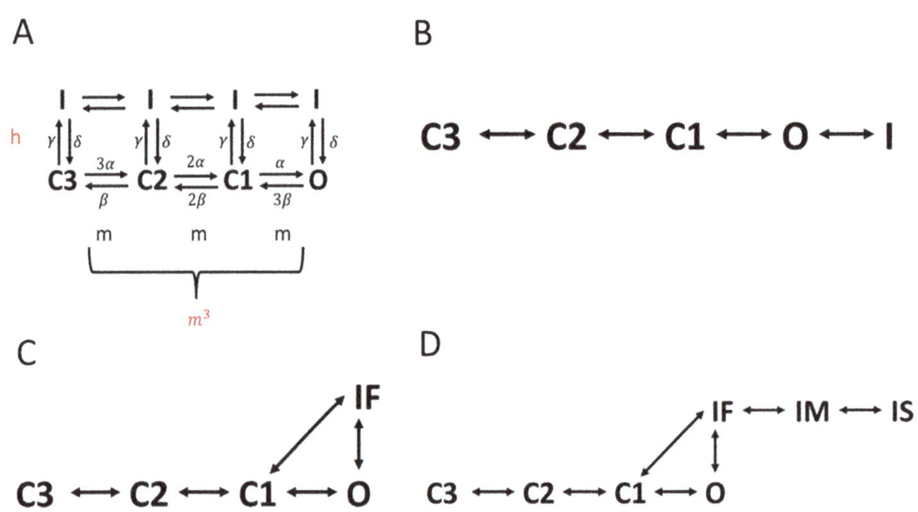

Figure 2.3. Sodium channel Markov model development. (a) Hodgkin and Huxley's model in the Markov model formalism. Multiple closed states (C1–C3) resemble the m^3 activation formalism. Rates α and β designate forward and backward transitions, respectively. Inactivated states (I) are reachable from all C3 to O states with rates γ, forward, and δ, backward, to represent independence of the onset of inactivation ($m^3 * h$). (b) Model structure accounting for the state-dependence of activation and inactivation. (C1–C3) are multiple closed states to represent various stages of VSD activation. Inactivation (I) only occurs once the channel is open (O). (c) Addition of a connection between a closed state (C1) and a fast-inactivated state (IF) to control bursting. (d) Sample model of today with two additional inactivated states: intermediate inactivated (IM) and slow (deep) inactivated (IS).

state-dependence of activation and inactivation. Multiple closed states reflect the VSD transitions, and the inactivated state is only reachable from the open state (figure 2.3(B)). This model structure implies 'bursting', however, when deactivating through the open state back to closed. Because bursting is not generally seen physiologically, an additional connection is added between the closed and inactivated states (figure 2.3(C)). Modelers now also account for the extent of channel inactivation [24], so figure 2.2(D) depicts two additional states for intermediate (IM) and deep inactivated states (IS) of the channel.

Models of today reflect this experimental knowledge and recapitulate a range of voltage clamp protocols used to characterize ion channel behavior. Models may be trained from voltage clamp experiments using optimization by minimizing differences between model and data points [25]. Colquhoun and Hawkes created the first Markov model of an ion channel in 1977 [26]. Years later, Clancy *et al* created models trained on single-channel sodium and potassium currents [27, 28], which pioneered reproducing a single-channel phenotype at simulated cellular levels. The Markov models were parameterized based on single-channel and macroscopic wild type and mutant channel data. These models were then substituted into a ventricular action potential model [29, 30] to predict cellular excitability consequences. Figure 2.4 highlights this modeling routine with the mutation ΔKPQ in the sodium channel, which impairs inactivation and results in greater late sodium current compared to wild type [27]. Figure 2.4(A) and (B) depict the Markov model structure for wild type and ΔKPQ sodium channels. Inclusion of these modeled sodium currents into the ventricular action potential model resulted in a proarrhythmic, longer action potential duration for the ΔKPQ sodium channel compared to wild type as seen phenotypically (figure 2.4(C) and (D)). Other notable examples of Markov models attempted to include channel temperature dependence [31] and, more recently, antiarrhythmic drug binding kinetic activity [32, 33].

2.4 Inclusion of molecular level detail in Markov models

With the successes in modeling ion channel gating, researchers wished to take the next step by including structural data along with the functional current data in the Markov models. In Silva *et al* 2009, a Markov model was created of four VSDs in a voltage-gated potassium channel $K_V7.1$ (KCNQ1). Each of the VSDs could occupy three states: down, intermediately up, and fully up [34]. The structure was based on a prior proposed model of $K_V7.1$ gating in Silva and Rudy [34] that was based on Zagotta *et al* [7] for *Shaker* K^+ channels (figure 2.5(A)). The model assumes that the four VSDs of $K_V7.1$ are identical (like HH's n^4), so that only once all four VSDS are fully activated may the channel conduct. The Markov model is drawn so that all permutations of the two-step VSD activation process are represented along the way to full channel opening. The rate parameters were parameterized based on $K_V7.1$ activation gating data. This Markov VSD model then included an AP model to predict how the VSD state changed over the course of the action potential.

The experimental technique of VCF has provided modelers with structurally related data quantifying VSD movement. A technique called voltage clamp

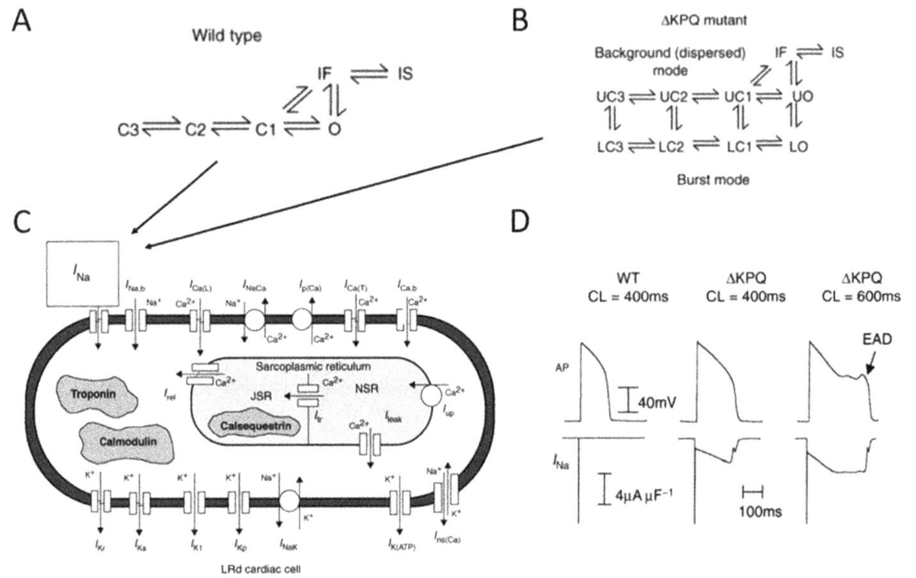

Figure 2.4. Connecting sodium channel single-channel dynamics to the cardiac action potential. Figure adapted by permission from Springer Nature [27], Copyright 1999. (A) Markov model structure of wild type (WT) sodium channel. (B) Markov model structure of the mutant ΔKPQ sodium channel with an additional 'burst mode' to represent the late sodium current. The ΔKPQ sodium channel has impaired inactivation resulting in the late sodium current. (C) Schematic of the Luo–Rudy dynamic (LRd) [29, 30] model cell depicting the inclusion of the Markov modeled sodium channel current from A and B. (D) Top: Simulated action potentials for WT and ΔKPQ modeled sodium currents at cycle lengths of (CL) 400 and 600 ms. Bottom: Corresponding simulated sodium current traces for each action potential above. WT shows a shorter action potential duration compared to ΔKPQ because of the late sodium current. A simulated ΔKPQ action potential at 600 ms displays an early after depolarization (EAD) resulting from the excess sodium current. For an in-depth reading about the relationship between late sodium current and impaired inactivation see reference [24].

fluorometry (VCF) [35] allows for tracking the VSD movements through the membrane during channel activation and deactivation [36]. Briefly, a cysteine is introduced into the S4 of each VSD so that a sulfhydryl fluorophore may be attached. Upon channel activation, the VSDs move to open the channel. This movement can be tracked via fluorescence changes from environmental changes the fluorophore experiences along the activation trajectory. In the past ten years, this technique has been widely adopted to study the VSD movement of cardiac ion channels. The fluorescence changes from the fluorophores on each VSD moving upon activation may be normalized to depict a fluorescence voltage curve.

Osteen *et al* used VCF to quantify the VSD movements in $K_v7.1$ (KCNQ1) [37]. They parameterized Markov models of VSD subunit activation and subsequent channel opening with the fluorescence kinetics (figure 2.5(A)). After rate parameter optimization on current and fluorescence experimental data, the model was able to recapitulate the structural, kinetic relationship between the VSDs and channel

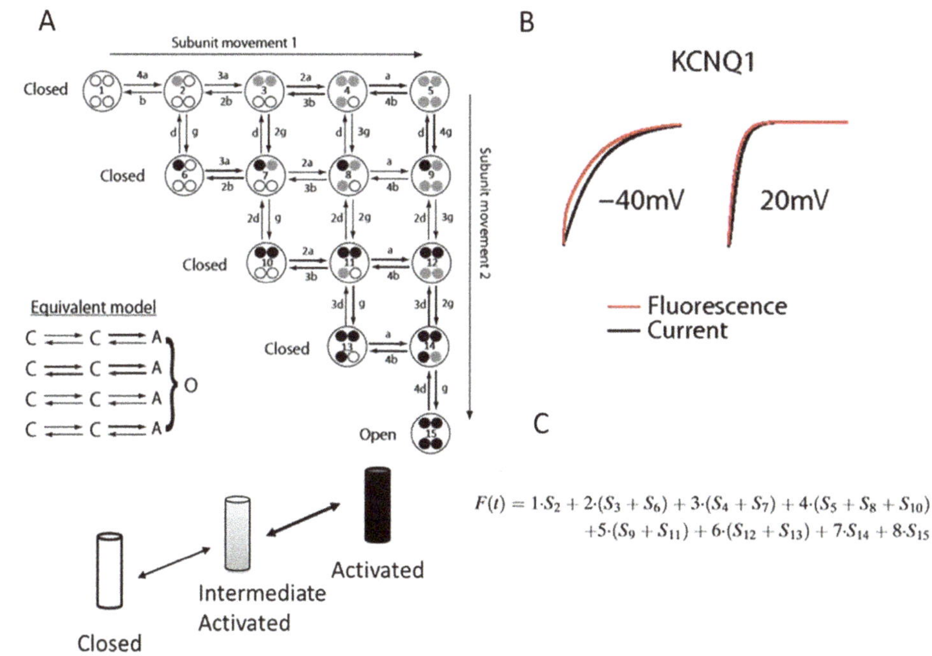

$$F(t) = 1{\cdot}S_2 + 2{\cdot}(S_3 + S_6) + 3{\cdot}(S_4 + S_7) + 4{\cdot}(S_5 + S_8 + S_{10}) + 5{\cdot}(S_9 + S_{11}) + 6{\cdot}(S_{12} + S_{13}) + 7{\cdot}S_{14} + 8{\cdot}S_{15}$$

Figure 2.5. Incorporation of VSD movement into a Markov model of $K_V7.1$ (KCNQ1). Figure adapted from reference [37] with permission. Copyright 2010 National Academy of Sciences. (A) Model structure of KCNQ1 voltage sensor movement. Channel conductance is calculated from the open probability (State 15). (B) Simulated fluorescence and current traces after rate parameter optimization. Fluorescence precedes current activation. (C) Equation for calculating fluorescence intensity based on state occupancy.

opening. Fluorescence activation preceded current activation, and the steady state kinetics of fluorescence and current was reproduced well (figure 2.5(B)).

Finally, a prior experimental study showed how far the DIII VSD of the sodium channel translocated through the membrane was key to the antiarrhythmic drugs' efficacy [38]. Moreno *et al* created a model that included the extent of DIII VSD activation in a cardiac sodium channel with the simulated antiarrhythmic drug [39]. As in the previously described modeling studies, the Markov model included clusters of states representing the extent of DIII VSD activation parametrized by VCF kinetic data. The model successfully recapitulated drug efficacy in tissue models of patients with mutations affecting the extent of DIII VSD movement. This model allows for unprecedented tracking of a key VSD during an action potential and demonstrating whether the drug could rescue arrythmia in patients harboring mutations that affect the extent of DIII VSD activation.

2.5 The power of molecular dynamics

Markov models represent channel dynamics in a set of discrete, descriptive hypothetical states. While these states may describe the general ion channel gating and limited molecular level detail (VCF), recent work focuses on modeling specifically a channel's VSD and pore domain dynamics at the molecular level. In order to model channels at the molecular level, one must start the structure of the channel of interest. This structural information is stored in the Protein Data Bank (PDB). Each file contains the sequence of amino acid residues in the channel along with the associated atoms making up each amino acid. The spatial (x,y,z) coordinates of each atom are listed so that one may rebuild the protein as depicted in the crystal structure resolved by solution NMR, cryo-electron microscopy, or x-ray diffraction. To prepare for simulation, any missing hydrogen atoms are added along with the solvent representation. This conformation of the molecule and solvent is then used as the initial condition for the molecular dynamics (MD) simulation [40].

Starting from this initial condition, the positions of the individual atoms are 'stepped forward' in time by calculating the forces on each atom. These forces result from the neighboring covalently or noncovalently bound atoms [41]. The potential energy (V in units of μV) of the system can be written in terms of the covalently or noncovalently bond neighbors of atom i:

$$V = \sum_{i=1}^{N} V_{\text{bond}}(i) + V_{\text{angle}}(i) + V_{\text{dihedral}}(i) + V_{\text{non-bonded}}(i). \tag{2.9}$$

Each of the terms for the total potential energy on atom i has a simple formulation for efficient calculation at each time step [42]. The potential energy of atom i due to covalent bonds and angles can be represented by springs while periodic functions represent potential energy from rotations (dihedral angles). For non-bonded potential energies, the Lennard-Jones formulation represents van der Waals interactions, which are local, transient dipoles between atoms. Coulomb's law accounts for electrostatic forces between two charged particles inversely related to distance. Taking the derivative of equation (2.9) with respect to position for each atom i yields $F(x_i)$, which is the total forces acting on atom i. Using Newton's second law of motion we can write a formulation for how each atom is 'stepped forward' in time during the simulation:

$$\frac{d^2x_i}{dt^2} = \frac{F(x_i)}{m_i}, \tag{2.10}$$

where $\frac{d^2x_i}{dt^2}$ is the acceleration of atom i, and m_i is the mass of atom i. Starting from the configuration given by the crystal structure as the initial condition, the next time step is prescribed by solving this second-order ordinary differential equation. In practice, formulations for potential energy are given in common packages such as CHARMM and AMBER to simplify the simulation process.

Of course, these simulations would not be complete without considering the effect of the solvent on the protein's dynamics. There are two ways to model the solvent: explicitly or implicitly. Modeling the solvent explicitly means that interactions between all atoms of the channel and the solvent are tracked. Calculating all these potential energies is well-defined in equation (2.9) but time consuming. This long computational time is compounded by the fact that very small timesteps must be taken to ensure stability in the atomic trajectories. In fact, a timestep should be no longer than the fastest atomistic movements [42], which limits the biggest time steps to the femtosecond (fs) scale. Ion channel dynamics typically occur on the milliseconds to seconds timescales, which means there is a great disparity between achievable, realistic simulation timescales (microseconds) and the desired timescales. This disparity remains the main limitation in (MD) simulations despite advances in computing power and resources. Thus, the field has proposed various approaches to reduce the computational complexity.

Treating the solvent implicitly and course graining allow for such simplifications. Course graining reduces the simulation time by clustering similar atoms in a protein into groups. These groups then are treated as the individual particles to be tracked in the MD simulation to reduce computational steps at each time step.

Implicit methods essentially 'average' the interactions between other ions in the solvent, the solvent, and the protein [43]. The solvent is treated as being continuously polarizable. The charges of the other ions in vicinity of the protein are averaged into a mean electric field that can be represented as a 'cloud' of charge surrounding the protein. Numerically solving the Poisson–Boltzmann [44] equation allows one to calculate the electrostatic potential energy of a given channel conformation by accounting for the 'average' behavior of the solvent. The electrostatic potential energy contribution may be added to equation (2.9), which describes other potential energy of the protein conformation to step the MD simulation forward.

2.6 MD simulations of ion channels

The first MD simulations began in the 1980s by simulating ion solvation in gramicidin A to simulate channel conduction [45, 46]. In the 1990s after the release of the structure of KcsA, a two transmembrane segment potassium channel from *Streptomyces lividans*, allowed for more sophisticated simulations. One simulation explained the selectivity of KcsA channel for potassium over sodium: the potassium ions could be completely solvated in the pore while sodium ions were incompletely solvated [47]. A later study in KcsA channels focused on simulating potassium ions moving down through the selectivity filter. A careful balance between attractive and repulsive forces with potassium ions and the channel filter and the ions themselves, respectively, drives the conduction [48].

After the release of the *Shaker* voltage-gated potassium channel structures in the 2000s [49, 50], MD simulations were able to focus on simulating VSD dynamics rather than just pore dynamics. A question plaguing the ion channel community was how a highly charged S4 segment could energetically reside in the hydrophobic lipid bilayer [51]. Early simulations showed that water reaches the inner membrane and is thus able to hydrate the VSD while the outermost gating charges on S4 interact with the polar lipid headgroups extracellularly when the channel is open [52]. Later simulations elucidated the salt bridge interactions between the other gating charges on S4 and the S1–S3 segments of the VSD [51, 53]. Another open question in the community was exactly how did S4 move through the membrane. Was the S4 movement purely rotational, translational, or some combination of both? Grabe *et al* used the Poisson–Boltzmann equation to test various activation pathways. They found S4 could not interact with the inner lipid membrane and there needed to be some rotational and translational movement based on electrostatics [54]. With these questions in mind, the next step was to derive the closed state structure and its associated dynamics to provide a complete structural representation of the channel across all states.

Researchers applied a variety of methods to infer the structure of the closed state. They included inferring structure from mutagenesis experimental constraints and homology modeling with previously solved non-voltage-gated potassium channels. The most promising approach, however, has been to artificially apply a very large hyperpolarizing potential to the solved structure to force closure into the closed state. Specifically, Jensen *et al* applied hyperpolarizing voltages between −375 and −750 mV to simulate closure of the $K_V1.2$ structure in the timeframe of an MD simulation [55]. These voltages are obviously beyond the physiological range used to characterize voltage-gated ion channels but provide key insight into the true deactivation mechanism. Simulations showed the deactivation mechanism includes the S4 segments of the VSDs moving inward by 15 Å and rotating by approximately 120°. This inward movement and rotation relax the VSD to pore connection and allows for the gate to close. Activation requires the opposite movement: an outward motion of S4 by 15 Å and rotation to pull the S4–S5 linker to open the inner pore. Figure 2.6(A) illustrates this activation and deactivation process. However, a limitation of this approach is that a limited number of trajectories were identified due to the immense computational power that was required.

Delemotte *et al* thoroughly investigated the free energy of the $K_V1.2$ VSD conformational pathway during channel deactivation by applying a deep hyperpolarizing pulse [56]. They observed sequential 'zipper-like' salt-bridges between the S4 charges, neighboring charges and lipid headgroups. They found three distinct intermediate energetic states that the VSD passes through dictated by the salt bridges with S4 residues (figure 2.6(B)). The simulated total gating charge transfer from the model agreed with values found experimentally during deactivation. When using the closed channel structure to infer an activation pathway, the hypothesized pathway agreed with VSD activation occurring in sequential steps with the associated simulated gating current in the range of experimental values.

Later, in 2015, Delmotte *et al* focused on simulating free energy changes of the VSDs during the hypothesized activation process [57]. In order to most efficiently explore the simulation space, they used 'well-tempered metadynamics.' This process adds additional energy to conformations already visited in the simulation to encourage maximum exploration. When simulating the free energy transition between the closed channel and first hypothesized intermediate state (see figure 2.6(B)), they found the VSD activation free energy landscape is 'rough': translocation of the VSD through the membrane has many momentary energetically favorable or unfavorable states in the activation trajectory. To reduce the complexity of the free energy landscape, they presented the free energies in terms of gating charge. This coarse graining technique remarkably showed that the VSD transition contains one prominent transition state. This result supports the Eyring rate parameter formulation assuming a single transitional state in Markov models to describe channel gating (figure 2.5(A)) [7].

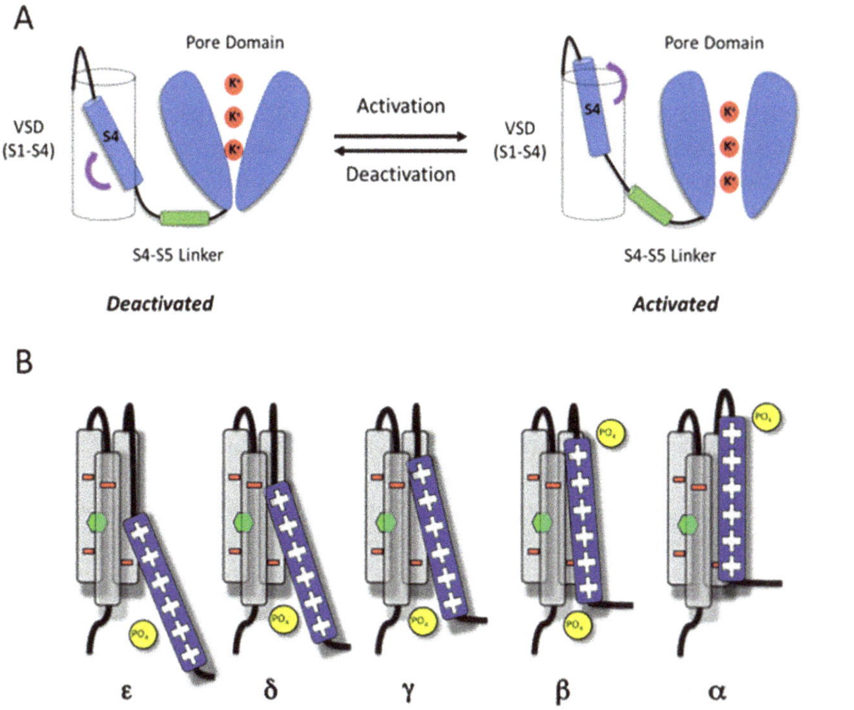

Figure 2.6. VSD movement in $K_V1.2$ from MD simulation. (A) A cartoon of the excitatory cycle from MD simulations of $K_V1.2$. Starting from the deactivated state, the outward movement of S4 by 15 Å and rotation by 120° creates tension on the S4–S5 linker open the pore. Deactivation requires the opposite motion of S4 to relax the S4–S5 linker to close the pore reference [55]. (B) States of activation from a free energy simulation focused on S4 movement. The S4 segment moves from the resting (ϵ) to fully activated (α) and passes through a series of intermediate states dictated by S4 interacting with negative charges (red) in a 'ratchet-like' motion. Figure adapted from reference [57] with permission. Copyright 2015 National Academy of Sciences.

MD simulations are invaluable to understand the fundamental structural dynamics underlying the channel activation and deactivation. This power can also be harnessed to better understand antiarrhythmic drug interactions with ion channels. As mentioned before in the context of Markov models, exactly how anti-arrhythmic drugs specifically modulate ion channels remains poorly understood. In comparison to the latest Markov models including effects of antiarrhythmic drugs to preliminarily study tissue level excitatory effects, MD simulations provide an opportunity to elucidate modulation at the molecular level. Nguyen *et al* modeled the human cardiac sodium channel with various antiarrhythmic drugs [58]. The simulations supported experimental data showing that DIVS6 is an ideal location for drug binding in the channel's inactivated state. Smaller drugs, such as lidocaine, showed concentrated interactions near the DIVS6 while larger drugs, such as flecainide and ranolazine, interacted up to the fenestration between DIII and DIV and down into the selectivity filter. This wide expanse of interaction surface area agrees with the wide variety of mutagenesis studies in these regions that affect drug binding efficacy. Simulation of lidocaine interactions specifically demonstrated two pathways for accessing the channel pore: a hydrophilic pathway inside the pore behind the inactivation gate and a hydrophobic route located near the DIII and DIV fenestration with the associated membrane lipids [59]. These molecular level insights may allow researchers to develop more effective antiarrhythmic drugs by highlighting two potential pathways to modulate the channel.

2.7 Intermediate modeling techniques

MD simulations provide the necessary molecular level scale detail needed for some ion channel gating studies, but the complexity of tracking force fields for all atoms, along with the effects of the solvent, is very computationally expensive. As mentioned in the introduction to MD simulations, the simulations are limited to taking femtosecond (fs) timesteps with a total simulation time of around a few microseconds. Ion channel conformation dynamics within the scope of the cardiac action potential, however, take on the order of milliseconds. This great disparity in time scales needed to simulate relevant channel conformations suggests the need for certain simplifications in MD protocols to relate structural dynamics to ion channel function. Nekouzadeh *et al* (2016) used a novel method [60] that removes high frequency atom vibrations to simulate the dynamics of the VSD's S3–S4 to the S4–S5 linker during $K_V7.1$ gating. Secondary structures were assumed constant along with limited degrees of the freedom so that the energetic calculations were greatly simplified. Greater time steps were taken allowing for a longer simulation so that channel conformations could be related to the macroscopic current. The simplified simulations still successfully recapitulated the mechanism from Jensen *et al* [55]. with the VSD twisting and translating through the membrane to pull the S4–S5 linker away from the pore to cause activation.

Ramasubramanian *et al* went a step further and used machine learning to associate energies to the many conformations of $K_V7.1$ during its excitatory cycle. The various conformations included all degrees of freedom and high frequency atom vibrations for an unprecedently detailed simulation. After using a subset of these conformations with their associated energies chosen for training, the model allowed for complete traversal of the conformational landscape [61]. The conformational landscape was then mapped with respect to two key components of $K_V7.1$ structure during gating: S4 movement and the pore diameter. The predicted S4 movement showed two steps of activation, which excitingly agrees with previous VCF data collected experimentally. The S4 translocation and pore diameter map was then used to predict gating and thus subconductance at various membrane potentials. These subconductances were then transformed to simulate experimental single-channel and macroscopic currents with good fidelity (figure 2.7). This use of machine learning in ion channel MD simulations serves as a basis for including maximum molecular details into simulations while reducing computational time. Transforming the calculated subconductances into single-channel and macroscopic currents throughout the channel's excitatory cycle allows for an extraordinary connection between the molecular and tissue levels through cellular action potential models.

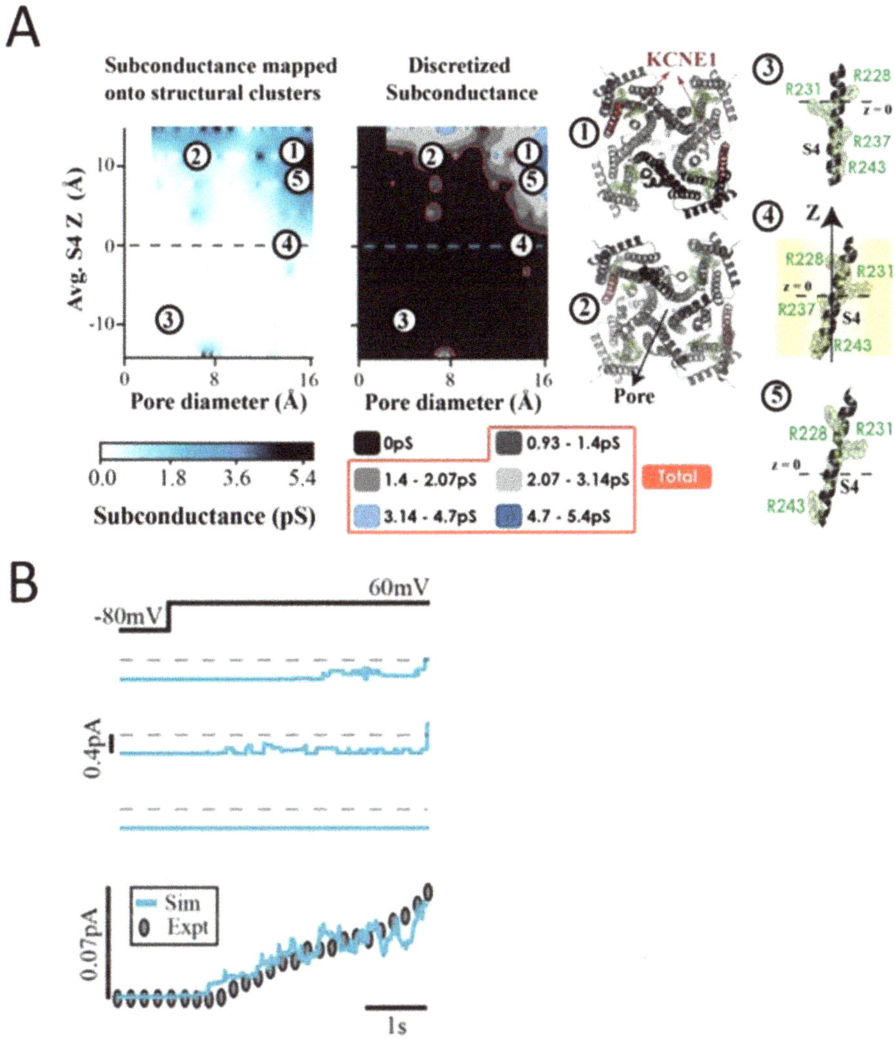

Figure 2.7. Use of machine learning to predict channel gating directly from a channel's excitatory cycle at the molecular scale. Figure adapted from reference [61] with permission from Elsevier. (A) Map of the two key components of $K_V7.1$ structure during gating, S4 movement and the pore diameter, and their relationship to subconductance of the channel. The left panel depicts the raw subconductance map while the middle shows the map transformed into five discrete subconductance levels used for calculations. The right panel illustrates structure conformations labeled in the left and middle panels. S4 charges are labeled in green. (B) Transformation of the discrete subconductance levels to simulate single-channel currents. For a step depolarization from −80 mV to 60 mV, three representative single-channel current traces are shown along with macroscopic current.

2.8 Concluding remarks

Ion channel modeling has clearly come a long way since the early days of Hodgkin and Huxley's work even though it remarkably still serves as a framework for modeling studies today. Ion channel modelers have access to a wealth of options to recapitulate channel dynamics from the microscopic to the macroscopic levels. Markov models provided the first opportunity to capture ion channel dynamics at the single-channel and macroscopic levels to predict cellular level excitability effects. As more ion channel structures were resolved more recently, these structures provided the opportunity to include microscopic, atomic level detail in channel models needed to elucidate precise channel mechanisms. This precise picture is becoming more invaluable today with the need to understand how drugs and other accessory subunits specifically modulate channel behavior in the context of drug development and disease.

The modeler must remain cognizant of the strengths and weaknesses of an approach along the spectrum of microscopic and macroscopic modeling techniques available. What is the goal of the project, and how much detail is required at the expense of simulation time to answer the question? Future ion channel modeling studies may try innovative approaches that attempt to combine the strengths of microscopic and macroscopic modeling. Singhal and Pande describe an approach that marries the use of Markov state modeling to MD simulations by taking a 'bottom up' approach to Markov model modeling rather than 'top down'. Clusters of similar structures during MD simulations may be assigned into states rather than assigning abstract kinetic states to macroscopic experimental data [62]. After assigning transition probabilities between these clusters of structures, or states, one may construct a Markov model of the main transitions in the MD simulation. This approach may be the future for longer channel simulations with drugs and accessory subunits [63]. Another especially promising approach is to use machine learning as in Ramasubramanian *et al* [61] to include the molecular level detail of channel structure into longer length simulations to connect up to tissue levels with cellular action potential models. With continued advances in computational power, machine learning methods and modeler ingenuity to recapitulate dynamics, the coming decades of ion channel modeling looks bright.

References

[1] Qu Z, Hu G, Garfinkel A and Weiss J N 2014 Nonlinear and stochastic dynamics in the heart *Phys. Rep.* **543** 61–162

[2] Hodgkin A L and Huxley A F 1952 A quantitative description of membrane current and its application to conduction and excitation in nerve *J. Physiol.* **117** 500–44

[3] Rudy Y and Silva J R 2006 Computational biology in the study of cardiac ion channels and cell electrophysiology *Q. Rev. Biophys.* **39** 57–116

[4] Plonsey R and Barr R C 2007 *Bioelectricity: A Quantitative Approach* (Berlin: Springer)

[5] Pallotta B S and Wagoner P K 1992 Voltage-dependent potassium channels since Hodgkin and Huxley *Physiol. Rev.* **72** S49–67

[6] Levy A 2013 What was Hodgkin and Huxley's achievement? *Br. J. Philos. Sci.* **65** 469–92

[7] Zagotta W N, Hoshi T and Aldrich R W 1994 Shaker potassium channel gating. III: Evaluation of kinetic models for activation *J. Gen. Physiol.* **103** 321–62

[8] Chanda B and Bezanilla F 2002 Tracking voltage-dependent conformational changes in skeletal muscle sodium channel during activation *J. Gen. Physiol.* **120** 629–45

[9] Goldschen-Ohm M P, Capes D L, Oelstrom K M and Chanda B 2013 Multiple pore conformations driven by asynchronous movements of voltage sensors in a eukaryotic sodium channel *Nat. Commun.* **4** 1350

[10] Hsu E J, Zhu W, Schubert A R, Voelker T, Varga Z and Silva J R 2017 Regulation of Na+ channel inactivation by the DIII and DIV voltage-sensing domains *J. Gen. Physiol.* **149** 389–403

[11] Capes D L, Goldschen-Ohm M P, Arcisio-Miranda M, Bezanilla F and Chanda B 2013 Domain IV voltage-sensor movement is both sufficient and rate limiting for fast inactivation in sodium channels *J. Gen. Physiol.* **142** 101–12

[12] Ahern C A, Payandeh J, Bosmans F and Chanda B 2016 The Hitchhiker's guide to the voltage-gated sodium channel galaxy *J. Gen. Physiol.* **147** 1–24

[13] Fox R F and Lu Y 1994 Emergent collective behavior in large numbers of globally coupled independently stochastic ion channels *Phys. Rev.* E **49** 3421

[14] Neher E and Sakmann B 1976 Single-channel currents recorded from membrane of denervated frog muscle fibres *Nature* **260** 799–802

[15] Goldwyn J H, Imennov N S, Famulare M and Shea-Brown E 2011 Stochastic differential equation models for ion channel noise in Hodgkin–Huxley neurons *Phys. Rev.* E **83** 41908

[16] Bezanilla F and Armstrong C M 1977 Inactivation of the sodium channel. I. Sodium current experiments *J. Gen. Physiol.* **70** 549–66

[17] Armstrong C M and Bezanilla F 1977 Inactivation of the sodium channel. II. Gating current experiments *J. Gen. Physiol.* **70** 567–90

[18] Horn R and Lange K 1983 Estimating kinetic constants from single channel data *Biophys. J.* **43** 207–23

[19] Sigg D 2014 Modeling ion channels: past, present, and future *J. Gen. Physiol.* **144** 7–26

[20] Fink M and Noble D 2009 Markov models for ion channels: versatility versus identifiability and speed *Philos. Trans. R. Soc. A Math. Phys. Eng. Sci.* **367** 2161–79

[21] Linaro D and Giugliano M 2015 *Markov Models of Ion Channels BT - Encyclopedia of Computational Neuroscience* ed D Jaeger and R Jung (New York: Springer) pp 1649–60

[22] Teed Z R and Silva J R 2016 A computationally efficient algorithm for fitting ion channel parameters *MethodsX* **3** 577–88

[23] Patlak J 1991 Molecular kinetics of voltage-dependent Na+ channels *Physiol. Rev.* **71** 1047–80

[24] Mangold K E, Brumback B D, Angsutararux P, Voelker T L, Zhu W, Kang P W, Moreno J D and Silva J R 2017 Mechanisms and models of cardiac sodium channel inactivation *Channels* **11** 517–33

[25] Clerx M, Beattie K A, Gavaghan D J and Mirams G R 2019 Four ways to fit an ion channel model *Biophys. J.* **117** 2420–37

[26] Colquhoun D and Hawkes A G 1977 Relaxation and fluctuations of membrane currents that flow through drug-operated channels *Proc. R. Soc. Lond. Ser. B Biol. Sci.* **199** 231–62

[27] Clancy C E and Rudy Y 1999 Linking a genetic defect to its cellular phenotype in a cardiac arrhythmia *Nature* **400** 566–9

[28] Clancy C E and Rudy Y 2001 Cellular consequences of HERG mutations in the long QT syndrome: precursors to sudden cardiac death *Cardiovasc. Res.* **50** 301–13

[29] Luo C-H and Rudy Y 1994 A dynamic model of the cardiac ventricular action potential. II. After depolarizations, triggered activity, and potentiation *Circ. Res.* **74** 1097–113

[30] Luo C and Rudy Y 1994 A dynamic model of the cardiac ventricular action potential. I. Simulations of ionic currents and concentration changes *Circ. Res.* **74** 1071–96

[31] Irvine L A, Saleet Jafri M and Winslow R L 1999 Cardiac sodium channel markov model with temperature dependence and recovery from inactivation *Biophys. J.* **76** 1868–85
[32] Moreno J D, Zhu Z I, Yang P-C, Bankston J R, Jeng M-T, Kang C, Wang L, Bayer J D, Christini D J and Trayanova N A 2011 A computational model to predict the effects of class I anti-arrhythmic drugs on ventricular rhythms *Sci. Transl. Med.* **3** 98ra83
[33] Moreno J D, Lewis T J and Clancy C E 2016 Parameterization for in-silico modeling of ion channel interactions with drugs *PLoS One* **11** e0150761
[34] Silva J and Rudy Y 2005 Subunit interaction determines I Ks participation in cardiac repolarization and repolarization reserve *Circulation* **112** 1384–91
[35] Mannuzzu L M, Moronne M M and Isacoff E Y 1996 Direct physical measure of conformational rearrangement underlying potassium channel gating *Science* **271** 213
[36] Zhu W, Varga Z and Silva J R 2016 Molecular motions that shape the cardiac action potential: Insights from voltage clamp fluorometry *Prog. Biophys. Mol. Biol.* **120** 3–17
[37] Osteen J D, Gonzalez C, Sampson K J, Iyer V, Rebolledo S, Larsson H P and Kass R S 2010 KCNE1 alters the voltage sensor movements necessary to open the KCNQ1 channel gate *Proc. Natl Acad. Sci. USA* **107** 22710–5
[38] Zhu W, Mazzanti A, Voelker T L, Hou P, Moreno J D, Angsutararux P, Naegle K M, Priori S G and Silva J R 2019 Predicting patient response to the antiarrhythmic mexiletine based on genetic variation: personalized medicine for long QT syndrome *Circ. Res.* **124** 539–52
[39] Moreno J D, Zhu W, Mangold K, Woenho C and Silva J R 2019 A molecularly detailed Nav1.5 model reveals a new class I antiarrhythmic target *JACC Basic Transl. Sci.* **4** 736–51
[40] Lindahl E 2015 *Molecular Dynamics Simulations BT - Molecular Modeling of Proteins* ed A Kukol (New York: Springer) pp 3–26
[41] Nurisso A, Daina A and Walker R C 2012 *A Practical Introduction to Molecular Dynamics Simulations: Applications to Homology Modeling BT - Homology Modeling: Methods and Protocols* ed A J W Orry and R Abagyan (Totowa, NJ: Humana) pp 137–73
[42] Hospital A, Goñi J R, Orozco M and Gelpí J L 2015 Molecular dynamics simulations: advances and applications *Adv. Appl. Bioinform. Chem.* **8** 37–47
[43] Baker N A 2004 Poisson–Boltzmann methods for biomolecular electrostatics *Numerical Computer Methods, Part D* (New York: Academic) pp 94–118
[44] Wang J, Tan C, Tan Y-H, Lu Q and Luo R 2008 Poisson-Boltzmann solvents in molecular dynamics simulations *Commun. Comput. Phys.* **3** 1010–31
[45] Fischer W, Brickmann J and Läuger P 1981 Molecular dynamics study of ion transport in transmembrane protein channels *Biophys. Chem.* **13** 105–16
[46] Mackay D H, Berens P H, Wilson K R and Hagler A T 1984 Structure and dynamics of ion transport through gramicidin A *Biophys. J.* **46** 229–48
[47] Allen T W, Kuyucak S and Chung S-H 1999 Molecular dynamics study of the KcsA potassium channel *Biophys. J.* **77** 2502–16
[48] Bernèche S and Roux B 2001 Energetics of ion conduction through the K+ channel *Nature* **414** 73–7
[49] Sokolova O, Kolmakova-Partensky L and Grigorieff N 2001 Three-dimensional structure of a voltage-gated potassium channel at 2.5 nm resolution *Structure* **9** 215–20
[50] Long S B, Campbell E B and Mackinnon R 2005 Crystal structure of a mammalian voltage-dependent Shaker family K+ channel *Science* **309** 897–903

[51] Delemotte L, Klein M and Tarek M 2012 Molecular dynamics simulations of voltage-gated cation channels: insights on voltage-sensor domain function and modulation *Front. Pharmacol.* **3** 97

[52] Sands Z A and Sansom M S P 2007 How does a voltage sensor interact with a lipid bilayer? Simulations of a potassium channel domain *Structure* **15** 235–44

[53] Jogini V and Roux B 2007 Dynamics of the Kv1.2 voltage-gated K+ channel in a membrane environment *Biophys. J.* **93** 3070–82

[54] Grabe M, Lecar H, Jan Y N and Jan L Y 2004 A quantitative assessment of models for voltage-dependent gating of ion channels *Proc. Natl Acad. Sci. USA* **101** 17640–5

[55] Jensen M O, Jogini V, Borhani D W, Leffler A E, Dror R O and Shaw D E 2012 Mechanism of voltage gating in potassium channels *Science* **336** 229–33

[56] Delemotte L, Tarek M, Klein M L, Amaral C and Treptow W 2011 Intermediate states of the Kv1.2 voltage sensor from atomistic molecular dynamics simulations *Proc. Natl Acad. Sci.* **108** 6109–14

[57] Delemotte L, Kasimova M A, Klein M L, Tarek M and Carnevale V 2015 Free-energy landscape of ion-channel voltage-sensor–domain activation *Proc. Natl Acad. Sci.* **112** 124–9

[58] Nguyen P T, DeMarco K R, Vorobyov I, Clancy C E and Yarov-Yarovoy V 2019 Structural basis for antiarrhythmic drug interactions with the human cardiac sodium channel *Proc. Natl Acad. Sci.* **116** 2945–54

[59] Hille B 1977 Local anesthetics: hydrophilic and hydrophobic pathways for the drug-receptor reaction *J. Gen. Physiol.* **69** 497–515

[60] Nekouzadeh A and Rudy Y 2011 Continuum molecular simulation of large conformational changes during ion-channel gating *PLoS One* **6** e20186

[61] Ramasubramanian S and Rudy Y 2018 The structural basis of IKs ion-channel activation: mechanistic insights from molecular simulations *Biophys. J.* **114** 2584–94

[62] Singhal N and Pande V S 2005 Error analysis and efficient sampling in Markovian state models for molecular dynamics *J. Chem. Phys.* **123** 204909

[63] Silva J R 2018 How to connect cardiac excitation to the atomic interactions of ion channels *Biophys. J.* **114** 259–66

[64] Colquhoun D, Dowsland K A, Beato M and Plested A J R 2004 How to impose microscopic reversibility in complex reaction mechanisms *Biophys. J.* **86** 3510–8

Chapter 3

Modeling cardiac calcium signaling, regulation, and control

Andrew G Edwards and Eleonora Grandi

Calcium is the third most common metal on Earth, and the most prevalent in the human body. It is also nature's most ubiquitous intracellular messenger. Its unique coordination chemistry has provided evolution with a remarkably flexible binding partner for constructing an intricate and diverse signaling machinery [10]. In the heart, interactions between Ca^{2+} and its various binding partners control processes as immediate as ion channel gating and myofilament contraction [3] and as prolonged as the transcriptional regulation that determines myocardial development [66], growth [55], and in certain contexts, death [30]. The biophysical mechanisms that allow Ca^{2+} to control these diverse processes rely upon differences in the intrinsic binding affinities, kinetics and actions of each Ca^{2+}-binding protein. In this chapter we will use these characteristics to reconstruct the transport mechanisms that allow electrical activation to drive contraction in cardiac myocytes—a process termed cardiac excitation–contraction coupling (ECC). Cyclical influx and extrusion of Ca^{2+} are fundamental to this coupling, and these fluxes both shape and are shaped by the cardiac action potential (AP). Of course, there are limits to which this dynamic system can remain stable, and our second objective will be to provide a quantitative framework for describing those limits. We will particularly focus on understanding how the structure of the system endows cardiac muscle with a unique ability to balance contractile plasticity and electromechanical stability.

3.1 Calcium in cardiac physiology

3.1.1 Fundamentals of cardiac ECC and Ca^{2+} homeostasis

The critical triggering event in cardiac ECC is the influx of Ca^{2+} that accompanies voltage-dependent opening of L-type Ca^{2+} channels (LCCs). This influx causes Ca^{2+}-dependent activation of Ryanodine receptors (RyRs) located in the closely apposed sarcoplasmic reticulum (SR) membrane. RyRs then open and allow Ca^{2+}

doi:10.1088/978-0-7503-2064-1ch3

to flow down its very large concentration gradient (~ 3–4 orders of magnitude) from the SR into the cytosol. This process initiates contraction of cardiac muscle and is termed Ca^{2+}-induced Ca^{2+} release (CICR) [21]. In ventricular myocytes projections of the cell's surface membrane extend transversely into the centre of the cell (figure 3.1(A)). This creates an array of transverse tubules (TTs), which allow the Ca^{2+}-dependent communication between LCCs and intracellular RyRs to occur synchronously at the cell periphery and interior. When observed at the whole cell level, this synchronous activation causes the change in cytosolic Ca^{2+} to exhibit a rapid but smooth rise, followed by a prolonged decay over several hundred milliseconds as Ca^{2+} is re-sequestered into the SR and extruded from the cell (figure 3.1(B)).

3.1.2 The role of models in the study of cardiac Ca^{2+} signaling

Computational and theoretical approaches have been central to establishing what we know of the fundamental structure-function relationships in cardiac ECC. In part, this is because the Ca^{2+} signaling that drives cardiac SR Ca^{2+} release occurs in spaces that are an order of magnitude smaller than the diffraction limit of light. Hence, the Ca^{2+} gradients and molecular events that fuel these dynamics cannot be

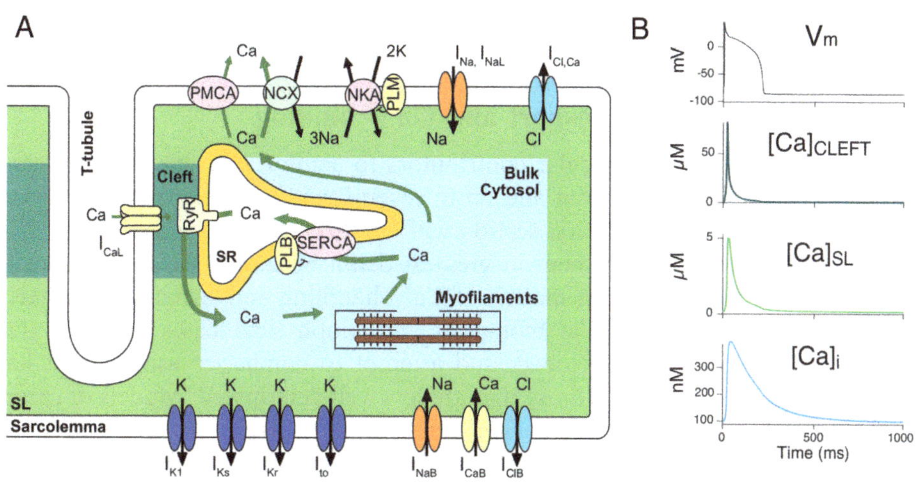

Figure 3.1. (A) Schematic representation of a cardiac myocyte model (in this case a human ventricular myocyte [31]). This model is an example of a low-dimension local control model, where the processes governing CICR occur in a small dyadic cleft volume (subscript CLEFT, dark green) housing 90% of LCCs and all RyRs. Ca^{2+} entering this compartment either from the extracellular space (via LCCs) or the SR (via RyRs) diffuses first into the sub-sarcolemmal space (SL, light green), and then the remaining intracellular space (i, blue). All other major surface membrane transporters are evenly distributed throughout the surface membrane, meaning that 89% of their flux is distributed to the SL compartment, and 11% into the dyadic cleft. Unlike the cleft, which can be visualized and reconstructed from electron micrographs, the SL space is not physically observable. Instead, it is a phenomenological implementation that is used to simulate the gradients in Ca^{2+} generated during the AP, which importantly modulate the function of Ca^{2+}-sensitive membrane transporters. These gradients are shown for Ca^{2+} in (B), where they appear as the discrete differences in $[Ca^{2+}]$ between the three compartments at any point in time.

observed by light microscopy, or any other direct method. The general approach to modeling these processes has been to attempt to reconcile measurable single protein function and membrane structure, with observable changes in cytosolic and SR [Ca^{2+}] that occur in larger surrounding volumes. These efforts have given rise to a large family of models that include some representation of the gradients in intracellular Ca^{2+} concentration occurring during a normal cardiac cycle. One such model is shown schematically in figure 3.1. The defining characteristic of these 'local-control' models is that they distinguish the intracellular Ca^{2+} pool that coordinates CICR from the pool that participates in myofilament Ca^{2+} signaling and SR Ca^{2+} reuptake. This property is enormously important for physiologic control of cardiac ECC.

Theoretical approaches in cardiac ECC have also filled a role to first integrate and then quantitatively abstract existing knowledge, for the purpose of distilling the complexity of the system to its essential dynamical properties. For example, elegant early theoretical work was key for establishing that Ca^{2+} gradients are necessary to simultaneously allow high ECC gain (a small LCC influx can trigger a large SR Ca^{2+} release), while also permitting stable Ca^{2+} release [84]. A variety of analytic and numerical approaches have also been used to establish a number of basic dynamical properties of cardiac ECC. In particular, they have provided a formal quantitative framework to understand different types of Ca^{2+} instabilities, and the contexts in which they are important.

3.2 Models of Ca^{2+} transport and Ca^{2+} binding

In the heart, the major molecular contributors to acute Ca^{2+}-dependent processes (those occurring in a single beat and up to several minutes of cardiac activity) are now well-characterized. Our first section will describe the most important of these proteins. We will deal with them in greatest detail because these models are the foundation for all higher-level models of Ca^{2+}-handling and signaling. We particularly focus on describing the biological and biophysical bases for how these transporter models have developed, and attempt to emphasize aspects that have relatively high certainty from those that are less well determined. Models of cardiac Ca^{2+} signaling occurring over longer time scales, particularly those involving regulation of transcription and translation (sometimes termed excitation–transcription coupling), are much less well developed to date, and so we will leave them to other texts.

3.2.1 Models of cardiac Ca^{2+} channels

In this section we focus on representing the function of LCCs, and the cardiac RyR. As well as coordinating the essential dynamics of CICR, these two proteins are critical sites of autonomic and intrinsic regulation of cardiac Ca^{2+} handling and contractile force [3].

L-type Ca^{2+} *channel*

The LCC is the most important sarcolemmal Ca^{2+} channel in the heart, and is particularly prominent in ventricular myocytes where other Ca^{2+} channel subtypes are trivially expressed [2]. LCC single-channel properties are unique among the major cardiac ion channels, and explain the 'L-type' nomenclature. They exhibit a *large* unitary conductance, relatively *long* openings, and require *large* depolarizations to activate. LCC opening events are characterized by major changes in gating kinetics that shift between three different prototypical 'modes' of behavior. Channels in mode 0 remain quiescent or exhibit infrequent and very brief (<0.1 ms) openings during a depolarization step. Channels in mode 1 undergo irregular bursts of frequent (~1–2 ms) openings, and mode 2 gating is characterized by opening events of 10s to 100s of ms, and relatively brief closures (figure 3.2(D)). Mode 2 dynamics are rarely observed under normal conditions, but are strongly facilitated by channel phosphorylation (by protein kinase A, PKA; and Ca^{2+}-calmodulin dependent protein kinase II, CaMKII) [19, 98], or binding of agonist dihydropiridines

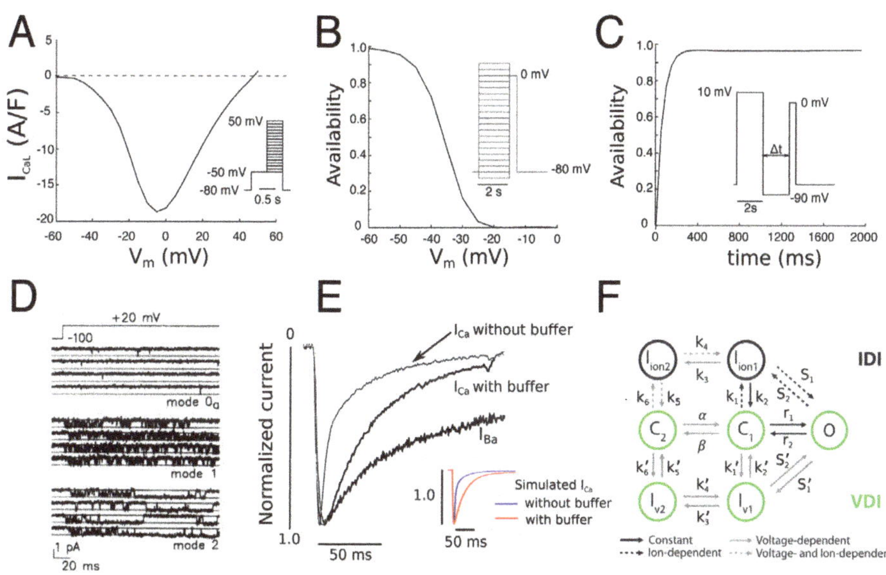

Figure 3.2. Characteristics of LCC gating and permeation. The three whole-cell properties commonly applied to build and fit LCC models are shown in panels A–C. They are the voltage dependencies of peak current activation (A) and inactivation (B), and the kinetics of I_{CaL} recovery from inactivation (C). These are fundamental properties for all I_{CaL} models, and are typically fit to data specific to the species and experimental or disease conditions of interest. Panels D–F show some of the unique properties of LCCs and I_{CaL}. (D) The three characteristic modes of LCC gating (with permission from [98], copyright 1993 National Academy of Sciences) with clearly distinguishable transition kinetics. Downward current deflections represent channel opening events. (E) Separating the influence of VDI and CDI on overall I_{CaL} inactivation is often done by applying heavy intracellular buffering and substituting Ba^{2+} in place of extracellular Ca^{2+} to prevent CICR. These traces are redrawn with permission from [59]. Copyright 2012 John Wiley & Sons Ltd. (F) These two major mechanisms of LCC inactivation (Ca^{2+}- or ion-dependent inactivation, and VDI) are generally modeled as alternative parallel pathways for channel inactivation, as in this schematic modification of the model originally from Mahajan *et al* [54, 59].

(e.g. Bay K 8644). LCCs are also potently inhibited by intracellular Ca^{2+}, a process known as Ca^{2+}-dependent inactivation (CDI) [97]. In this mechanism, a calmodulin tethered to the channel's C-terminus promotes channel inactivation upon Ca^{2+} binding. This is the major mechanism of I_{CaL} inactivation in the heart, and allows increases in SR Ca^{2+} release to inhibit further Ca^{2+} influx. Voltage-dependent inactivation (VDI) also contributes I_{CaL} decay during the ventricular AP. While it is generally less prominent than CDI, deficits in VDI underlying Timothy Syndrome (Long QT subtype 8) have clearly established that this process is important in human ventricular repolarization. Both steady-state activation and VDI are right-shifted (relatively positive) in LCCs compared to other voltage-activated Ca^{2+} channels (VGCs). Lastly, like other VGCs, LCCs are permeable to varying extents by divalent cations (other than Mg^{2+}), and the major monovalents (Na^+ and K^+) [2]. Because channel permeability is much higher for Ca^{2+} (~3000-fold), the compound I_{CaL} is carried by Ca^{2+} for much of the AP, but when V_m approaches the reversal potential for multi-ion permeation, there will still be net Ca^{2+} entry that is opposed by an equal charge efflux of K^+.

All of the properties mentioned above feature in modern models of cardiac I_{CaL} and its regulation by intracellular signaling. We will introduce two representative models of rabbit I_{CaL}, which have also been re-parameterized for other species including human, and been integrated with models of β-adrenergic and CaMKII regulation. The first is an HH framework that extends the gating frameworks you have been introduced to in prior chapters to include CDI [74]. The second is a Markovian representation specifically developed to incorporate aspects of our biophysical understanding of CDI and account for the frequency-dependence of changes to CDI and VDI [54].

The Shannon *et al*, rabbit myocyte model was one of the first to incorporate a comprehensive Ca^{2+} handling machinery [74]. The overall I_{CaL} equation is in the Goldmann–Hodgkin–Katz current form to explicitly represent the nonlinear multi-ion permeability. Each LCC conducts currents carried by the three permeant ions (IP_{Ca}, IP_{Na}, and IP_{K}):

$$I_{CaL} = F_X(IP_{\mathrm{Ca}} + IP_{\mathrm{Na}} + IP_{\mathrm{K}}), \tag{3.1}$$

Here, F_X represents the fraction of channels located in compartment X. Figure 3.1(A) depicts these compartments, such as the dyad, and the subsarcolemmal space (SL). The cardiac dyad is an observable structure, and named for its appearance as two closely associated membrane components (t-tubular membrane and junctional SR membrane) in longitudinal EM sections. The space between these closely apposed (~12–15 nm) membranes is the dyadic (junctional) cleft, which is home to the LCCs and RyRs that coordinate CICR. The SL is a phenomenological compartment, which cannot be directly observed, but is implemented to represent the intermediate calcium concentration (i.e. between the high dyadic concentration and much lower average cytosolic calcium concentration) sensed by membrane calcium transporters. Most LCCs are thought to reside within or very close to dyads, and in the Shannon model (figure 3.1) 90% are located in the junctional space and

10% in the SL membrane. This is an important characteristic for determining the average local [Ca^{2+}] experienced by LCCs, and hence the [Ca^{2+}] that drives CDI. Each of the ionic currents carried by the open fraction of LCCs ($IP_{Ca}...IP_K$) is then calculated by a more familiar looking formulation:

$$IP_S = dff_{Ca}\bar{I}_S, \tag{3.2}$$

In Ca^{2+} channels it is conventional to use d and f to represent steady-state activation and VDI, respectively, and f_{Ca} is inversely proportional to CDI. Together, these three components (activation, VDI and CDI) are the critical pieces describing gating in virtually all modern models of I_{CaL}. $\bar{I}_S$ is the maximal current calculated for each permeant ionic species (S) via the Goldmann–Hodgkin–Katz current equation:

$$\bar{I}_S = P_S z_S^2 V_m \frac{F^2}{RT} \frac{[S]_X \exp\left(\frac{z_S V_m F}{RT}\right) - [S]_e}{\exp\left(\frac{z_S V_m F}{RT}\right) - 1}, \tag{3.3}$$

where X is again the relevant intracellular compartment.

The evolution of d and f in time is given by:

$$d' = \frac{(\bar{d} - d)}{\tau_d}, \quad f' = \frac{(\bar{f} - f)}{\tau_f}, \tag{3.4}$$

where $\bar{d}$ and $\bar{f}$ are the steady-state activation and inactivation functions (figures 3.2(A) and (B)), and τ_d and τ_f are the time-constants of activation and inactivation, respectively:

$$\tau_d = \frac{1}{\alpha_d + \beta_d}, \quad \tau_f = \frac{1}{\alpha_f + \gamma_f}, \tag{3.5}$$

In general, these voltage dependent components are fit simultaneously to corresponding datasets for steady-state activation and inactivation, and recovery from inactivation (figures 3.2(A)–(C)). Importantly, because SR Ca^{2+} release varies with voltage, a range of manipulations have been used to separate both the voltage dependencies and kinetics of VDI from the effects of CDI. Early work decomposed the inactivation time course into two exponential components: (1) a large and fast component attributed to CDI, and (2) a smaller slower component to VDI. To do this well requires good control of SR Ca^{2+} load. More commonly now, CDI is eliminated through heavy chelation of internal Ca^{2+} (with BAPTA) and use of barium as the charge carrier (figure 3.2(E)). This has its own drawbacks because barium still appears to have some calmodulin affinity and ability to induce mild CDI-like inactivation (ion-dependent inactivation).

f_{Ca} is the new gating component in equation (3.2) to represent the fraction of channels not inactivated via CDI. The gating process for CDI in this model is implemented as a simple mass action relationship for Ca^{2+} association and dissociation:

$$\frac{df_{\text{CaB}}}{dt} = k_{on}[\text{Ca}^{2+}]_{cleft}(1 - f_{\text{CaB}}) - k_{off}f_{\text{CaB}}, \quad f_{\text{Ca}} = 1 - f_{\text{CaB}}, \tag{3.6}$$

where f_{CaB} is the fraction of channels with Ca^{2+} bound, $[\text{Ca}]_{\text{cleft}}$ is the Ca^{2+} concentration in the junctional cleft and k_{on} and k_{off} are rate constants for Ca^{2+} association and dissociation, respectively. Importantly, for all LCC models these constants are poorly defined by experiments and in the model they interact with the many factors that determine $[\text{Ca}]_{\text{cleft}}$ (RyR release current, cleft volume, local buffering and diffusion) to dictate I_{CaL} inactivation. For this reason they are often highly model-dependent. In the original Shannon model [74] the ratio $k_{\text{off}}/k_{\text{on}}$ gives an equilibrium Ca^{2+} dissociation constant (K_d) of 7 μM, which is close to our best estimates of half-maximal CDI (on the order of ~10 μM).

More detailed models of CDI have attempted to explore the interaction of VDI and CDI in cardiac I_{CaL}. An early Markovian representation of ventricular I_{CaL} was developed by Jafri *et al* [43], and then later condensed and modified for the rabbit [31, 54, 59] (figure 3.2(F) [59]). These models have attempted to account for biophysical evidence that Ca^{2+} binding to the C-terminal calmodulin simply facilitates the VDI gating event, and reconcile this with the measurable I_{CaL} across physiologic frequencies and with different charge carriers. Finally, a recent iteration of this model has been applied in stochastic simulations to interrogate coupled gating of LCCs [70]. This concept suggests that physically associated LCCs may interact to allow them to open and close together [18], which may provide a part of the explanation for how low peak open probability (P_o) of single LCCs is capable of eliciting RyR openings with high probability in intact myocytes (a property termed high ECC fidelity).

Together, these three components (activation, VDI and CDI) are the critical pieces in virtually all modern models of I_{CaL}.

Ryanodine receptor

This massive tetrameric protein (2.26 MDa) is named for its affinity and regulation by the plant alkaloid Ryanodine, which was key for isolating the muscle isoforms. Because RyRs are embedded in the SR membrane they cannot be assayed *in situ* via conventional microelectrode techniques, and our knowledge of RyR function has developed largely from single channel measurements made in bilayer preparations. Because RyRs in these preparations experience conditions that are far from those in a cardiac dyad (e.g. ionic composition, molecular partners), it has proven challenging to extend single protein behavior to *in vivo* function. Model-based extrapolation has therefore been crucial for reconciling the properties of isolated RyRs with evidence of Ca^{2+} release in isolated SR microsomes (vesicles), and local and whole-cell Ca^{2+} release in intact myocytes [14].

Permeation of open RyRs exhibits two characteristics that are crucial for understanding SR Ca^{2+} release. First, they have a very large unitary conductance (~100–200 pS for divalents and up to 700 pS for K^+) [79]. Second, their low selectivity ($P_{\text{Ca}}/P_{\text{K,Na}} \sim 6$) is closer to non-specific cation channels than voltage-gated Ca^{2+} channels like the LCC [79]. For these reasons, RyRs are probably capable of conducting their own countercurrent during SR Ca^{2+} release [28], although this is likely to be unnecessary *in vivo* as TRIC (trimeric intracellular cation) channels are also available for that countercurrent [27, 102].

Moving on to RyR gating, as mentioned earlier, Ca^{2+}-dependent activation of RyRs is the central process underlying cardiac CICR. This, and the processes that cause RyR adaptation (similar to inactivation), have been addressed by a broad range of experimental and computational studies. An important first point to make about these models is that many have been built to replicate either whole-cell Ca^{2+} release (e.g. figure 3.1), or the discrete Ca^{2+} release events that result from the activation of RyRs in a single junctional terminal. These smaller isolated events can be measured, and have been coined Ca^{2+} 'sparks'. As a result, most of the whole-cell models and many spark models are some hybrids of the mechanisms of single RyR gating, and the release dynamics that arise from interaction of 10s to many 1000s of receptors in larger intracellular volumes. In this section we will cover the properties and models of single RyR channels. These models have typically employed two Ca^{2+}-dependent processes, activation and inactivation (or adaptation).

RyR activation depends on both cytosolic (cleft) and SR lumenal $[Ca^{2+}]$. Channel equilibrium P_o begins to increase when cleft $[Ca^{2+}]$ is in the high nM range, and continues to a peak in the hundreds of μM (figure 3.3(A)). This sensitivity is regulated by many factors, but the most important *in vivo* are desensitization due to free Mg^{2+} and calmodulin, and sensitization by ATP. A range of post-translational modifications and binding interactions also regulate RyRs, but most remain at least somewhat controversial, and we will not discuss them further here. The cytosolic Ca^{2+} binding sites are not yet established, and it is not agreed how many Ca^{2+} ions must bind to open the channel (figure 3.4). Estimates have suggested as few as two [94]

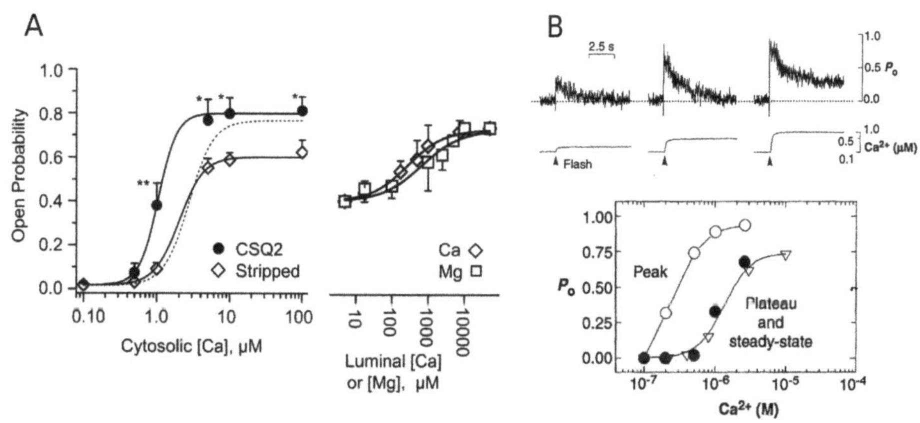

Figure 3.3. Essential properties of RyR gating. (A) Steady-state Ca^{2+} dependencies for single rat RyRs reconstituted in bilayers, and either with CSQ stripped through high-salt wash, or subsequently re-added recombinant CSQ. Note that steady-state P_o declines at higher $[Ca^{2+}]_i$ (1 mM) not shown here, but which is typically described as absorbing $[Ca^{2+}]_i$-dependent inactivation. Data reproduced from [67], copyright Rockfeller University Press. (B) Adaptation is the relatively slow decline from peak P_o accompanying acute increases in $[Ca^{2+}]_i$. Here those increases are applied by flash photolysis of caged Ca^{2+}. Data reproduced from [35], with permission from AAAS.

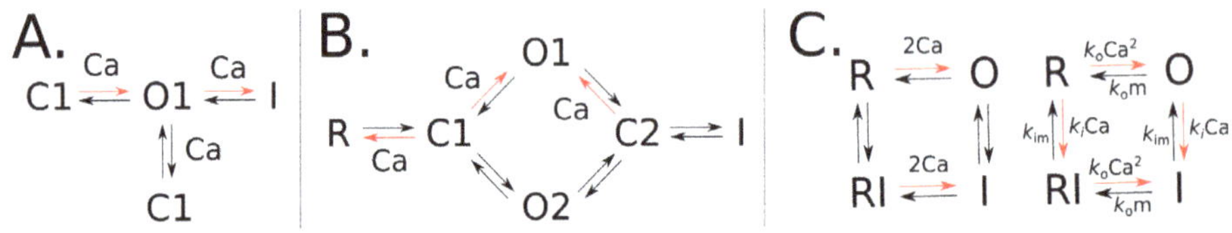

Figure 3.4. Early models of RyR gating. For all models, Ca^{2+}-dependent transitions are indicated by red arrows. R indicates the resting state, C states are closed, O are open, and I states are inactive (adapted). The models shown in (A) and (B) are built directly from single RyR2 recordings providing Ca^{2+} dependencies and best estimates of kinetics for the processes of activation and inactivation. The model in (B) attempts to account for high- and low-activity modes observed as bursts of RyR activity [69, 99]. Like many early models built from single channel recordings, both of these models fail to terminate when incorporated into stochastic models of dyadic CICR. The two models shown in (C) are phenomenologic and permit rapid inactivation to an extent not supported by single-channel recordings, but which also allows stable and high-gain CICR [85].

or at least four [69, 100]. What is clear, is that activation results from a marked decrease in the closed lifetime, with very little impact on the open duration [8, 69]. The role of lumenal Ca^{2+} is also clear, although two mechanistic aspects have seen vigorous debate [26, 67]: (1) is the $[Ca^{2+}]_{SR}$ dependence due to direct intra-SR regulation of the channel, or simply due to the ability for higher single channel Ca^{2+} flux to increase the local Ca^{2+} concentration at the cytosolic surface? (2) If direct intra-SR regulation does occur, does it involve accessory proteins (e.g. calsequestrin, CSQ) or direct binding to a lumenal site/s on RyR? At this stage the available data suggest that intra-SR regulation is the dominant mechanism [22, 67], and that it occurs via both direct and indirect interactions between RyR and lumenal Ca^{2+} [12, 13, 67].

RyR inactivation and adaptation are processes that result in channel closure in response to increased cytosolic Ca^{2+}. The terminology describes two different characteristics of the regulation, but the underlying mechanism may be the same [36]. Inactivation refers to the effect of very high $[Ca^{2+}]_i$ (upwards of several mM) to reduce steady-state P_o (figure 3.3). Adaptation refers to a time-dependent decrease in RyR P_o (at constant lumenal $[Ca^{2+}]$) back to the steady-state value after a higher peak P_o during rapid application of Ca^{2+}. Adaptation is observed with rapid Ca^{2+} applications across physiologic $[Ca^{2+}]_i$, whereas inactivation is only observed at supra-physiologic $[Ca^{2+}]_i$. While the mechanisms of this (these) process(es) are poorly understood, adaptation features as an important mechanism of RyR closure and CICR termination in many models. A key point of contention is that the time course of adaptation measured in single channels is usually upwards of 100 ms. So the degree to which it can contribute to halting CICR has been treated with reasonable skepticism. Without it, however, many early models of CICR fail to terminate, as we discuss below.

Finally, an important constraint for model function is that only a small fraction of channels are required to open simultaneously to achieve peak whole-cell Ca^{2+} release flux in cardiac myocytes. This means that peak cell-wide RyR P_o during EC-coupling is probably only on the order of 0.01–0.1 [2, 74]. This again has important implications for modeling local versus whole-cell RyR function. The basic formulation for release flux J_{rel} from most RyR models is analogous to Ohmic expressions of ion channel currents, recalling that the SR membrane is assumed not to develop a meaningful membrane potential during release:

$$J_{rel} = k_{rel}P_o([Ca^{2+}]_{SR} - [Ca^{2+}]_{cleft}), \tag{3.7}$$

In this equation, the dynamics determining the fraction of channels in the open state (P_o) are where much of the detail and debate has lived. The parameter k_{rel} is often provided as a constant, and assumed not to vary with the conditions of the simulated experiment, but can be calculated as $k_{rel} = \frac{i_{RyR}}{2FV}$, where i_{RyR} is the unitary current, which can vary with ionic composition or voltage.

Reliable termination of Ca^{2+} release was a major challenge for early models. This problem has given rise to many phenomenologic approaches to enforce termination, and is well summarized by Stern *et al* [85, 86]. Briefly, the dynamic problem is that an ensemble of ten or more clustered RyRs should be stable at rest and exhibit high CICR gain during activation. That is, the ensemble should exhibit sufficiently low P_o to remain inactive at diastolic $[Ca^{2+}]_{cleft}$ and $[Ca^{2+}]_{SR}$, but upon injection of a small Ca^{2+} influx from one or perhaps two juxtaposed LCCs, this system should transition to a highly non-linear regenerative regime in which the majority of the RyRs activate. Typically, the total Ca^{2+} released via the RyR ensemble should be at least ~10 times that entering through the local LCCs (high-gain CICR). To construct a model capable of both characteristics, which also reliably terminates Ca^{2+} release, proved remarkably challenging. Figure 3.4 shows two (A and B) early models of Ca^{2+}-dependent RyR activation that exhibit a stable resting state and regenerative release, but fail to terminate in spark simulations. Figure 3.4(C) then shows two phenomenological gating schemes with a rapid inactivation (adaptation) process that permits termination.

More recent RyR models have been able to overcome these prior challenges by incorporating updated measures of RyR bilayer function. The key characteristics have been: (1) sensitivity of RyR activation to SR lumenal Ca^{2+} (not a feature of most early models), (2) improved estimates of both cytosolic and lumenal Ca^{2+} sensitivity in physiologic conditions (realistic ATP, Mg^{2+}), and (3) accurate definition of RyR unitary current under physiologic conditions [34]. With RyR unitary currents and $[Ca^{2+}]_i$ sensitivities that reflected physiologic [ATP] and $[Mg^{2+}]$, two models published almost simultaneously in 2013 described essentially the same mechanism of termination via SR Ca^{2+} depletion. They have been coined the 'pernicious attrition' [29], and 'induction decay' [8, 48] models. In both cases, termination results from the inability of reduced RyR current (as a result of local SR Ca^{2+} depletion) to maintain regenerative release. Similar results have been achieved by applying the updated RyR properties to other widely-used models [93].

In the interest of covering the RyR and LCC in detail, we will not devote further space to the many Ca^{2+}-dependent processes in gating and regulation of Ca^{2+} non-selective channels, although there are a number in the heart. We have also chosen not to describe models of the T-type calcium channel. These channels are not highly expressed in the heart, and where they are, the model construction process is essentially as it would be for an LCC without CDI.

3.2.2 Models of cardiac Ca^{2+} pumps and exchangers

To this point, all of the transporters and models we have discussed have been passive. For that reason the forms of most models we have discussed have been relatively similar. They have comprised an overall conductance or permeability term (e.g. g_X), a term (O) or product of terms (e.g. dff_{Ca}) to indicate the fraction of channels in the open state, and some measure of the electrochemical driving force for the permeant ion ($V_m - E_x$). In this section we move on to discuss the major pumps involved in active Ca^{2+} transport, as well as the passive but directly coupled co-transport performed by the Na^+–Ca^{2+} exchanger (NCX). Some of the quantities here are analogous to those used for ion channel transport, but in all cases the binding interactions between the transporter and ions are more explicitly dealt with.

Sarco-endoplasmic reticulum Ca^{2+}-ATPase (SERCA)

The most important Ca^{2+} transporting ATPase in the heart is SERCA, and it is a member of the P-type superfamily of ion pumps. Its role is to faithfully return Ca^{2+} to the SR after it has been released as a result of CICR. Of course, the pump cannot distinguish between Ca^{2+} released by RyR and Ca^{2+} entering through LCCs, but under conditions of steady-state Ca^{2+} cycling, SERCA returns the same quantity of Ca^{2+} as is released by RyR (~ 60%–70% of the Ca^{2+} participating in contraction in humans). To achieve this level of transport, SERCA is abundantly expressed in cardiac muscle.

For each ATP hydrolyzed, SERCA transports two Ca^{2+} ions from the cytosol into the SR, and returns protons (most likely two of them) from the SR to the cytosol [88]. The directionality of the transport is determined by the difference in affinity of the Ca^{2+} binding sites when the pump is open to the cytosol (the E_1 conformation) compared to their affinity when exposed to the SR lumen (E_2, figure 3.5(A)). This difference, which is intrinsic to the change in conformation between E_1 and E_2, is approximately four orders of magnitude. However, this is a complex multistep process involving various MgATP, Ca^{2+} and H^+ binding and dissociation events, as well as ATP hydrolysis, and slower long-range conformational changes associated with ion trapping and release. The integration of these effects is a Ca^{2+} transport mechanism that is capable of generating the very large SR-cytosol Ca^{2+} gradient, but which also has an important physiologic dependency on pH [57], and to a much lesser extent on physiologic variations in Mg^{2+} and ATP.

The first step for constructing a SERCA transport model is to define the rate of transport as a function of $[Ca^{2+}]_i$, and conditions for the equilibrium $[Ca^{2+}]_{SR}:[Ca^{2+}]_i$ relationship. Early models for SERCA formulated SR Ca^{2+} reuptake via Michaelis–Menten or Hill relationships. For example, one of the first and most widely used ventricular myocyte models (the Luo–Rudy guinea pig myocyte, LRd [52]) employed pure Michaelis–Menten dynamics:

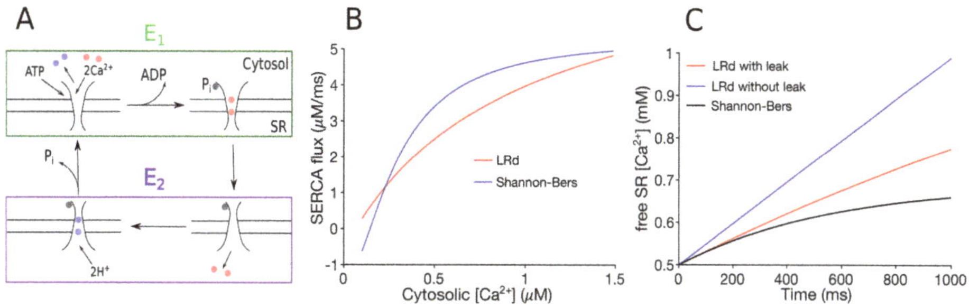

Figure 3.5. The SERCA catalytic cycle and the functional consequences of two popular SERCA models. (A) A schematic of the SERCA catalytic cycle. The two major conformations, E1 and E2 are open to ion binding from the cytosolic and lumenal surfaces, respectively. The decrease in Ca^{2+} affinity from the E1 to E2 conformation provides the thermodynamic basis for transport. (B) With fixed free $[Ca^{2+}]_{SR} = 1$ mM, the transport rates of the LRd and Shannon–Bers SERCA models are similarly sensitive to $[Ca^{2+}]_i$, although the cooperativity of the Shannon–Bers model can be seen in its steeper rise. (C) With fixed $[Ca^{2+}]_i = 100$ nM the initial refill rates of these two models allow similar early SR filling, but the asymptotic bidirectional relationship built in to the Shannon–Bers model more rapidly limits the upper limit. We present here the LRd function with (red) and without (blue) the impact of the leak counterflux.

$$J_{\text{up}} = V_{\max}\frac{[Ca^{2+}]_i}{K_m + [Ca^{2+}]_i}, \tag{3.8}$$

This relationship assumes completely uncooperative binding of the two Ca^{2+} ions, and has the typical hyperbolic relationship between transport rate and substrate concentration (figure 3.5(A)), where $V_{\max}$ is the saturated (maximal) pump velocity, and K_m is half-maximal $[Ca^{2+}]_i$. Importantly, there is no implicit dependency on $[Ca^{2+}]_{\text{SR}}$, which means that this SERCA formulation can continue pumping so long as Ca^{2+} is present in the cytosol. Given enough time this can permit unrealistic values for $[Ca^{2+}]_{\text{SR}}$. This is an important point, because we know that the ATPase activity of SERCA provides the energy to pump Ca^{2+} against the very high concentration gradient across the SR membrane. As such, the thermodynamic limit for the gradient in *free* Ca^{2+} between the SR and cytosol is the free energy of ATP hydrolysis (ΔG_{ATP}). Formally, for net forward transport:

$$-\eta\Delta G_{\text{ATP}} > \Delta G_{\text{SR-cyto}} = 2\text{RT}\,ln\left(\frac{[Ca^{2+}]_{\text{SR}}}{[Ca^{2+}]_i}\right), \tag{3.9}$$

Where η is SERCA mechanical efficiency (~75% [2]). For the LRd SERCA model (equation (3.8)), reuptake is not constrained to this limit. Instead it is balanced by an SR Ca^{2+} leak flux (J_{leak}) at equilibrium SR Ca^{2+} load. Few models have thermodynamically coupled J_{leak} to forward SERCA pumping, for example, two common formulations for the phenomenological SR Ca^{2+} leak are:

$$J_{\text{leak}} = k_{\text{leak}}([Ca^{2+}]_{\text{SR}} - [Ca^{2+}]_i), \tag{3.10}$$

$$J_{\text{leak}} = J_{\text{up}} \frac{[\text{Ca}^{2+}]_{\text{SR}}}{k_{\text{SRmax}}}, \tag{3.11}$$

In the first, J_{leak} simply reflects a passive and constant release mechanism scaled by k_{leak} (mM ms^{-1}), and depending on the SR Ca^{2+} gradient. In the second, J_{leak} approaches J_{up} as $[Ca^{2+}]_{SR}$ approaches a constant maximum concentration (k_{SRmax}), but again k_{SRmax} is not thermodynamically defined. An alternative approach to applying a thermodynamic constraint was developed by Shannon *et al* [74], and sets the net SERCA flux as a combination of the forward and reverse Hill-type processes:

$$J_{\text{up}} = \frac{V_{\max,f}\left(\frac{[\text{Ca}^{2+}]_i}{K_{m,f}}\right)^2 - V_{\max,r}\left(\frac{[\text{Ca}^{2+}]_{\text{SR}}}{K_{m,r}}\right)^2}{1 + \left(\frac{[\text{Ca}^{2+}]_i}{K_{m,f}}\right)^2 + \left(\frac{[\text{Ca}^{2+}]_{\text{SR}}}{K_{m,r}}\right)^2}, \tag{3.12}$$

In this case, because $V_{\max}$ is a function of SERCA number and thought to be independent of direction, $V_{\max,f} = V_{\max,r}$. As a result, the equilibrium condition ($J_{up} = 0$) implies $\frac{[\text{Ca}^{2+}]_{\text{SR}}}{[\text{Ca}^{2+}]_i} = \frac{K_{m,r}}{K_{m,f}}$. Also, note that each transport mode has Hill coefficients of 2 to represent cooperativity in Ca^{2+} binding. For some time there was some debate about the implication of the equilibrium condition in this model, i.e. that most SR Ca^{2+} leak at equilibrium was due to reverse SERCA flux. In recent years it has generally been accepted that most diastolic SR Ca^{2+} leak is due to spontaneous stochastic openings of RyR. In part this is because SR load can be markedly increased by RyR blockade, and in part because we now know much of this spontaneous RyR activity was not visible in earlier datasets. Practically, this means that while the Shannon model (3.12) presents a nicely compact and thermodynamically faithful description of SERCA function, it should be combined with a SERCA-independent SR Ca^{2+} leak to replicate the balance of fluxes determining maximum SR Ca^{2+} content *in vivo*.

All of these formulations only implement SERCA in terms of its dependencies on SR and cytosolic $[Ca^{2+}]$. As mentioned earlier there are other important determinants of SERCA transport. Phospholamban (PLN) is one of the most physiologically important. Its dissociation from SERCA is driven by protein kinase A (PKA) phosphorylation at the Ser16 residue of PLN, which itself is secondary to β-adrenergic stimulation. The outcome for SERCA function is a marked increase in cytosolic Ca^{2+} affinity, which reduces the overall K_m and increases transport for any $[Ca^{2+}]_i$. Fortunately, it has been relatively straightforward to implement this effect in most SERCA models because K_m is a fundamental parameter in all of them (e.g. [72, 83]). Finally, the pH-sensitivity, and to a lesser extent ATP-sensitivity described above causes SERCA transport to be subject to metabolic challenge. While relatively few models have examined the physiologic outcomes of these effects specifically, the detailed Markov scheme presented by Tran *et al* [88] provides a framework for doing so.

Sarcolemmal Ca^{2+} pump (PMCA)

This pump is embedded in the sarcolemmal membrane, and its ATPase activity is used to extrude Ca^{2+} from the cytosol into the extracellular space. Unlike SERCA, PMCA stoichiometry allows only a single Ca^{2+} to be transported for each ATP consumed. Most importantly for our discussion, overall Ca^{2+} transport by PMCA makes a minor contribution to Ca^{2+} removal from the cytosol during a beat (<2%, [2]), and so most cardiac models treat it with a simple Michaelis–Menten scheme:

$$J_{\mathrm{PMCA}} = V_{\max}\frac{[\mathrm{Ca}^{2+}]_i}{K_m + [\mathrm{Ca}^{2+}]_i}, \tag{3.13}$$

Sarcolemmal Na^{+}–Ca^{2+} exchanger (NCX)

NCX is responsible for removing between ~10 and 40% of cytosolic Ca^{2+} during cardiac relaxation (SERCA contributes the vast majority of the remainder). At a constant pacing rate (steady state), the total Ca^{2+} extruded to the extracellular space via NCX in each beat is approximately equal to the amount that enters via I_{CaL}, much in the same way that total SERCA reuptake balances RyR Ca^{2+} release from the SR. The transport process for NCX is a coupled antiport action between 3 Na^{+} ions and a single Ca^{2+} ion, thus it generates current in the direction of Na^{+} transport, and is influenced by membrane potential. At resting potential and under physiologic Na^{+} and Ca^{2+} gradients, that direction is inward and often called forward-mode exchange. The reversal potential can be calculated by the stoichiometry of charge movement and the Nernst equilibrium potentials for the two ion species:

$$E_{\mathrm{NCX}} = 3E_{\mathrm{Na}^{2+}} - 2E_{\mathrm{Ca}^{2+}}, \tag{3.14}$$

Whenever $V_m < E_{\mathrm{NCX}}$ the exchanger will function in forward mode, and vice versa. The velocity of NCX cycling is, of course, dependent on the rapid changes in $[\mathrm{Ca}^{2+}]_i$ occurring during a beat, and slower changes in $[\mathrm{Na}^{+}]_i$. Additionally, there is an important form of allosteric regulation by cytosolic $[\mathrm{Ca}^{2+}]$ that occurs through cooperative binding of Ca^{2+} to at least two high affinity sites [45, 49]. Allosteric inactivation due to $[\mathrm{Na}^{+}]_i$ also occurs, and its inhibition by allosteric Ca^{2+} binding contributes to activation by $[\mathrm{Ca}^{2+}]_i$, but because it is observed at very high $[\mathrm{Na}^{+}]_i$ (> 30 μM) its independent role is not clear [2]. Together, this range of factors creates an extraordinarily complex regulatory scheme, and one which depends on rapid variations in V_m and multiple nonlinear effects of the local Na^{+} and Ca^{2+} concentrations.

Most models of these dynamics have broken I_{NCX} into three components: (1) the maximal exchange velocity under fully activating conditions ($V_{\max}$), (2) allosteric activation by $[Ca^{2+}]_i$ (A_{Ca}), and (3) a thermodynamic component involving the full voltage- and ion-binding-dependencies of the transport mechanism (ΔE):

$$I_{NCX} = V_{max} A_{Ca} \Delta E, \tag{3.15}$$

The thermodynamic component is a complex formulation that describes the intrinsic dependencies of cycling velocity on the ion binding and translocation steps in this bi–bi cyclic ping–pong reaction. Reactions of this type are coupled for two reactants ('bi–bi') but the two components of the reaction occur consecutively ('ping–pong'), rather than through an intermediate state with both reactants bound to the exchanger. The formulation below is from Weber *et al* [91]:

$$\Delta E = \frac{([Na^+]_i^3[Ca^{2+}]_o e^{\eta Vk} - [Na^+]_o^3[Ca^{2+}]_i e^{(\eta-1)Vk}}{\left\{\begin{array}{l} K_{Cao}[Na^+]_i^3 + K_{Nao}^3[Ca^{2+}]_i + K_{Nai}^3[Ca^{2+}]_o\left(1 + \dfrac{[Ca^{2+}]_i}{K_{Cai}}\right) + \\ K_{Cai}[Na^+]_o^3\left(1 + \left(\dfrac{[Na^+]_i}{K_{Nai}}\right)^3\right) + [Na^+]_i^3[Ca^{2+}]_o + [Na^+]_o^3[Ca^{2+}]_i \end{array}\right\} \{1 + k_{sat} e^{(\eta-1)Vk},\}} \tag{3.16}$$

While there are many components to this representation, it can be dissected intuitively. In the numerator the overall velocity is set up as the difference of the rates of exchange in the two directions (outward current–inward). These are scaled by exponential terms that impart the voltage dependence that we have independently calculated in equation (3.14). The parameter η, the partition factor, is borrowed from rate theory as the position of the translocation energy barrier within the membrane, and thus is a measure of how the membrane field impacts charge movement in either direction. It controls the asymmetry of the voltage-dependence for the two exchange reactions. k is the constant F/RT. A separate voltage dependence is also implemented in the denominator (right term) as a phenomenologic control to ensure saturation at very negative membrane potentials. The left term of the denominator collects the combined effects of the various ion-binding affinities on exchange velocity. The K_{mSX} terms are the dissociation constants for binding of ion S in compartment X to the transport sites, and here are assumed to be equivalent to the concentration at half-maximal transport.

The major aspects of the thermodynamic component of I_{NCX} have not been altered much in recent years, but models of allosteric regulation continue to evolve to reflect new experimental findings. The original Weber model (equations (3.15) and (3.16)) accounted for Ca^{2+} allostery with an instantaneous Hill formulation:

$$A_{Ca} = \frac{1}{1 + \left(\dfrac{K_m}{[Ca^{2+}]_i}\right)^n}, \tag{3.17}$$

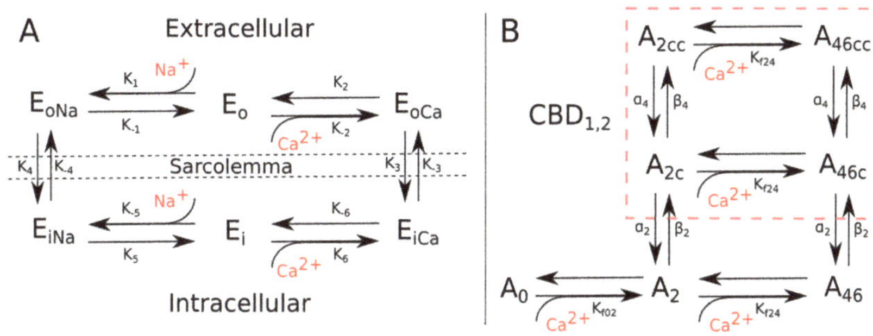

Figure 3.6. A state-transition (Markov) model of the bi–bi cyclic ping–pong NCX reaction, and its modulation by allosteric Ca^{2+} binding at the intracellular binding domains. (A) The thermodynamic module carries Ca^{2+} through coupled antiport, where the parameters K_3, K_{-3}, K_4, and K_{-4} carry the voltage dependence due to energy barrier crossing in each transport direction. The Ca^{2+} and Na^+ binding affinities, and their differences on each side of the membrane are then implemented through mass action dynamics with the remaining rate constants. (B) The allosteric module involves two Ca^{2+} binding steps A_0 to A_2 and A_2 to A_4, each of which can involve Ca^{2+} binding to one of three sites (not shown). Once Ca^{2+} is bound Ca^{2+}-independent transitions allow the model to enter Ca^{2+}-activated states (in red outline). The transitions to and between these states (α_2 and α_4) are fast and slow, respectively, to replicate the two kinetic components of Ca^{2+}-activation.

where K_m and n are the concentration at half maximal allosteric activation and the Hill coefficient, respectively. Both the $[Ca^{2+}]_i$-dependence and kinetics of this activation have been difficult to measure, particularly in intact myocytes. K_m estimates range from several tens to a few hundred nM and activation time constants have varied from 100 ms to 10 s [15]. As mentioned above, allosteric activation was originally found to involve two high-affinity binding sites in the large intracellular loop [45, 49], and chymotrypsin cleavage of this loop has been key to separating the allosteric and thermodynamic sensitivities to $[Ca^{2+}]_i$ [56]. More recent studies involving structural and mutagenic approaches have identified up to six sites in two Ca^{2+} binding domains, and shown that the interaction of these sites determines the properties of allosteric Ca^{2+} activation [4, 11, 62, 63].

Chu *et al* [15] have proposed a detailed Markov model that provides the full kinetic scheme for thermodynamic cycling (figure 3.6(A)), as well as a module for allosteric activation involving interaction within and between $[Ca^{2+}]$-binding domains and incorporating Ca^{2+}-independent steps in activation. This model suggested that allosteric modulation is an important factor for allowing NCX transporters located relatively far from SR Ca^{2+} release sites to maintain high activity and effectively extrude Ca^{2+} through diastole.

3.2.3 Cardiac Ca^{2+} buffers and cytosolic bulk transport

During cardiac ECC the free $[Ca^{2+}]_i$ measured by cytosolic fluorophores reflects only ~1% of the total Ca^{2+} in the cytosol. The remaining 99% is associated with a variety of Ca^{2+}-binding molecules. Most of these are proteins, and many are responsible for transducing the Ca^{2+}signal into some form of molecular control. Myofilament activation and contraction being the most obvious and important. Negatively charged membrane phospholipids are also meaningful contributors [2].

In the SR the most important buffer is calsequestrin. It is essential to have a realistic implementation of this buffer because it greatly influences the releasable SR Ca^{2+} store by preventing lumenal free Ca^{2+} from limiting SERCA transport. In the cytosol, the myofilament binding sites (primarily troponin C) make the largest single contribution, but SERCA, sarcolemmal membrane sites, and calmodulin also absorb meaningful fractions during the cytosolic transient [2].

In this section we present the buffering formalisms and reductions that are common in electrophysiologic models. Of course it should be remembered that many of these buffers (e.g. myofilaments) have complex interactions with the cytosolic Ca^{2+} pool that vary with influences other than time (e.g. contractile force or velocity), and are themselves the topic of large bodies of literature. The basic mass action association and dissociation of a single Ca^{2+} with a buffer (B) is given by:

$$\mathrm{Ca}^{2+}+\mathrm{B} \underset{k_{\mathrm{off}}}{\overset{k_{\mathrm{on}}}{\rightleftharpoons}} \mathrm{CaB}, \tag{3.18}$$

$$\frac{d\mathrm{CaB}}{dt} = k_{\mathrm{on}}[\mathrm{Ca}^{2+}]([\mathrm{B}_{\mathrm{tot}}] - [\mathrm{CaB}]) - k_{\mathrm{off}}[\mathrm{CaB}], \tag{3.19}$$

where k_{on} and k_{off} have their usual meanings, and B_{tot} is the total concentration (B + CaB) of the buffer with those binding rates. For many buffers, the specific values for k_{on} and k_{off} have often been inherited through many model generations and lineages, and been applied disregarding major differences in species and temperature [80]. Wherever possible, these rates should be verified for the model conditions being studied.

Until now, we have described protein-level models (figures 3.2–3.6) and whole-cell models (figure 3.1) with several well-mixed intracellular compartments. For the whole-cell models it is typical to implement equation (3.19) for each of the important buffer species in the relevant compartment. This only introduces a few ordinary differential equations (ODEs) to the model, and so does not markedly alter computation time given that most modern whole-cell ODE models already include at least 20 and up to 100 or more equations. In the following section we begin discussing models that involve detailed spatial representations of $[Ca^{2+}]$, and the role of intracellular $[Ca^{2+}]$ gradients in EC coupling. Of course, these models require partial differential equation (PDE) representations to capture Ca^{2+} transport, and buffers markedly alter the effective or apparent rate of Ca^{2+} diffusion. To account for these effects we consider two reaction–diffusion equations. One for free Ca^{2+} and one for Ca^{2+} bound to each (jth) species of buffer (CaB_j):

$$\frac{\delta \mathrm{Ca}^{2+}}{\delta t} = D_{\mathrm{Ca}}\nabla^2[\mathrm{Ca}^{2+}] + f(\mathrm{Ca}) - f(\mathrm{CaB}), \tag{3.20}$$

$$\frac{\delta \mathrm{CaB}_j}{\delta t} = D_{\mathrm{CaB}_j}\nabla^2[\mathrm{CaB}_j] + f(\mathrm{CaB}_j), \tag{3.21}$$

where D_{Ca} is the isotropic diffusion coefficient for free Ca^{2+}, $DCaB_j$ is the corresponding coefficient for Ca^{2+} bound to the jth buffer, ∇ is the Laplacian, $f(Ca)$ is a reaction term describing the sum of the local Ca^{2+} transporter fluxes (e.g. RyR, SERCA, or I_{CaL}), and $f(CaB)$ represents the sum of the various local buffering reactions $f(CaB_j)$, each given by equation (3.19):

$$f(\mathrm{CaB}) = \sum_j f(\mathrm{CaB}_j). \tag{3.22}$$

Ca^{2+} bound to mobile buffers, such as ATP and fluorophores (3.21), usually diffuses considerably more slowly than free Ca^{2+} ($D_{CaB_j} < D_{Ca}$). For stationary buffers (tethered proteins and lipids), $D_{CaB_j} = 0$, so the dynamics reduce to equation (3.19). It is also worth noting that we have assumed that all buffers are uniformly distributed. This can be tailored by introducing B_{tot} as some function of space, but usually this only applies to stationary buffers with known distributions. Due to computational cost, relatively few spatial cardiac models have implemented all different forms (with corresponding kinetics) of mobile and stationary buffers, although there are some good examples simulating in small domains [37, 46].

An important and common reduction for many spatial models is to assume that the dynamics of buffering occur much more rapidly than diffusion, and thus are at equilibrium with respect to transport of Ca^{2+}. Setting equation (3.19) = 0, gives:

$$\mathrm{CaB} = \frac{B_{tot}[\mathrm{Ca}^{2+}]}{K_d + [\mathrm{Ca}^{2+}]}. \tag{3.23}$$

This is commonly called the rapid buffering approximation, and is applied in electrophysiologic models of all types. It is particularly useful in spatial models of whole cardiac myocytes where it is often only troponin C binding that is simulated in the full kinetic detail of equation (3.19), e.g. [68]. For models in which the buffering is assumed to be both rapid and spatially invariant, the analytic solution for Ca^{2+} is an advection–diffusion relationship [90], although this relatively rarely applied in cardiac models:

$$\begin{aligned}\frac{\delta \mathrm{Ca}^{2+}}{\delta t} &= \frac{D_c + D_b\theta}{1+\theta}\nabla^2[\mathrm{Ca}^{2+}] \\ &\quad - \frac{D_b\theta}{(K_d + [\mathrm{Ca}^{2+}])(1+\theta)}|\nabla[\mathrm{Ca}^{2+}]|^2 + \frac{f(\mathrm{Ca})}{1+\theta}.\end{aligned} \tag{3.24}$$

Depending on the nature of the model being studied, and the degree of detail required for representing intracellular $[Ca^{2+}]$ gradients, these are the various strategies that are applied to represent Ca^{2+} buffering and transport. Next we will discuss how the ion channel and transporter models described above are combined with these models of diffusion and buffering to build models of Ca^{2+} signaling at various spatial scales.

3.3 Multiscale cardiac ECC and Ca^{2+} homeostasis

The macroscopic properties of cardiac ECC (such as CICR [21]) were measurable well before the underlying unitary events (Ca^{2+} sparks) could be visualized and studied [14]. In fact, the structures involved in both cardiac and skeletal muscle ECC were observed via electron microscopy (EM) [17, 95] several decades before those first sparks were recorded. Over the years, models and experimental measures of function have developed together to fill in this middle ground linking macroscopic function and microscopic structure. In this section we will discuss characteristics of models at each scale from single-dyad models of Ca^{2+} sparks, to subcellular Ca^{2+} handling involving hundreds to tens of thousands of dyads, to whole-cell models of cardiac ECC which compress these details into a few representative Ca^{2+} pools. Because we have already covered the transporter models in detail, we will focus here on describing how their spatial arrangement influences ECC properties at the various levels. A key objective is to highlight how behaviors at each level influence those at the other levels, and which dynamics are deemed 'dispensable' when stepping up to larger spatial scales.

3.3.1 Structural details and protein localization

The basic structures and localisation of proteins in cardiac ECC are shown schematically in figure 3.1, but of course the true nanoscale architecture is vastly more intricate. For the purposes of focusing on the structure–function relationships, we will discuss the important structures at three functional scales.

- **The nanoscale: dyads, couplons, clusters and CRUs.** The cardiac dyad is the crucial site of CICR and home to the LCCs and RyRs that coordinate this process. The terms dyad and RyR cluster are *structural* definitions. Dyad refers to the two membrane structures of the TT and juxtaposed 'junctional' region of the SR (jSR), which are separated by the ~12–15 nm cleft. RyR clusters are areas of densely packed RyR tetramers (most often but not always within dyads), and are defined as regions with contiguous RyR antibody labeling in super-resolution light microscopy (SRM) [1, 46]. There are also two commonly-used *functional* definitions. The first is the 'Ca^{2+} release unit' (CRU), which includes the local jSR terminal and the ensemble of RyRs that operate together as an all-or-none unit during a spark. The second is the 'couplon', which includes the CRU, the dyad, and the local LCCs that can trigger a spark. Whichever definition is used, the important aspect here is to understand the size of the dyadic domain and the number of LCCs and RyRs it contains, because these characteristics fundamentally determine CICR. EM tomography [40] and SRM [44, 75] have been used to measure dyadic geometries and the number of RyRs in (dyadic) clusters, respectively. The *average* RyR exists in a dyad of $\sim 5 \times 10^5$ nm^3 volume, and with between ~20 and 45 other RyR tetramers depending on the imaging technique and definition of the dyadic border. Perhaps most importantly, the dimensions other than dyadic width (cleft separation) vary greatly and the size distributions are steep exponentials, with the majority of dyads being very small (< 100 000 nm^3, <8 RyRs) and only a few being very large (> 2×10^6 nm^3, and > 150 RyRs) [1, 40, 46, 75]. The number of LCCs responsible for controlling the average dyadic cleft is poorly known, although LCCs do cluster in TTs [73], and on a whole-cell basis the ratio of LCCs:RyRs is ~ 1:10. Thus 1–10 LCCs is likely reasonable for most dyads. Several other proteins are certainly located very near or within the cleft (e.g. FKBP12.6, calmodulin, CaMKII, PKA), but we caution that this space has become vastly overcrowded by the number of regulators proposed to occupy such a small Ca^{2+} nanodomain.

- **The microscale: t-tubules and sarcomeres.** The next functionally important scale is at the transition between CRUs or couplons. The mean centre-to-centre distance between neighboring jSR terminals in the longitudinal axis closely reflects the length of sarcomeres and spacing of TTs (~1.8 μm), and the variation is skewed [40]. Spacing in the transverse axes is much closer, more variable, and somewhat debated. The highest resolution datasets suggest that the edge-to-edge spacing between axially adjacent jSR terminals is as little as 19nm on average [40], while SRM suggests ~50% of clusters exist within 100 nm of their nearest neighbour and ~1/3 are within 50 nm [1]. These data were all collected in rodents, and it may be that the small inter-dyadic distances along TTs are not representative of spacing in larger mammals, but it is almost certain that they are considerably shorter than the longitudinal edge-to-edge separation. The variability and spacing of CRUs, and particularly of TTs also differ markedly between atria and ventricle [7, 23, 24].
- **The macroscale: whole cell ECC.** So far we have only discussed the structural characteristics of SR Ca^{2+} release sites and their component RyRs and LCCs. The remaining Ca^{2+} transporters and binding proteins present in myocytes must also interact with the [Ca^{2+}] gradients generated during ECC, and their localization dictates those interactions. Like LCCs and RyRs, NCX exhibits a punctate distribution in the myocyte sarcolemma [44], although it is less pronounced than for either of the dyadic channels. While NCX may exist in the dyad it is unlikely to markedly alter CICR, although it is very clear that it senses a higher [Ca^{2+}] than cytosolic fluorophores, at least early in the cytosolic transient. This has been one of the major reasons for introducing a sub-sarcolemmal compartment in a number of whole-cell ECC models (e.g. figure 3.1). SERCA is likely to be a component of the SR membrane very close to the dyad, but its maximal turnover rate is extremely low compared to maximal RyR flux, and it can do relatively little to shape whole-cell SR release, although it may have some involvement in isolated spark kinetics and stability. Instead SERCA's primary role during ECC is to compete with the myofilaments for free cytosolic Ca^{2+}, and so it is generally modeled that SERCA responds to the bulk $[Ca^{2+}]_i$.

3.3.2 Models of Ca^{2+} sparks, couplons, and local control

Before describing important characteristics of spark and couplon models, an important historical note is that these models followed a defining theoretical study of CICR by Michael Stern [84]. This classic paper identified a quantitative necessity for CICR to respond to the Ca^{2+} concentration in the immediate vicinity of RyRs, rather than the bulk cytosolic $[Ca^{2+}]$ (or at least different to the $[Ca^{2+}]$ driving SR reuptake) [84], and laid the foundation for quantitative approaches in local control. In detailing the development of RyR modeling, we have already mentioned many of the most important models and properties of local control, but to reiterate, those key properties are: (1) a stable resting state; (2) strongly regenerative release and high CICR gain; (3) reliable termination of release (at ~10%–20% of the resting local jSR Ca^{2+} load); and (4) adherence to bilayer-measured RyR single channel properties.

The most realistic spark models necessarily involve stochastic operation of a local ensemble of RyR proteins, performed with some form of Monte Carlo formalism. These stochastic differential equations are solved as part of the reaction component, $f(\mathrm{Ca})$, of equation (3.20). An important modifier of Ca^{2+} diffusion in such a small domain, so close to the sarcolemma, is the influence of fields generated by membrane surface charge and transmembrane potential. Soeller and Cannell introduced these effects to the conventional reaction–diffusion formulation to simulate anisotropic intra-dyadic electrodiffusion [82], where diffusion in the radial (r) plane parallel to the membrane is conventional, and electrodiffusion occurs across the cleft in the orthogonal (z) dimension:

$$\frac{\delta \mathrm{Ca}^{2+}}{\delta t} = -\vec{\nabla}\vec{J} + f(\mathrm{Ca}) - f(\mathrm{CaB}), \tag{3.25}$$

where $\vec{J}$ includes the two components (r, z),

$$J_r = -D_r\frac{\delta \mathrm{Ca}^{2+}}{\delta r}, \quad J_z = -D_z\left(\frac{\delta \mathrm{Ca}^{2+}}{\delta z} + 2[\mathrm{Ca}^{2+}]\frac{d\Phi}{dz}\right), \tag{3.26}$$

and Φ represents the dynamic electric field.

Early models of this type involved stereotyped or idealized geometries reflecting what was considered the average dyad (figure 3.7(A)). Because the diffusion in this small domain is extraordinarily rapid, but must still dictate the gating of local RyRs and LCCs, the spatial (0.1 nm in z) and temporal ($\sim 10^{-14}$ s) resolution used for numerical solution were extremely fine, and the simulations predictably expensive.

As detailed above, these early RyR models struggled to observe reliable release termination through exclusively Ca^{2+}-dependent coupling of RyRs. Sobie *et al*, surmised that the critical closure dynamics depend on inter-RyR interactions that are not directly Ca^{2+} dependent. Instead, these interactions occurred due to physical interaction among the densely packed RyRs, with accessory proteins connecting adjacent tetramers. They proposed that mechanically coupled gating could allow RyRs to open and close together in a way that diverges from single channel measurements. The model built to reflect these ideas is known as the 'sticky cluster' [81].

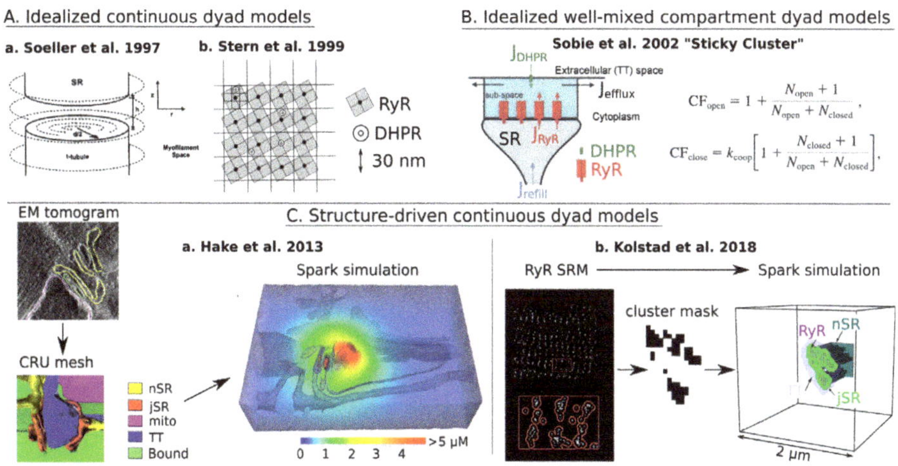

Figure 3.7. Key developments in modeling local control of cardiac CICR. (A) Models developed by Soeller, Cannell, and Stern [9, 82, 85], copyright Rockfeller University Press, were among important early computational studies of dyadic local control, and implemented detailed intradyadic Ca^{2+} transport including electrodiffusive effects and and local buffering. These principles have been incorporated in later models, and been central to defining the characteristics of RyR gating *in vivo*. (B) The 'sticky cluster' model from Sobie *et al* [81] offered a computationally compact dyad model that used phenomenologic coupling between RyRs to stabilize high-gain CICR. As experimental RyR data have refined RyR gating models, this Ca^{2+}-independent coupling has become unnecessary. (C) Recent structure-driven models have explored the importance of realistic variations in structure. On the left we show the model of Hake *et al* [37], which required hand segmentation to reconstruct a single CRU from EM tomography. On the right is an alternative approach based on automated reconstruction of CRU geometries from SRM localization of individual RyRs, and which has allowed a more systematic approach to studying the affect of RyR cluster dissociation in disease [46].

Because it employed a single spatially homogeneous Ca^{2+} concentration in the dyad it was much more computationally efficient, while still remaining phenomenologically reliable. These characteristics have allowed it to become an important base both for questions related to dyadic remodeling in disease [51, 65], and for multiscale extensions [93].

Because modern RyR gatung models can reliably replicate spark dynamics exclusively through Ca^{2+}-dependent coupling, recent models have expanded efforts away from idealized geometries and towards simulating CICR in structures reconstructed from data. The eventual objective being to understand how the very large variation in these structures impacts CRU and couplon function. A first effort in these data-driven approaches applied a simple RyR gating scheme in a very detailed geometry reconstructed from EM tomography [37]. This study asked whether assumptions surrounding the constraints (volumes, SR Ca^{2+} content, detailed buffering, and rate of depletion or refilling) in idealized geometries were reasonable with respect to a real CRU structure. In large part that work supported the earlier assumptions, and these structural questions have since moved on to address the impact of disease-associated degradation of CRUs with fully stochastic models built from SRM datasets [46].

3.3.3 Spatial models of subcellular Ca^{2+} signaling

Extending the principles of spark and couplon modeling up to thousands of CRUs allows us to understand how nanoscale CICR dynamics determine inter-CRU interactions that in turn manifest in whole-cell ECC. These properties both permit the Ca^{2+} handling system some plasticity to respond to physiologic cues (e.g. neurohormonal modulation), but also impose limits to stability, and the resulting instabilities are major contributors to the most important cellular mechanisms of cardiac arrhythmia.

A first important point about these multi-CRU models is that they are forced to condense intra-dyadic dynamics to maintain computational tractability. With $N \sim$ 20 000 CRUs per myocyte and tens of RyRs per CRU, it is not realistic to track stochastic operation of individual RyRs. Instead, it is typical to treat the couplons as stochastic units, within which the RyR dynamics can be calculated deterministically, so long as each local $[Ca^{2+}]$ is known. With these 1000s of local $[Ca^{2+}]$'s it is possible to calculate, for each couplon, the fraction of channels in each modeled channel state by sampling the probability of transition over some time interval (Δt). This can then be multiplied by the number of local channels to give the local Ca^{2+} release flux, and LCC influx. In this way, it becomes possible to reduce the number of ODEs being tracked in each dyad to just two, the local $[Ca^{2+}]_{\text{cleft}}$ and $[Ca^{2+}]_{\text{jSR}}$, which can be formulated in simple terms:

$$\frac{d[Ca^{2+}]^n_{\text{cleft}}}{dt} = \frac{\beta^n_{\text{cleft}}}{\lambda_{\text{cleft}}}(J_{\text{LCC}} + J_{\text{RyR}} - J_{\text{efflux}}), \tag{3.27}$$

$$\frac{d[Ca^{2+}]^n_{\text{jSR}}}{dt} = \frac{\beta^n_{\text{jSR}}}{\lambda_{\text{jSR}}}(J_{\text{refill}} - J_{\text{RyR}}), \tag{3.28}$$

where in the nth CRU, β^n_{jSR} and β^n_{cleft} are the rapid equilibrium buffer fractions in the jSR and cleft, respectively. J_{tr} is the nSR to jSR transfer flux, J_{RyR} is the RyR release flux, J_{LCC} is LCC influx, and diffusion of Ca^{2+} from the cleft to the bulk cytosol is J_{efflux}. The λ terms represent the volume ratios between compartments for mass conservation. This formulation still yields a system in excess of $2N$ ODEs to be solved, and the first models of this type were very costly [33]. One of the key observations carried over from the detailed early dyadic models was that the Ca^{2+} efflux from the dyad is extremely rapid (on the order of 10 μs), and much faster than channel gating. This means that the evolution of each $[Ca^{2+}]^n_{\text{cleft}}$ is history-independent, and can be approximated algebraically:

$$[Ca^{2+}]^n_{\text{cleft}} \sim [Ca^{2+}]_i + \frac{J^n_{\text{LCC}} + J^n_{\text{RyR}}}{J^n_{\text{efflux}}}, \tag{3.29}$$

This approximation was first described by Hinch *et al* [41, 42], and halves the number of ODEs to be solved. As a result it has become a general feature of models of this type, which have been further developed to address questions that span the scale between stochastic operation of single couplons and macroscopic behaviors.

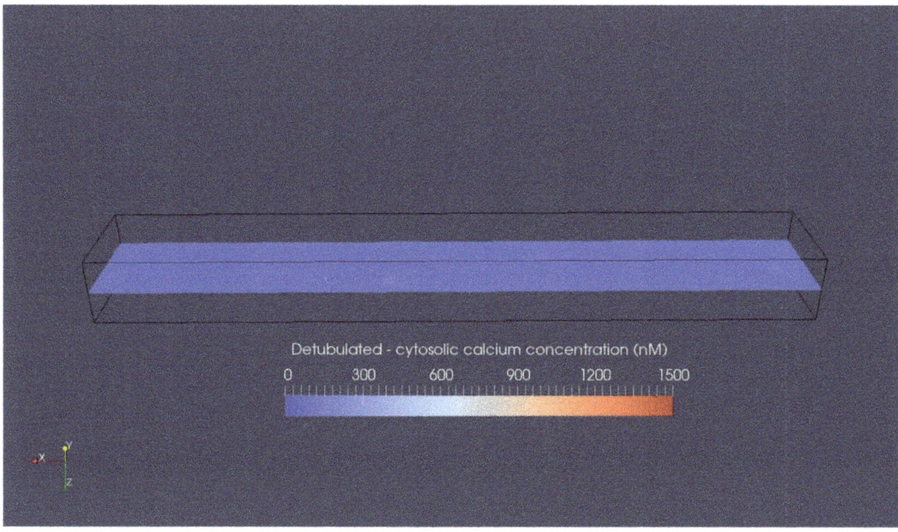

Figure 3.8. Video 1. A cardiac cycle simulated in the Restrepo rabbit ventricular model [68] under action potential clamp. Notice the stochastic occurrence of sparks in the diastolic periods before and after the global Ca^{2+} transient subsides. These represent the probabilistic firing of individual CRUS due to spontaneous RyR openings. The AP then clearly coordinates the synchronous opening of CRU throughout the cell, thus initiating the whole-cell transient in a rapid and uniform manner, as is characteristic of healthy ventricular myocytes of all species. Video available at https://iopscience.iop.org/book/978-0-7503-2064-1.

One of the most widely used and extended of these models is the Restrepo rabbit ventricular myocyte [60, 68, 76]. Simulations of the original Restrepo model driven by a clamped AP waveform are shown in figure 3.8. A modified version of that model (figure 3.9), in which all TTs have been removed to simulate structural characteristics known to occur in the atria, and certain ventricular diseases. In both cases, spontaneous sparks can be observed due to stochastic channel behaviors. During the action potential we observe very different behaviors. In the original model, the interaction of all modeled CRUs with LCCs in the TT membrane allows synchronous, rapid Ca^{2+} release. Conversely, in the detubulated cell, CICR propagates slowly from membrane coupled CRUs at the cell surface towards the interior. This propagation occurs by a fire–diffuse–fire interaction among the CRUs that are decoupled from LCCs, but which can be reached and activated by Ca^{2+} diffusing in from adjacent CRUs. This type of propagation is termed a Ca^{2+} wave, and is also often observed traversing the length of fully tubulated cells that have become overloaded with Ca^{2+}. This is one of the key forms of Ca^{2+} instability.

Separate to the line of models extended from Restrepo *et al*, the recent updates to lumenal and cytosolic Ca^{2+} sensitivity of RyR gating has allowed the hypothetical 'stickyness' of the sticky cluster to be removed. This in turn has been leveraged to develop a new line of idealized but detailed multiscale Ca^{2+} models representing mouse ventricular Ca^{2+} dynamics and electrophysiology [93].

Most recently, models of these subcellular dynamics have moved towards applying serial sectioning scanning EM datasets, which can reconstruct large

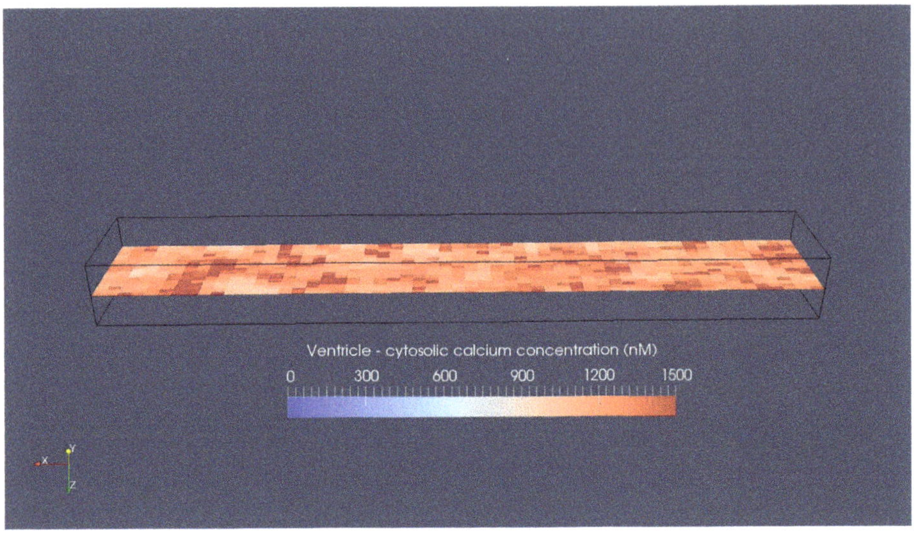

Figure 3.9. Video 2. As in Video 1, a cardiac cycle simulated in AP-clamp conditions in the Restrepo rabbit ventricular model [68]. In this case the simulation involves nearly complete removal of the TT network. Thus, CICR proceeds as a wave from the cell periphery (surface) to its interior. As Ca^{2+} released by CRUs near the cell surfaces diffuses towards the centre of the cell, it triggers internal CRUs along the way. As noted in the text, a similar behavior in which waves traverse the cell in the longitudinal direction occurs spontaneously under conditions of Ca^{2+} overload. Normally individual sparks remain isolated events, but SR Ca^{2+} overload can allow these spontaneous activations to to propagate to adjacent release sites, thereby reaching a critical point in the Ca^{2+} release dynamics that is thought to contribute to arrhythmia. Video available at https://iopscience.iop.org/book/978-0-7503-2064-1.

volumes of myocyte membrane structures at very high resolution (~15 nm) [64]. The modeling strategies that can leverage the simultaneous resolution and scale offered by these datasets are still being established, but initial efforts suggest that hybrid approaches are applicable (figure 3.10(B)) [16].

3.3.4 Zero-dimensional models of myocyte Ca^{2+} handling

Most of the commonly-used whole-cell models of myocyte electrophysiology and Ca^{2+}-handling are often called zero-dimensional (0D) because they exclusively involve time-varying ODEs to represent transport and diffusion of Ca^{2+} between several well-mixed intracellular compartments (e.g. figure 3.1). Each of these compartments represents the spatially-averaged (but time-varying) concentration that acts upon the Ca^{2+}-sensitive transporters in that compartment. The focus of this family of models is to reflect the interaction of Ca^{2+} with these transporters and voltage dynamics in a way that simultaneously permits efficient computation and accurate coupling between the AP and Ca^{2+} handling. This in turn allows the modeler to ask questions about dynamics that require longer simulation time-courses, and permits them to be incorporated into tissue electrophysiology and mechanics simulations. Of course, in moving up in scale some of the important considerations of the lower scales are left behind, as they were in moving from spark

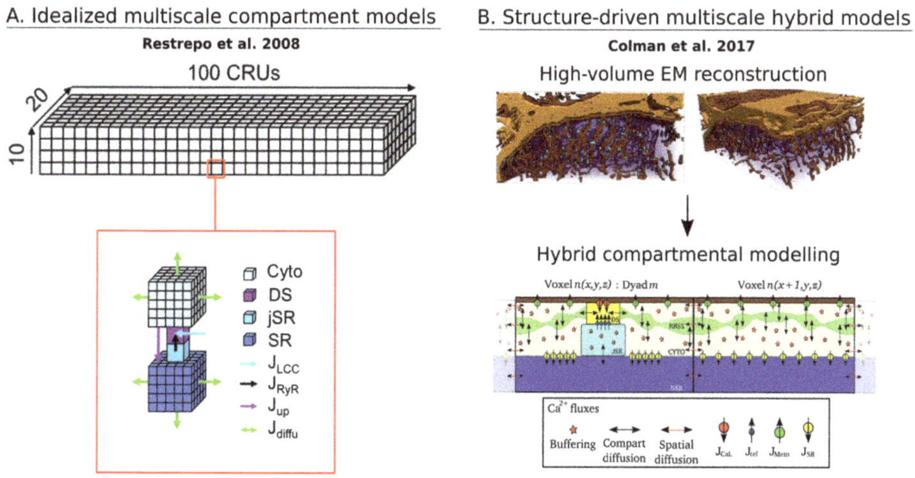

Figure 3.10. Multiscale models of CICR and EC coupling. Popular multiscale models of ECC approximate CRU structure and diffusion dynamics to permit tractable computation in a whole myocyte. The model from Restrepo *et al*, was one of the first of this type, and it separates each CRU into cytosolic (Cyto), dyadic (DS), jSR, and network SR (SR) volumes, coupled by fluxes due to LCCs (J_{LCC}), RyRs (J_{RyR}), SERCA (J_{up}), and compartmental diffusion (J_{diffu}) [60, 68]. The most recent multiscale approach from Colman *et al*, reconstructed CRUS from a serially-sectioned (block-face) scanning EM dataset [16]. Each CRU contained cytosolic, SR, dyadic, and jSR compartments, and they applied a mixture of compartmental and continuous diffusion, and incorporated a phenomenologic restricted buffering domain (RBSS) to allow rapid diffusion between CRUs. This model constitutes a first step towards developing hybrid methods that will be useful for simulating in geometries arising from these new imaging methods, which are both very large and very detailed.

to subcellular models. For example, virtually all 0D models ignore concerns about phenomenologic termination of SR Ca^{2+} release, because phenomenological mechanisms are much more stable, and markedly improve model utility [5, 6, 52, 74]. In this section we will focus on highlighting the dynamic characteristics of whole-cell Ca^{2+} models that often define their utility for simulations requiring the more compact 0D framework.

- **Ca^{2+} flux balance.** As mentioned earlier, during steady-state pacing SR Ca^{2+} reuptake by SERCA is equal and opposite to RyR Ca^{2+} release, and NCX-mediated Ca^{2+} extrusion is balanced by I_{CaL} influx. These relationships are fundamental to cardiac myocyte Ca^{2+} dynamics and vary with species and physiologic conditions. They set a key foundation for many aspects of whole-cell Ca^{2+} homeostasis [20]. When perturbations are applied, for example by altering pacing frequency, autonomic control, or pharmacologic intervention, it is important for the model to both reach a correct new steady-state and do so over an approximately correct time course. Changes in the expression or function of SERCA, NCX, LCCs and RyR alter the characteristics of both the eventual steady-state and the kinetics of the response.
- **Ca^{2+}-dependent autoregulation.** One of the most compact demonstrations of the role that Ca^{2+} flux balance plays in cardiac ECC is a dynamic property named Ca^{2+}-dependent autoregulation. The principle of autoregulation states

that acute changes that alter SR Ca^{2+} release *without* changing the fundamental balance relationships, will be dynamically compensated, such that Ca^{2+} release and the cytosolic Ca^{2+}transient amplitude return to their level prior to the perturbation [20]. This property has been studied by sensitizing RyRs to cleft Ca^{2+} with low dose caffeine. The result is an immediate increase in Ca^{2+} release, and Ca^{2+} transient amplitude. However, that large Ca^{2+} release event reduces I_{CaL} influx by inactivating LCCs, promotes NCX extrusion during diastole, and thereby reduces SR Ca^{2+} load at the subsequent beat. This in turn depresses the subsequent release and over a few cycles returns Ca^{2+} release to the pre-caffeine level, even though the new steady state SR Ca^{2+} load is reduced. The reverse process occurs if RyR is partially blocked.

- **Ca^{2+} frequency-dependence.** While there are many factors that influence the frequency-dependence of myocyte Ca^{2+} cycling, the most important rely on how changes in frequency modify Ca^{2+} flux balance. To understand these dynamic relationships there are two important preliminary facts. First, as pacing frequency increases the duration of the AP decreases (steady-state AP restitution), but if the system is stable the AP duration (APD) does not decrease as much as the diastolic interval. Second, NCX-mediated Ca^{2+} extrusion occurs at negative potentials, whereas I_{CaL} influx occurs maximally near 0 mV. Thus, there is a phase-selectivity such that Ca^{2+} entry is favored during the AP and extrusion during the diastolic interval. When the diastolic interval shrinks to a greater extent than the AP shortens, Ca^{2+} influx is favored, and the new steady state exhibits higher whole-cell and SR Ca^{2+} load. This is a characteristic frequency-dependent effect in the heart from rodents through to large mammals. In large mammals this increase in SR load is the basis for the characteristic positive force- and Ca^{2+}-frequency relationships, whereas in rodents it is balanced (probably by refractory RyRs) to result in relatively flat relationships for both.
- **Neurohormonal and phosphoregulation.** Over the last fifteen years, whole-cell models have been extended from their electrophysiologic and ionic roots to incorporate regulation by intracellular signaling cascades, particularly phosphoregulation. The best established of these models include β-adrenergic [72, 83] and CaMKII-dependent regulation, which itself is Ca^{2+}-dependent [38, 39, 71]. These two kinases again alter whole-cell Ca^{2+} balance on a seconds-to-minutes time scale, and together with the independent and interactive effects of frequency, create a rich set of frequency-dependent characteristics in many of these models. In some cases, CaMKII regulation is a major feature of the frequency-dependence ([61]).

Given this range of important macroscopic properties of cardiac Ca^{2+} handling, how does the modeler ensure that they are incorporated into their 0D model? There are two basic answers: (1) by inheriting many of the properties from a parent model which has previously been rigorously validated; and (2) by optimizing the model parameters to a broad range of data sources specific to the species, disease condition, and questions being investigated. Of course there are mutual exclusivities in such an optimization process, and often there are unforeseen outcomes when the optimization goals are so broad. Sensible and physically justified constraints, as described throughout this chapter, are *critical*. The Shannon–Bers rabbit myocyte was one of the earliest models to successfully combine relatively realistic SR Ca^{2+} release, with these whole-cell Ca^{2+} balance considerations. The foundation of that model has been inherited and extended by a long list of subsequent models incorporating species- [58], disease- [31], tissue- [32], and phosphoregulatory-specific [83] details. There are of course many other model lineages worth mentioning, for example those beginning with the LRd guinea-pig [52] and Winslow canine models [96].

3.3.5 Emergent instabilities in multiscale cardiac Ca^{2+} signaling

One of the major goals of modeling in cardiac electrophysiology is to capture basic physiologic mechanisms with sufficient accuracy to predict emergent properties, particularly the emergent instabilities that underlie arrhythmia. In the case of Ca^{2+}-driven arrhythmia, these instabilities can be observed at the cellular level. This is because Ca^{2+} is not transferred as readily or rapidly between cells as other charge carriers, so the essential arrhythmogenic behavior generally does not rely on tissue-level dynamics, such as reentry. We will discuss two arrhythmogenic cellular behaviors thought to involve major contributions from unstable Ca^{2+} cycling: (1) Ca^{2+} waves and delayed afterdepolarizations (DADs), and (2) Ca^{2+} transient amplitude and APD alternans.

- **Ca^{2+} waves and DADs.** We have already seen (in Video 1, figure 3.8) the basic dynamics of Ca^{2+} waves, where activation of Ca^{2+} release at a CRU (or local collection of CRUs) is large enough to propagate in a fire–diffuse–fire regime through the cell. In that video, the wave is triggered at the surface membrane by depolarization, and propagates towards the centre of the cell. When cardiac myocytes become overloaded by Ca^{2+}, the spontaneous sparks that occur anywhere in the cell due to random RyR openings no longer remain localized events, and instead they spread via this propagation mechanism throughout the cell. Because this occurs in diastole when the cell is at resting membrane potential, the large spontaneous Ca^{2+} wave can drive inward I_{NCX} and depolarize the sarcolemma, a DAD.

  Ca^{2+} wave initiation depends on the factors that promote transition from sparks to a propagating wave (spark-wave criticality). Experimental studies established quite early a central role for SR Ca^{2+} content in this criticality. Intuitively, the parameters that most influence this transition are those that influence Ca^{2+} flux balance, steady state $[Ca^{2+}]_{SR}$ (SERCA V_{max} and Ca^{2+} affinity, and NCX V_{max}), and SR Ca^{2+} release flux (k_{rel}) [87, 89]. The interpretation is that increasing $[Ca^{2+}]_{SR}$, or increasing CRU release flux through some other method, allows the spatial profile of the Ca^{2+} spark to become broader and overcome the normally isolating diffusion distance between CRUs. Nivala *et al*, applied the Restrepo model to formalize and extend this intuitive explanation, and showed that by increasing the effective coupling between CRUs, for example by reducing intracellular buffering or increasing RyR release flux, the distribution of spark sizes changes from an exponential to a power rule [60]. This has important implications in that it describes how Ca^{2+} wave criticality is just one critical point the dynamical space. At high enough $[Ca^{2+}]_{SR}$, or under other conditions of strong coupling, the Ca^{2+} cycling system becomes an oscillator and can drive autorhythmicity [47].

- **Alternans.** Cellular cardiac alternans is a period-2 instability, in which amplitude of the Ca^{2+} transient and the duration of the AP increase and then decrease in alternation. This alternation can be in-phase (long APD = large Ca^{2+} transient), called temporally concordant alternans, or out of phase (temporally discordant alternans). The dynamic basis for this behavior is complex, and involves multiple dynamical regimes for which either unstable Ca^{2+} cycling or unstable AP dynamics can act as the driving mechanisms. All of these mechanisms revolve around the bidirectional coupling between $[Ca^{2+}]_i$ and membrane voltage V_m. While a full treatment of these dynamics would require a separate chapter, it is increasingly well accepted that Ca^{2+} instabilities probably underlie the majority of alternans in large mammals. We will give a cursory explanation of the core principles and have included several important references for interested readers.

  At the whole-cell level, both AP and Ca^{2+} driven alternans rely on the steepness with which each variable (APD and Ca^{2+} transient amplitude) changes with respect to a coupled variable. In the case of the AP-driven alternans, the coupled variable is the diastolic interval (DI). If a random perturbation causes APD to lengthen, the subsequent DI becomes shorter (ΔDI), and as described above, when DI becomes shorter so does the subsequent APD (ΔAPD) due to APD restitution. If $\Delta APD > \Delta DI$ (i.e. APD restitution > 1), alternans will ensue. A similar relationship occurs for Ca^{2+} transient amplitude (CaT) and SR Ca^{2+} load ($[Ca^{2+}]_{SR}$), where the interplay occurs between the quantity of Ca^{2+} reuptaken by SERCA on each beat and the steepness by which RyR release depends on $[Ca^{2+}]_{SR}$ [78, 92].

  As the unique dynamics of alternans have been further explored, the role for subcellular heterogeneity in Ca^{2+} handling has become a feature of the discussion. One of the most interesting sequences of studies began with theoretical work suggesting that regional subcellular Ca^{2+} alternans could arise from a Turing instability, which itself results from the very different rates at which Ca^{2+} diffuses and voltage propagates in cells [77]. This prediction has since been supported by recordings where an alternans control pacing scheme was imposed upon a myocyte undergoing whole-cell alternans, and this uncovered the underlying subcellular spatial alternation in Ca^{2+} release (see [25]).

3.4 The future of models in cardiac Ca^{2+} signaling

The continued expansion of computational power has opened new theoretical approaches to understanding cardiac Ca^{2+} signaling. However, unlike many other disciplines where heavily parallelized architectures and machine learning are at the forefront of those approaches, advances in structural biology are at least as important in shaping current perspectives of the future of cardiac Ca^{2+} research. Recent developments in cryo-electron microscopy are already providing new insights to long held questions in Ca^{2+} biology [13, 101]. Serial sectioning scanning EM is providing a basis for developing multiscale structure-function models in a way that was not foreseeable 10 years ago. Super resolution fluorescence approaches continue to improve rapidly [50, 53]. These techniques are permitting basic scientists to vigorously dissect the relationships between protein localisation and regulation, and will require sophisticated new mathematical approaches and computational pipelines.

Acknowledgements

We wish to thank Xianwei Zhang for his careful help in preparing figures and videos, and Drs Stefano Morotti and Haibo Ni for their critical reading of the manuscript. This work was supported by the National Institutes of Health grant R01-HL131517 and the American Heart Association grant 15SDG24910015 (EG)

References

[1] Baddeley D, Jayasinghe I D, Lam L, Rossberger S, Cannell M B and Soeller C 2009 Optical single-channel resolution imaging of the Ryanodine receptor distribution in rat cardiac myocytes *Proc. Natl Acad. Sci.* **106** 1–6

[2] Bers D M 2001 *Excitation-Contraction Coupling and Cardiac Contractile Force* (Berlin: Springer)

[3] Bers D M 2008 Calcium cycling and signaling in cardiac myocytes *Annu. Rev. Physiol.* **70** 23–49

[4] Besserer G M, Ottolia M, Nicoll D A, Chaptal V, Cascio D, Philipson K D and Abramson J 2007 The second Ca2+-binding domain of the Na^+Ca^{2+} exchanger is essential for regulation: Crystal structures and mutational analysis *Proc. Natl Acad. Sci.* **104** 18467–72

[5] Bondarenko V E, Bett G C L and Rasmusson R L 2004 A model of graded calcium release and L-type Ca^{2+} channel inactivation in cardiac muscle *Am. J. Physiol.-Heart Circ. Physiol.* **286** H1154–69

[6] Bondarenko V E and Rasmusson R L 2010 Transmural heterogeneity of repolarization and Ca^{2+} handling in a model of mouse ventricular tissue *Am. J. Physiol.-Heart Circ. Physiol.* **299** H454–69

[7] Brandenburg S *et al* 2018 Axial tubule junctions activate atrial Ca^{2+} release across species *Front. Physiol.* **9** 1227

[8] Cannell M B, Kong C H T, Imtiaz M S and Laver D R 2013 Control of sarcoplasmic reticulum Ca^{2+} release by stochastic RyR gating within a 3D model of the cardiac dyad and importance of induction decay for CICR termination *Biophys. J.* **104** 2149–59

[9] Cannell M B and Soeller C 1997 Numerical analysis of ryanodine receptor activation by L-type channel activity in the cardiac muscle diad *Biophys. J.* **73** 112–22

[10] Carafoli E and Krebs J 2016 Why calcium? How calcium became the best communicator *J. Biol. Chem.* **291** 20849–57

[11] Chaptal V, Ottolia M, Mercado-Besserer G, Nicoll D A, Philipson K D and Abramson J 2009 Structure and functional analysis of a Ca^{2+} sensor mutant of the Na^{+}/Ca^{2+} exchanger *J. Biol. Chem.* **284** 14688–92

[12] Chen H, Valle G, Furlan S, Nani A, Gyorke S, Fill M and Volpe P 2013 Mechanism of calsequestrin regulation of single cardiac ryanodine receptor in normal and pathological conditions *J. Gen. Physiol.* **142** 127–36

[13] Chen W *et al* 2014 The ryanodine receptor store-sensing gate controls Ca^{2+} waves and Ca^{2+}-triggered arrhythmias *Nat. Med.* **20** nm.3440

[14] Cheng H, Lederer W J and Cannell M B 1993 Calcium sparks: elementary events underlying excitation-contraction coupling in heart muscle *Science* **262** 740–4

[15] Chu L, Greenstein J L and Winslow R L 2016 Modeling Na^{+}–Ca^{2+} exchange in the heart: allosteric activation, spatial localization, sparks and excitation-contraction coupling *J. Mol. Cell. Cardiol.* **99** 174–87

[16] Colman M A, Pinali C, Trafford A W, Zhang H and Kitmitto A 2017 A computational model of spatio-temporal cardiac intracellular calcium handling with realistic structure and spatial flux distribution from sarcoplasmic reticulum and t-tubule reconstructions *PLoS Comput. Biol.* **13** e1005714

[17] Constantin L L, Franzini-Armstrong C and Podolsky R J 1965 Localization of calcium-accumulating structures in striated muscle fibers *Science* **147** 158–60

[18] Dixon R E, Moreno C M, Yuan C, Opitz-Araya X, Binder M D, Navedo M F and Santana L F 2015 Graded Ca^{2+}/calmodulin-dependent coupling of voltage-gated CaV1.2 channels *eLife* **4** e05608

[19] Dzhura I, Wu Y, Colbran R J, Balser J R and Anderson M E 2000 Calmodulin kinase determines calcium-dependent facilitation of L-type calcium channels *Nat. Cell Biol.* **2** 173–7

[20] Eisner D A, Caldwell J L, Kistams K and Trafford A W 2017 Calcium and excitation-contraction coupling in the heart *Circ. Res.* **121** 181–95

[21] Fabiato A and Fabiato F 1975 Contractions induced by a calcium-triggered release of calcium from the sarcoplasmic reticulum of single skinned cardiac cells *J. Physiol.* **249** 469–95

[22] Fill M and Gillespie D 2018 Ryanodine receptor open times are determined in the closed state *Biophys. J.* **115** 1160–5

[23] Frisk M, Koivumki J T, Norseng P A, Maleckar M M, Sejersted O M and Louch W E 2014 Variable t-tubule organization and Ca^{2+} homeostasis across the atria *Am. J. Physiol.-Heart Circ. Physiol.* **307** H609–20

[24] Frisk M, Sejersted O M and Louch W E 2012 Variable T-tubule density and organization in rat atrial cardiomyocytes *Biophys. J.* **102** 139a

[25] Gaeta S A, Bub G, Abbott G W and Christini D J 2009 Dynamical mechanism for subcellular alternans in cardiac myocytes *Circ. Res.* **105** 335–42

[26] Gaur N and Rudy Y 2011 Multiscale modeling of calcium cycling in cardiac ventricular myocyte: macroscopic consequences of microscopic dyadic function *Biophys. J.* **100** 2904–12

[27] Gillespie D, Chen H and Fill M 2012 Is ryanodine receptor a calcium or magnesium channel? Roles of K^{+} and Mg^{2+} during Ca^{2+} release *Cell Calcium* **51** 427–33

[28] Gillespie D and Fill M 2008 Intracellular calcium release channels mediate their own countercurrent: the ryanodine receptor case study *Biophys. J.* **95** 3706–14

[29] Gillespie D and Fill M 2013 Pernicious attrition and inter-RyR2 CICR current control in cardiac muscle *J. Mol. Cell. Cardiol.* **58** 53–8

[30] Gorski P A, Ceholski D K and Hajjar R J 2015 Altered myocardial calcium cycling and energetics in heart failure—a rational approach for disease treatment *Cell Metab.* **21** 183–94

[31] Grandi E, Morotti S, Ginsburg K S, Severi S and Bers D M 2010 Interplay of voltage and Ca-dependent inactivation of L-type Ca current *Prog. Biophys. Mol. Biol.* **103** 44–50

[32] Grandi E, Pandit S V, Voigt N, Workman A J, Dobrev D, Jalife J and Bers D M 2011 Human atrial action potential and Ca^{2+} model *Circ. Res.* **109** 1055–66

[33] Greenstein J L and Winslow R L 2002 An integrative model of the cardiac ventricular myocyte incorporating local control of Ca^{2+} release *Biophys. J.* **83** 2918–45

[34] Guo T, Gillespie D and Fill M 2012 Ryanodine receptor current amplitude controls Ca^{2+} sparks in cardiac muscle *Circ. Res.* **111** 28–36

[35] Györke S and Fill M 1993 Ryanodine receptor adaptation: control mechanism of Ca(2+)-induced Ca2+ release in heart *Science* **260** 807–9

[36] Györke S 1999 Ca^{2+} spark termination inactivation and adaptation may be manifestations of the same mechanism *J. Gen. Physiol.* **114** 163–6

[37] Hake J, Edwards A G, Yu Z, Kekenes-Huskey P M, Michailova A P, McCammon J A, Holst M J, Hoshijima M and McCulloch A D 2012 Modelling cardiac calcium sparks in a three-dimensional reconstruction of a calcium release unit *J. Physiol.* **590** 4403–22

[38] Hashambhoy Y L, Greenstein J L and Winslow R L 2010 Role of CaMKII in RyR leak, EC coupling and action potential duration: a computational model *J. Mol. Cell. Cardiol.* **49** 617–24

[39] Hashambhoy Y L, Winslow R L and Greenstein J L 2009 CaMKII-induced shift in modal gating explains L-type Ca^{2+} current facilitation: a modeling study *Biophys. J.* **96** 1770–85

[40] Hayashi T, Martone M E, Yu Z, Thor A, Doi M, Holst M J, Ellisman M H and Hoshijima M 2009 Three-dimensional electron microscopy reveals new details of membrane systems for Ca^{2+} signaling in the heart *J. Cell Sci.* **122** 1005–13

[41] Hinch R 2004 A mathematical analysis of the generation and termination of calcium sparks *Biophys. J.* **86** 1293–307

[42] Hinch R, Greenstein J L, Tanskanen A J, Xu L and Winslow R L 2004 A simplified local control model of calcium-induced calcium release in cardiac ventricular myocytes *Biophys. J.* **87** 3723–36

[43] Jafri M S, Rice J J and Winslow R L 1998 Cardiac Ca^{2+} dynamics: the roles of Ryanodine receptor adaptation and sarcoplasmic reticulum load *Biophys. J.* **74** 1149–68

[44] Jayasinghe I D, Baddeley D, Kong C H T, Wehrens X H T, Cannell M B and Soeller C 2012 Nanoscale organization of Junctophilin-2 and Ryanodine receptors within peripheral couplings of rat ventricular cardiomyocytes *Biophys. J.* **102** L19–21

[45] Kimura J, Noma A and Irisawa H 1986 Na–Ca exchange current in mammalian heart cells *Nature* **319** 596–7

[46] Kolstad T R *et al* 2018 Ryanodine receptor dispersion disrupts Ca2+ release in failing cardiac myocytes *eLife* **7** e39427

[47] Lakatta E G, Maltsev V A and Vinogradova T M 2010 A coupled SYSTEM of intracellular Ca^{2+} clocks and surface membrane voltage clocks controls the timekeeping mechanism of the hearts pacemaker *Circ. Res.* **106** 659–73

[48] Laver D R, Kong C H T, Imtiaz M S and Cannell M B 2013 Termination of calcium-induced calcium release by induction decay: an emergent property of stochastic channel gating and molecular scale architecture *J. Mol. Cell. Cardiol.* **54** 98–100

[49] Levitsky D O, Fraysse B, Leoty C, Nicoll D A and Philipson K D 1996 Cooperative interaction between Ca^{2+} binding sites in the hydrophilic loop of the Na^{+}–Ca^{2+} exchanger *Mol. Cell. Biochem.* **160–161** 27–32

[50] Lin R, Clowsley A H, Lutz T, Baddeley D and Soeller C 2019 3D super-resolution microscopy performance and quantitative analysis assessment using DNA-PAINT and DNA origami test samples *Methods* **174** 56–71

[51] Louch W E, Hake J, Mrk H K, Hougen K, Skrbic B, Ursu D, Tnnessen T, Sjaastad I and Sejersted O M 2013 Slow Ca(2+) sparks de-synchronize Ca(2+) release in failing cardiomyocytes: Evidence for altered configuration of Ca(2+) release units? *J. Mol. Cell. Cardiol.* **58** 41–52

[52] Luo C H and Rudy Y 1994 A dynamic model of the cardiac ventricular action potential. I. Simulations of ionic currents and concentration changes *Circ. Res.* **74** 1071–96

[53] Lutz T, Clowsley A H, Lin R, Pagliara S, Di Michele L and Soeller C 2018 Versatile multiplexed super-resolution imaging of nanostructures by Quencher-Exchange-PAINT *Nano Res.* **11** 6141–54

[54] Mahajan A *et al* 2008 A rabbit ventricular action potential model replicating cardiac dynamics at rapid heart rates *Biophys. J.* **94** 392–410

[55] Marks A R 2003 Calcium and the heart: a question of life and death *J. Clin. Invest.* **111** 597–600

[56] Matsuoka S and Hilgemann D W 1992 Steady-state and dynamic properties of cardiac sodium-calcium exchange. Ion and voltage dependencies of the transport cycle *J. Gen. Physiol.* **100** 963–1001

[57] Mintz E, Mata A M, Forge V, Passafiume M and Guillain F 1995 The modulation of Ca^{2+} binding to sarcoplasmic reticulum ATPase by ATP analogues is pH-dependent *J. Biol. Chem.* **270** 27160–4

[58] Morotti S, Edwards A G, McCulloch A D, Bers D M and Grandi E 2014 A novel computational model of mouse myocyte electrophysiology to assess the synergy between Na^{+} loading and CaMKII *J. Physiol.* **592** 1181–97

[59] Morotti S, Grandi E, Summa A, Ginsburg K S and Bers D M 2012 Theoretical study of L-type Ca^{2+} current inactivation kinetics during action potential repolarization and early afterdepolarizations *J. Physiol.* **590** 4465–81

[60] Nivala M, Ko C Y, Nivala M, Weiss J N and Qu Z 2012 Criticality in intracellular calcium signaling in cardiac myocytes *Biophys. J.* **102** 2433–42

[61] O'Hara T, Virg L, Varro A and Rudy Y 2011 Simulation of the undiseased human cardiac ventricular action potential: model formulation and experimental validation *PLoS Comput. Biol.* **7** e1002061

[62] Ottolia M, Nicoll D A, John S and Philipson K 2010 Interactions between Ca^{2+} binding domains of the Na^{+}–Ca^{2+} exchanger and secondary regulation *Channels* **4** 159–62

[63] Ottolia M, Nicoll D A and Philipson K D 2009 Roles of Two Ca^{2+}-binding domains in regulation of the cardiac Na^{+}–Ca^{2+} exchanger *J. Biol. Chem.* **284** 32735–41

[64] Pinali C, Bennett H, Davenport J B, Trafford A W and Kitmitto A 2013 Three-dimensional reconstruction of cardiac sarcoplasmic reticulum reveals a continuous network linking transverse-tubules: this organization is perturbed in heart failure *Circ. Res.* **113** 1219–30

[65] Polkov E and Sobie E A 2013 Alterations in T-tubule and dyad structure in heart disease: challenges and opportunities for computational analyses *Cardiovasc. Res.* **98** 233–9

[66] Porter G A, Makuck R F and Rivkees S A 2003 Intracellular calcium plays an essential role in cardiac development *Dev. Dyn.* **227** 280–90

[67] Qin J, Valle G, Nani A, Nori A, Rizzi N, Priori S G, Volpe P and Fill M 2008 Luminal Ca^{2+} regulation of single cardiac Ryanodine receptors: insights provided by calsequestrin and its mutants *J. Gen. Physiol.* **131** 325–34

[68] Restrepo J G, Weiss J N and Karma A 2008 Calsequestrin-mediated mechanism for cellular calcium transient alternans *Biophys. J.* **95** 3767–89

[69] Saftenku E, Williams A J and Sitsapesan R 2001 Markovian models of low and high activity levels of cardiac Ryanodine receptors *Biophys. J.* **80** 2727–41

[70] Sato D, Dixon R E, Santana L F and Navedo M F 2018 A model for cooperative gating of L-type Ca^{2+} channels and its effects on cardiac alternans dynamics *PLoS Comput. Biol.* **14** e1005906

[71] Saucerman J J and Bers D M 2008 Calmodulin mediates differential sensitivity of CaMKII and calcineurin to local Ca^{2+} in cardiac myocytes *Biophys. J.* **95** 4597–612

[72] Saucerman J J, Brunton L L, Michailova A P and McCulloch A D 2003 Modeling β-adrenergic control of cardiac myocyte contractility in silico *J. Biol. Chem.* **278** 47997–8003

[73] Scriven D R, Dan P and Moore E D 2000 Distribution of proteins implicated in excitation-contraction coupling in rat ventricular myocytes *Biophys. J.* **79** 2682–91

[74] Shannon T R, Wang F, Puglisi J, Weber C and Bers D M 2004 A mathematical treatment of integrated Ca dynamics within the ventricular myocyte *Biophys. J.* **87** 3351–71

[75] Shen X *et al* 2019 3D dSTORM imaging reveals novel detail of Ryanodine receptor localization in rat cardiac myocytes *J. Physiol.* **597** 399–418

[76] Shiferaw Y, Aistrup G L and Wasserstrom J A 2017 Mechanism for triggered waves in atrial myocytes *Biophys. J.* **113** 656–70

[77] Shiferaw Y and Karma A 2006 Turing instability mediated by voltage and calcium diffusion in paced cardiac cells *Proc. Natl Acad. Sci. USA* **103** 5670–5

[78] Shiferaw Y, Sato D and Karma A 2005 Coupled dynamics of voltage and calcium in paced cardiac cells *Phys. Rev.* E **71** 021903

[79] Sitsapesan R, Williams A J and Williams A J 1998 *The Structure and Function of Ryanodine Receptors* (Singapore: World Scientific)

[80] Smith N P, Crampin E J, Niederer S A, Bassingthwaighte J B and Beard D A 2007 Computational biology of cardiac myocytes: proposed standards for the physiome *J. Exp. Biol.* **210** 1576–83

[81] Sobie E A, Dilly K W, dos Santos Cruz J, Lederer W J and Jafri M S 2002 Termination of cardiac Ca(2+) sparks: an investigative mathematical model of calcium-induced calcium release *Biophys. J.* **83** 59–78

[82] Soeller C and Cannell M B 1997 Numerical simulation of local calcium movements during L-type calcium channel gating in the cardiac diad *Biophys. J.* **73** 97–111

[83] Soltis A R and Saucerman J J 2010 Synergy between CaMKII substrates and β-adrenergic signaling inregulation of cardiac myocyte Ca^{2+} handling *Biophys. J.* **99** 2038–47

[84] Stern M D 1992 Theory of excitation-contraction coupling in cardiac muscle *Biophys. J.* **63** 497–517

[85] Stern M D, Song L S, Cheng H, Sham J S, Yang H T, Boheler K R and Ros E 1999 Local control models of cardiac excitation-contraction coupling. A possible role for allosteric interactions between Ryanodine receptors *J. Gen. Physiol.* **113** 469–89

[86] Stern M D and Cheng H 2004 Putting out the fire: what terminates calcium-induced calcium release in cardiac muscle? *Cell Calcium* **35** 591–601

[87] Swietach P, Spitzer K W and Vaughan-Jones R D 2010 Modeling calcium waves in cardiac myocytes: importance of calcium diffusion *Front. Biosci.* **15** 661–80

[88] Tran K, Smith N P, Loiselle D S and Crampin E J 2009 A thermodynamic model of the cardiac sarcoplasmic/endoplasmic Ca(2+) (SERCA) pump *Biophys. J.* **96** 2029–42

[89] Tveito A, Lines G T, Hake J and Edwards A G 2012 Instabilities of the resting state in a mathematical model of calcium handling in cardiac myocytes *Math. Biosci.* **236** 97–107

[90] Wagner J and Keizer J 1994 Effects of rapid buffers on Ca^{2+} diffusion and Ca^{2+} oscillations *Biophys. J.* **67** 447–56

[91] Weber C R, Ginsburg K S, Philipson K D, Shannon T R and Bers D M 2001 Allosteric regulation of Na/Ca exchange current by cytosolic Ca in intact cardiac myocytes *J. Gen. Physiol.* **117** 119–31

[92] Weiss J N, Karma A, Shiferaw Y, Chen P-S, Garfinkel A and Qu Z 2006 From pulsus to pulseless *Circ. Res.* **98** 1244–53

[93] Wescott A P, Jafri M S, Lederer W J and Williams G S B 2016 Ryanodine receptor sensitivity governs the stability and synchrony of local calcium release during cardiac excitation-contraction coupling *J. Mol. Cell. Cardiol.* **92** 82–92

[94] Wier W G 2007 Gain and cardiac E–C coupling: revisited and revised *Circ. Res.* **101** 533–5

[95] Winegrad S 1965 The location of muscle calcium with respect to the myofibrils *J. Gen. Physiol.* **48** 997–1002

[96] Winslow R L, Rice J, Jafri S, Marbán E and ORourke B 1999 Mechanisms of altered excitation-contraction coupling in canine tachycardia-induced heart failure, II *Circ. Res.* **84** 571–86

[97] Yue D T, Backx P H and Imredy J P 1990 Calcium-sensitive inactivation in the gating of single calcium channels *Science* **250** 1735–8

[98] Yue D T, Herzig S and Marban E 1990 Beta-adrenergic stimulation of calcium channels occurs by potentiation of high-activity gating modes *Proc. Natl Acad. Sci. USA* **87** 753–7

[99] Zahradnkov A and Zahradnk I 1996 A minimal gating model for the cardiac calcium release channel *Biophys. J.* **71** 2996–3012

[100] Zahradnkov A, Zahradnk I, Gyrke I and Gyrke S 1999 Rapid activation of the cardiac Ryanodine receptor by submillisecond calcium stimuli *J. Gen. Physiol.* **114** 787–98

[101] Zalk R, Clarke O B, des Georges A, Grassucci R A, Reiken S, Mancia F, Hendrickson W A, Frank J and Marks A R 2015 Structure of a mammalian Ryanodine receptor *Nature* **517** 44–9

[102] Zsolnay V, Fill M and Gillespie D 2018 Sarcoplasmic reticulum Ca^{2+} release uses acascading network of intra-SR and channel countercurrents *Biophys. J.* **114** 462–73

Chapter 4

Cardiac cell modeling

Claire Yanyan Ji, Elizabeth M Cherry and Flavio H Fenton

4.1 Introduction

To date, more than one hundred models have been developed to describe the action potential (AP) of a cardiac cell mathematically [1, 15]. For more than 50 years, these models have been created to represent cells from different regions of the heart, including atria, ventricles, sinoatrial node, atrioventricular node, and Purkinje fibers, and to describe electrophysiological differences across various animal species. The general form for describing an AP mathematically follows the charge conservation equation [25]:

$$C_m \frac{dV_m}{dt} = -i_{\text{ion}} + i_{\text{stim}}, \tag{4.1}$$

where C_m is the membrane capacitance in μF cm^{-2}, V_m is the membrane voltage in mV, and $-i_{\text{ion}}$ and i_{stim} are the total transmembrane ionic current and the stimulus current, respectively, in μA cm^{-2}. By convention, inward currents are negative. Thus, the negative sign for $-i_{\text{ion}}$ ensures that inward current depolarizes the membrane.

The complexity of the ionic current descriptions varies greatly across models. Phenomenological models, which usually are intended to be used in higher dimensional tissue simulations due to their simplicity, contain minimum numbers of ionic currents and gating variables to reproduce mesoscopic features of cell dynamics, such as action potential amplitude (APA), action potential duration (APD), diastolic interval (DI), and restitution curves [14, 32] among others. For more realistic models, i_{ion} is comprised of a sum of the individual transmembrane currents that represent the movement of ions through specific Na^+, Ca^{2+}, K^+, Cl^- channels, pump and exchangers, and are often inherited or modified from previous generations of models [33, 47, 53, 61]. Earlier models, known as the first generation [33, 53, 61], had fixed intracellular ion concentrations, following the classic Hodgkin–Huxley (HH) [25] type of equations based on experimental voltage-clamp data used

doi:10.1088/978-0-7503-2064-1ch4

to describe their ion channels [2, 38, 50]. Later models, classified as the second generation of models [33, 53, 61], improved on the first generation by allowing ion concentrations to change over the course of the AP [11, 24, 37]. Some more recent models, the third generation [33, 53, 61], included more detailed mathematical equations using Markov descriptions to express channels transitioning through different discrete states [41, 52, 58].

In this chapter, we briefly review different types of cardiac models and some of their properties. Then we provide eight examples of historically significant cardiac models that encompass most of the wide range of possible dynamics.

4.2 Types of cardiac models

In this section, we describe cardiac models based on four properties: the type of model for gating variables (Hodgkin–Huxley or Markov); the level of detail for the ionic currents (phenomenological or physiological); the region in the heart represented (such as the ventricles, atria, or Purkinje fibers); and the species described.

4.2.1 Hodgkin–Huxley type versus Markov type

As elaborated in previous chapters, in general, the ionic current depends on the conductance of the membrane and the driving force. The driving force is the difference between the transmembrane voltage (V_m) and the Nernst potential (E_{ion}). The membrane conductance is proportional to the maximal conductance ($\bar{g}_{\text{ion}}$) (which corresponds to the membrane conductance when all channels are open) and the probability of finding the ion channel in its open state (p). Thus, we can write a current generally following Ohm's law as

$$i_{\text{ion}} = \bar{g}_{\text{ion}} p(V_m - E_{\text{ion}}). \tag{4.2}$$

Based on how the open probability p is formulated, cardiac models can be classified as a Hodgkin–Huxley-type model or a Markov-type model.

For the Hodgkin–Huxley (HH) type of cardiac models, the ion channel is usually 'gated' by one or more gating variables, each of which takes a value from 0 to 1. The probability of finding the ion channel open is associated with the probability of finding each gating variable open, usually as the product of the probabilities of all gating variables, some of which may be raised to an integer power. For example, in the Beeler–Rueter model (BR) [2], for the Na^+ channel, $p = m^3hj$, with m, h, and j being three gates (see section 4.3.4 for details). Each gating variable has two states, open (O) and closed (C), with transition rates α (of opening) and β (of closing) that usually depend on the membrane potential V_m:

$$\text{O} \underset{\beta(V_m)}{\overset{\alpha(V_m)}{\rightleftharpoons}} \text{C}. \tag{4.3}$$

Some gating variables are associated with activation of the channel and others with inactivation. In a slight abuse of notation, a gating variable typically is used to refer both to the gate and, mathematically, to the probability that that gate is open. The rate of change of the probability that a representative gating variable s is open typically is described as $ds/dt = \alpha(V_m)(1 - s) - \beta(V_m)s$. Written in a slightly different form, we have the typical HH gating variable equation:

$$\frac{ds}{dt} = (s - s_\infty)/\tau_s, \tag{4.4}$$

$$s_\infty = \frac{\beta}{(\alpha + \beta)}, \tag{4.5}$$

$$\tau = \frac{1}{(\alpha + \beta)}. \tag{4.6}$$

Here, s_∞ is the steady-state value that the gate s eventually will take on if conditions are left unchanged and τ_s is the time constant that governs how s_∞ is reached. Because the transition rates α and β typically depend on voltage, the steady-state value and time constant also typically depend on voltage; for example, $s_\infty = s_\infty(V_m)$ is a function of voltage and represents the value the gate ultimately will approach after leaving the membrane fixed at a particular voltage value. Because there are typically thousands of ion channels of a particular type embedded in the cell membrane, a gating variable typically represents the collective behavior of all those channels rather than one individual channel and thus can be interpreted as the fraction of all channels that are in the open state. Note that with only two states represented, the fraction of channels in the closed state can be obtained easily as $1 - s$.

If it is important to represent more than two discrete configurations for an ion channel,as in cases when binding of particular drugs are studied, a Markov-type description is typically used. Where the channel can be in n states with transition rate a_{ij} associated with the transition from state i to state j. The probability of finding channel s in state i (s_i) follows the equation

$$\frac{ds_i}{dt} = \sum_{j=1,j\neq i}^{n} (\alpha_{ji}s_j - \alpha_{ij}s_i), \quad i = 1, 2, \dots, n. \tag{4.7}$$

Figure 4.1 illustrates an example of a Markov model for an ion channel with three states: open, closed, and inactive. Here $p = s_O$. The probability of finding the channel open, for example, is $ds_O/dt = \alpha_{CO}s_C + \alpha_{IO}s_I - (\alpha_{OC} + \alpha_{OI})s_O$. In practice, most Markov chain models used in cardiac models contain ~5–30 discrete states for each ion channel.

Markov models have been adopted for describing both ion channels and calcium dynamics in cardiac models. For example, the Iyer *et al* (2004) model [27], one of the

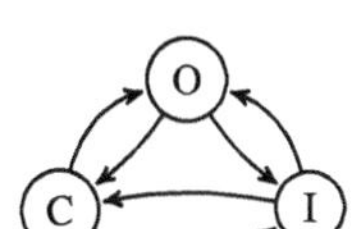

Figure 4.1. Example of a Markov chain model with three states: open ('O'), closed ('C'), and inactive ('I').

earliest full human ventricular cell models with Markov formulations for i_{Na}, i_{to}, i_{Kr}, and i_{CaL} has 67 variables, and all but one of its gate variables are described by a Markov-chain. The Shannon *et al* (2004) model [57] for rabbit ventricles, uses a four-state (resting, open, inactive, and resting inactive) Markov chain model to model the opening of the ryanodine receptors (RyRs) in its calcium dynamics.

The HH model is an empirical model established before technology allowed a detailed understanding of ion channel structures. The Markov chain model is closer to the true mechanisms associated with channel configurations. The HH-type ion channel with one gating variable can be seen as a special case for the Markov chain model when there are only two states ('C' and 'O') for the channel. While HH style models are usually simpler and easier to solve, Markov models usually contain many more variables and are numerically stiffer. Which may require semi-implicit methods [4].

4.2.2 Phenomenological models versus physiologically detailed models

Based on whether a cardiac model is highly summarized in describing the trans-membrane currents or not, cardiac models can be classified into phenomenological models and detailed models.

Phenomenological models do not intend to simulate each ion current specifically. Instead, they aim to describe the cardiac action potential in a highly summarized manner with a minimal amount of current detail. The intent is to capture some aspects of the fundamental dynamics of a cardiac cell sufficiently so that the model can be simulated efficiently in large high-dimensional tissue or to perform detailed mathematical analysis. For example, the Fenton–Karma model (FK) [16] has only three variables and three ion currents (fast inward, slow outward, and slow inward, approximating the total sodium, potassium, and calcium currents, respectively; see section 4.3.6). It does not include any ion concentrations. Nevertheless, it can reproduce essential action potential dynamics of several more detailed cardiac models by adjusting its parameters, and it can be used for studying the dynamics of spiral and scroll waves in 2D and 3D. Models like this can also be used to fit experimental [19] and clinical [36] data.

Realistic models, on the other hand, aim to describe the ion currents in detail. Such models usually divide the cell into several compartments, such as the bulk myoplasm (MYO), the junctional/network sarcoplasmic reticulum (JSR/NSR), and the subsarcolemmal space (SS). They include equations to describe ion currents, concentrations, pumps/exchangers, and buffers. Usually the currents are represented by a mixture of Hodgkin–Huxley and Markov models. As an example, a diagram of the O'Hara *et al* model [51] is shown in figure 4.2. It is a widely used realistic human ventricle model with 4 compartments, 48 variables, 11 ion currents, 5 ion pumps/exchangers, and 6 buffers. Some realistic models include discretization within a cell to incorporate the detailed locally decoupled calcium in sarcomeres [56], and some even include contraction [41, 54]. Contraction on many of these electrophysiological models can be implemented by coupling to the calcium dynamics [29].

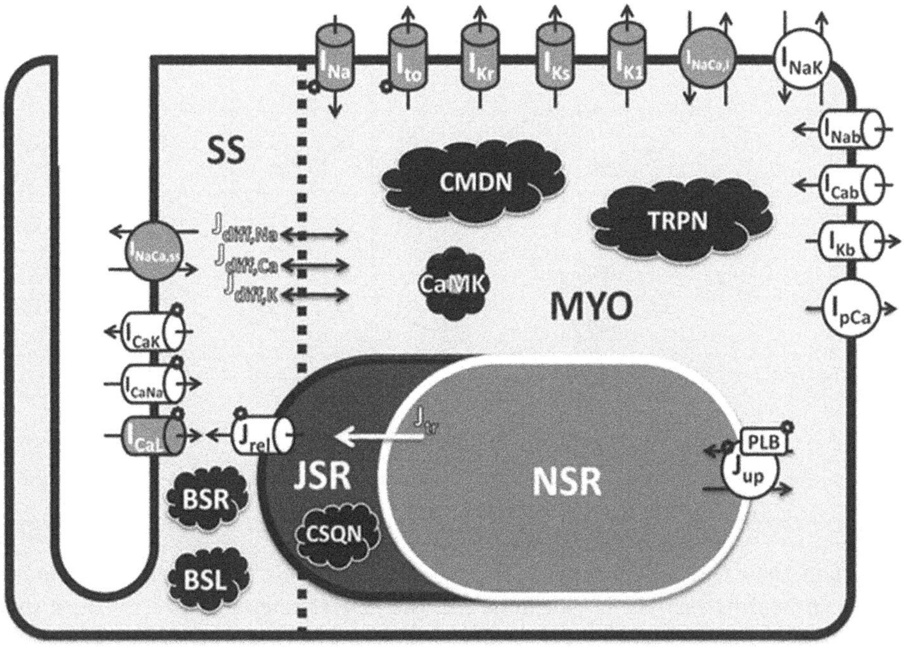

Figure 4.2. A diagram of the O'Hara *et al* model with 4 compartments, 48 variables, 11 ion currents, 5 ion pumps/exchangers, and 6 buffers. Reprinted from [51] with permission from PLoS One.

Realistic models can provide a large picture of cell dynamics. Many of their parameters are typically fitted directly from patch-clamp experimental data and while many can use data from a mix of animal species [46] most recent models are fitted specifically. However, as models become more complex they are often more difficult to solve numerically due to the large number of variables and the stiffness of their equations. Thus, it can be costly to implement them at the tissue and organ scale.

4.2.3 Action potentials for different regions of the heart

Electrical signals in the heart are generated from a small section of cells called the sinoatrial node (SA node) located in the upper right atrium (RA). The signals propagate from the SA node to the atria, resulting in the contraction of the atria. The signals continue passing through the atrioventricular node (AV node), which is the only electrical pathway connecting the atria and the ventricles in healthy hearts, to the His–Purkinje system. APs propagate with a high conduction velocity through the Purkinje fibers, which enables the action potential to propagate quickly to both ventricles, causing a synchronous ventricular contraction. The properties of APs in these different regions of the heart can be substantially different because of differences in channel expression and function. Cardiac models capture these differences through their formulations and parameterizations.

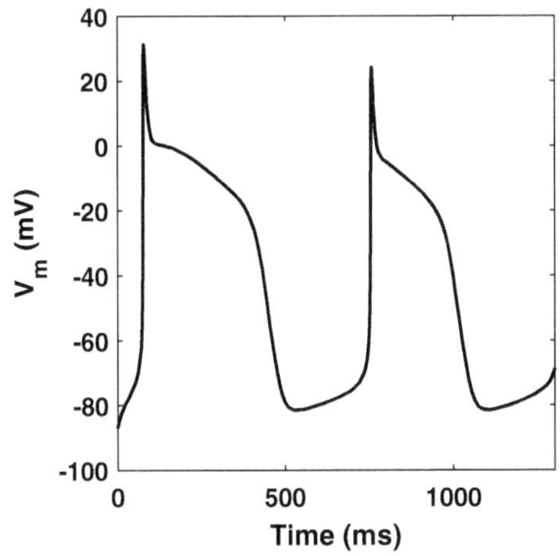

Figure 4.3. Action potential of a Purkinje model [50] showing the pacemaker potential and the oscillatory behavior.

Cells in the SA node, the AV node, and the Purkinje fibers all have pacemaking ability. In healthy hearts, the SA node has the fastest rate, and as a result, it drives the rest of the heart including the other pacemaking regions. Under pathological conditions when the SA node fails to serve as the main driving force, the AV node and the Purkinje fibers can serve as secondary and tertiary pacemakers. Therefore, models for the SA node, the AV node, and the Purkinje fibers are usually auto-oscillatory; they can produce APs without external stimuli. Figure 4.3 shows two APs generated from the Purkinje model [50]. After repolarization, instead of remaining at the resting membrane potential like atrial and ventricular cells, pacemaker cells slowly depolarize due to the decay of I_K and the presence of inward currents until a new AP is generated. This phase is called the pacemaker potential.

Cardiomyocytes in atria and ventricles are excitable and thus are characterized by a stable resting membrane potential (a stable fixed point) with the need for an external stimulus to produce an AP, a fast depolarization phase and a slow repolarization phase, and a refractory period during which a new excitation cannot be generated. Action potentials in atria and ventricles usually have different shapes. Figure 4.4(a) shows a comparison of human ventricular (black) and human atrial (red) APs generated by the ten Tusscher *et al* (2006) model [59] and the Morotti *et al* (2016) model [43], respectively. There are two important distinctions. First, the resting membrane potential is higher in atrial than in ventricular cells. Second, the atrial cell repolarizes at a slow, steady rate, while in ventricles, after depolarization, there is a clear plateau phase before a fast repolarization. Both types of behavior can be attributed to the inward-rectifier K^+ current (I_{K1}; see figure 4.4(b)). In atrial cells, I_{K1} is smaller and increases at a more moderate rate during repolarization. For ventricular cells, however, I_{K1} remains low until it spikes abruptly at the end of repolarization, which results in the differences in the resting potential and the slopes of repolarization [22].

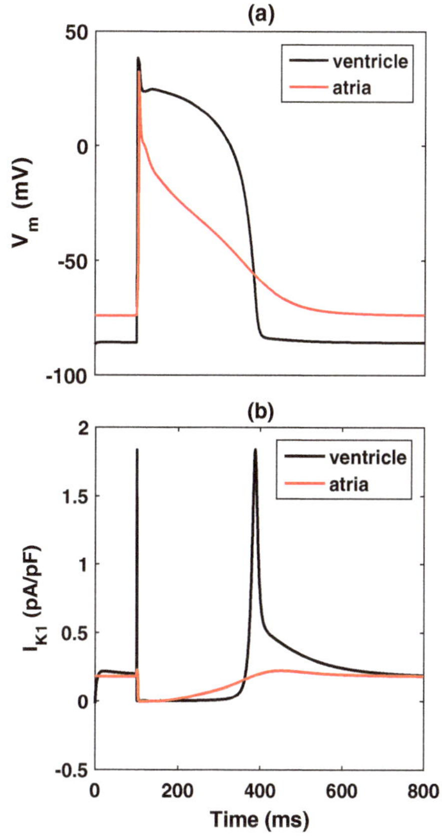

Figure 4.4. (a) Comparison of modeled human ventricular (black) and atrial (red) action potentials. The ten Tusscher *et al* (2006) model [59] and the Morotti *et al* (2016) model are used, respectively [43]. (b) Differences in the resting membrane potential and the slope of repolarization are attributed to differences in I_{K1}.

Action potentials in different regions of the atria and ventricles also can have distinct shapes. For example, figure 4.5(a) shows epicardial, midmyocardial (M cell), and endocardial APs generated with the ten Tusscher *et al* (2006) model [59]. There are two major differences. First, epicardial and midmyocardial APs have a 'spike-and-dome' shape, which is less pronounced in endocardial cells. In epicardial and midmyocardial cells, immediately after depolarization, a rapid repolarization occurs due to the activation of the transient outward K^+ current (I_{to}). After this repolarization, the membrane depolarizes again, forming the dome. This repolarization is less pronounced in endocardial cells because I_{to} is weakly expressed there (see figure 4.5(b)); therefore, the spike-and-dome shape in the AP is less evident. Second, midmyocardial cells have a longer APD compared with epicardial and endocardial cells because the slow delayed rectifier K^+ current (I_{Ks}) is weakly expressed (see figure 4.5(c)), so the repolarization is slower, causing a longer APD. In the ten Tusscher *et al* (2006) model [59], the maximal conductances of I_{to} and I_{Ks} are adjusted for epicardial, midmyocardial, and endocardial cells to reflect the correct morphology of the AP.

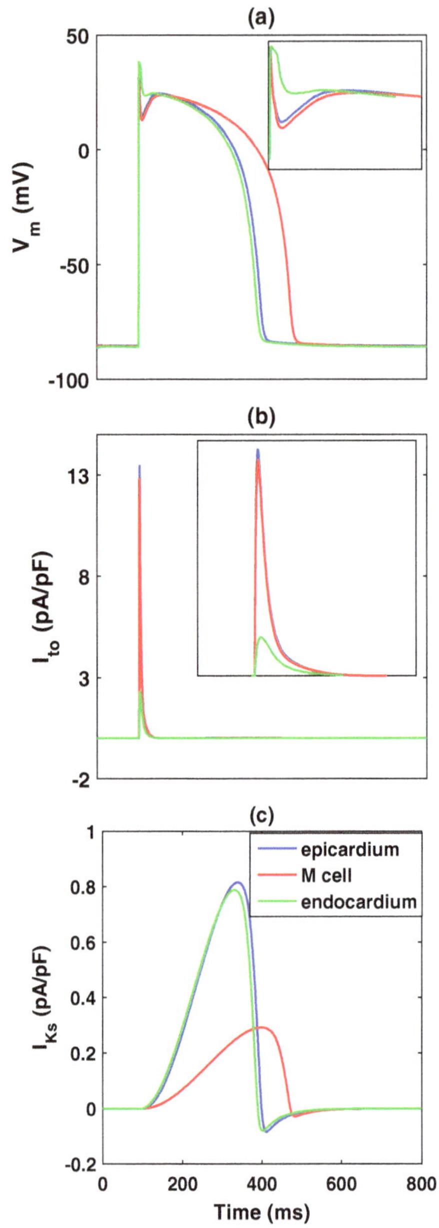

Figure 4.5. (a) Comparison of endocardial, midmyocardial, and epicardial action potentials in ventricles, demonstrated using the Ten Tusscher *et al* (2006) model [59]. (b) I_{to} contributes to the spike-and-dome morphology following the upstroke and differences in I_{Ks} expression (c) account for the difference in the repolarization.

4.2.4 Action potential across different species

Action potentials vary greatly from one species to another due to differences in ion channel expression and function as well as properties related to the concentrations of various ions. Figure 4.6 shows APs generated by models for rat [13], guinea pig [38], mouse [3], rabbit [40], canine [26], and human [59]. Rats and mice have triangular-shaped ventricular action potentials similar to those of atrial cells along with much shorter APDs (<100 ms) due to their overexpression of I_{to}. However, species size is not the only factor that influences APs. For example, guinea pigs, while similar to rats in size, have ventricular action potentials closer to larger mammals in both shape and duration. On the other hand, ectotherms, such as zebrafish and alligators, though varying greatly in sizes, can have ventricular APs that are similar to those of humans [23, 45].

4.3 Examples of cardiac models

In 1952, Hodgkin and Huxley published their ground-breaking paper describing quantitatively the membrane currents and the voltage in squid giant axons based on precise measurements of sodium and potassium currents [25]. For this work they received the Nobel Prize in Physiology/Medicine in 1963. The HH model has four variables: voltage (V), two sodium gates—one activation (m) and one inactivation (h)—and one potassium gate (n). Activation (inactivation) gates open (close) at more positive potentials. All the gating variables follow the HH type of equations from section 4.2.1, where their equations and α's and β's values, can be found in the original paper [25]. The model is composed of three currents—i_{Na}, i_K, and a leakage current i_l:

$$i_{Na} = \bar{g}_{Na} m^3 h(V - E_{Na}), \tag{4.8}$$

$$i_K = \bar{g}_K n^4 (V - E_K), \tag{4.9}$$

$$i_l = \bar{g}_l (V - E_K). \tag{4.10}$$

The mathematical formalism of the HH model was widely adopted by the cardiac modeling community over the next 60 years. Many review papers have provided a comprehensive picture of this history. Noble *et al* [49] presented a list of cardiac models that established and extended the field, as well as those that are potentially problematic. The Scholarpedia page of 'Models of cardiac cell' [15] discussed 45 models published through 2008 and summarized their properties with demonstrations of single-cell AP signals, bifurcation diagrams and alternans [17], and 2D spiral-wave dynamics using Java applets from [18]. In this section, we focus on only eight models that are historically important, with an emphasis on human models. For each model, we describe its significance and mathematical components and provide figures of its ion concentrations, currents, gating variables, and/or other variables as appropriate to the model. Finally, we show spiral-wave tip trajectories for all the cardiac models in 2D. For convenience, all currents plotted here are divided by C_m so that they are in units of $\mu A \mu F^{-1}$.

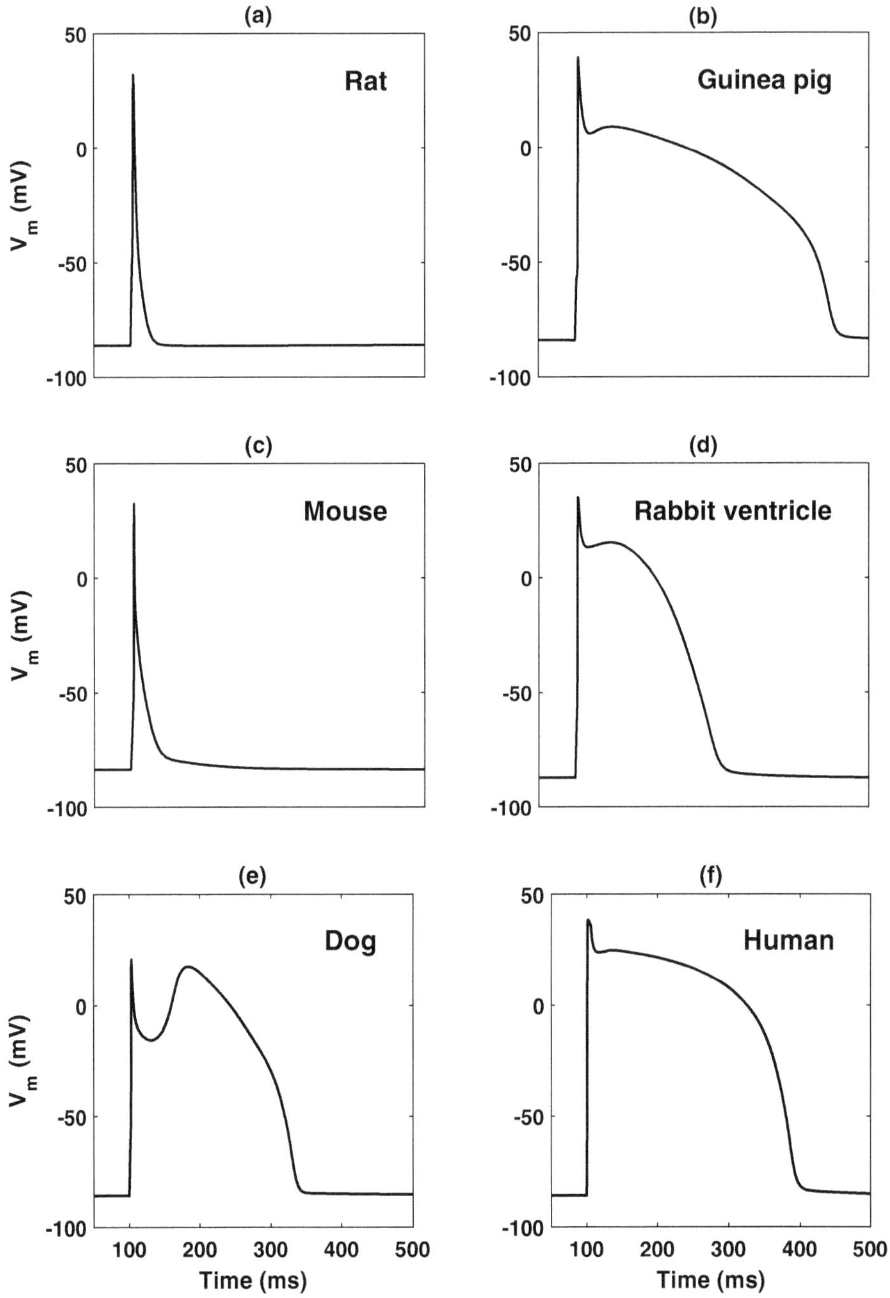

Figure 4.6. Action potentials in different species, generated by cardiac models.

4.3.1 The modified FitzHugh–Nagumo model

The FitzHugh–Nagumo (FHN) model is a widely-used two-variable model of a generic excitable medium. It was proposed first by FitzHugh in 1961 [20], followed

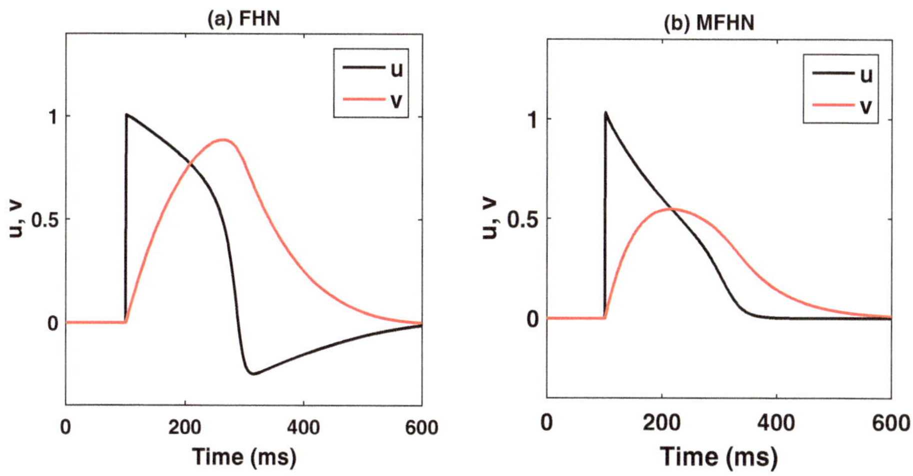

Figure 4.7. The fast (u) and the slow (v) variables of the FitzHugh–Nagumo model (FHN) and the modified FitzHugh–Nagumo model (MFHN). In the FHN model, the parameters are $a = 0.12$, $c_1 = 0.175$, $c_2 = 0.03$, $b = 0.011$, $d = 0.55$, $\delta = 0$. In the MFHN model, $a = 0.13$, $c_1 = 0.26$, $c_2 = 0.1$, $b = 0.013$, $d = 1$, $\delta = 0$.

by Nagumo *et al*'s equivalent circuit in 1962 [44]. The model is a simplification of the Hodgkin–Huxley model. Figure 4.7 shows the fast (black) and the slow (red) variables of the FHN model. The fast variable (u) can be seen as the dimensionless transmembrane potential, and the slow variable (v) is an inactivation gating variable similar to the n gate of the HH model. The model can be written [55] as

$$\frac{du}{dt} = c_1(a - u)(u - 1)u - c_2 v, \tag{4.11}$$

$$\frac{dv}{dt} = b(u - dv - \delta), \tag{4.12}$$

here c_1, c_2, a, b, d, and δ are constants. (Note that the model in this form is overparameterized, as every term on the right-hand side is multiplied by a constant.)

The FHN model can produce rich dynamics. By varying the value of δ, the model can be changed between oscillatory (like a pacemaker cell) and excitable (like a myocyte) regimes. It can also produce alternans and spiral waves. Alternans is a phenomena during which the action potential duration alternates in the pattern of long-short-long-short. In the original FHN model, $c_1 = c_2 = 1$. The AP has similar time constants for depolarization and repolarization, which is unphysiological for cardiac cells. By adjusting c_1 and c_2 we can adjust the depolarization and repolarization rates.

The modified FHN model (MFHN) [55] removes the hyperpolarization part of the AP by replacing $-c_2 v$ with $-c_2 uv$ in equation (4.11) so that it resembles APs in cardiomyocytes. Figure 4.7 shows the u and v variables for the MFHN model. Nevertheless, this model fails to reproduce some key features of cardiac cells. For example, in cardiac cells, the APD decreases as the DI is decreased but to a minimal value greater than zero; for smaller DIs there typically will be no AP generation, leading to block. The FHN and the MFHN models lack this property, as in both cases the DI can reach zero, as the excitability parameter ε goes to zero, which is the regime at which the FHN model goes from a neuronal model to a cardiac model, with large APDs.

All the figures of single-cell APs in this chapter are obtained at long cycle lengths (or from its intrinsic rate if it is an auto-oscillatory model). At fast rates, the APs will look different. For example, with the current parameters the FHN model will produce alternans with a fast pacing rate (200 ms).

4.3.2 The Noble model

The Noble 1962 model [50] is a generic Purkinje cell model that is historically significant because it is the first cardiac cell model. It is a modification of the HH model that incorporates the slow recovery of the potassium channel conductance. Like the HH model, it has four variables (u, m, h, n) and three currents (i_{Na}, i_{K} and i_{leak}). The expressions for i_{Na} and i_{K} are modified and an anionic leak current added as follows:

$$i_{\mathrm{Na}} = (\bar{g}_{\mathrm{Na}} m^3 h + 140.0)(V - E_{\mathrm{Na}}), \tag{4.13}$$

$$i_{\mathrm{K}} = (g_{\mathrm{K1}} + g_{\mathrm{K2}})(V - E_{\mathrm{K}}), \tag{4.14}$$

$$i_{\mathrm{leak}} = g_{\mathrm{leak}}(V - E_{\mathrm{leak}}), \tag{4.15}$$

$$g_{\mathrm{K1}} = 1200e^{-(u+90)/50} + 15e^{(u+90)/60}, \tag{4.16}$$

$$g_{\mathrm{K2}} = \bar{g}_{\mathrm{K}} n^4. \tag{4.17}$$

In i_{Na}, a constant component is added to the total conductance, independent of the gating variables. The constant sodium current both prolongs the AP and supports pacemaking activity, although the potassium current can be made large enough to suppress it. In the formulation of i_{K}, g_{K1} accounts for an outward potassium current that decreases at depolarization, instantaneously responding to the membrane potential. g_{K2}, which has the same form as the HH model, accounts for the slowly recovering potassium current after depolarization. Figure 4.8 shows the AP (a), the three gating variables (b), the conductance (c), and the three currents (d) of the Noble model. After repolarization, i_{Na} increases due to its constant component, and i_{K} increases but with a slower rate due to the decrease of both g_{K1} and g_{K2}. They result in a positive i_{sum} (here $i_{\mathrm{sum}} = -i_{\mathrm{ion}}$, see equation (4.1)) that slowly increases the membrane potential, leading to a spontaneous depolarization that captures the pacemaker behavior of the Purkinje fiber.

This model is also interesting because it was published before the discovery of calcium in cardiac cells [21], so only sodium and potassium currents contribute to the AP in this model. However, the Noble model can produce alternans in single cells and 1D fibers, consistent with behavior observed in Purkinje cells and fibers [8].

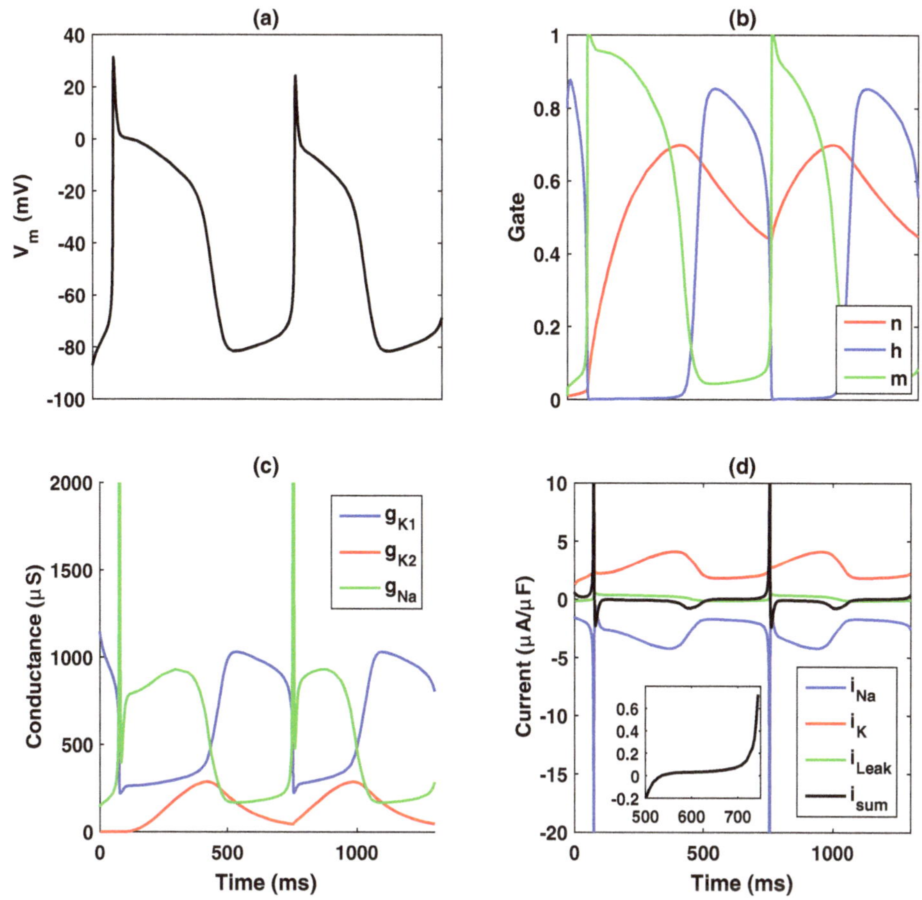

Figure 4.8. The AP (a), the gating variables (b), the conductance (c), and the ion currents (d) of the Noble 1962 model. Note that in (c) and (d), g_{Na}, i_{Na}, and i_{sum} exceed the y-axis limit.

In contrast, a later modification of this model that included the calcium current failed to produce alternans [11]. This example shows that it is possible for models to produce behavior consistent with experimental observations even when the model excludes important components, which underscores that mechanisms responsible for behavior in models may not be the same as those responsible for similar behavior in experiments. Therefore, models in general need to be well characterized before used to study a particular dynamics.

4.3.3 The Karma model

The Karma model [32] is a reduction of the Noble model with two adiabatically eliminated variables (the h and m). Therefore, it consists of only two variables: the fast variable that resembles the membrane potential (E) and the slow gating variable (n), which is equivalent to the n gate in the Noble model. The equations for E and n are as follows:

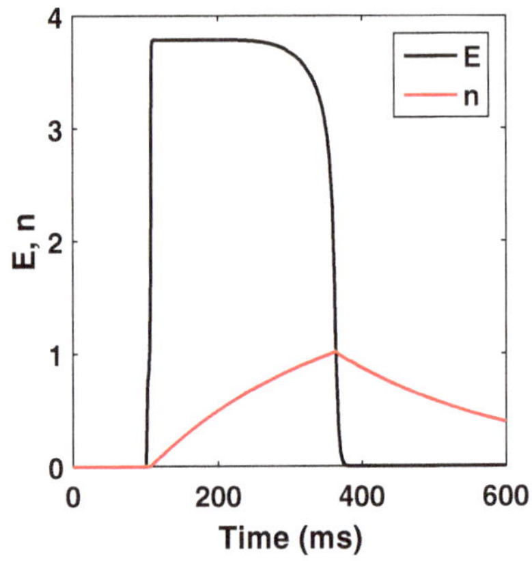

Figure 4.9. The fast (E) and the slow (n) variables of the Karma model.

$$\frac{dE}{dt} = \frac{1}{\tau_E}\left(-E + (E_s - \mathcal{D}(n))(1 - \tanh(E - E_h))\frac{E^2}{2}\right), \tag{4.18}$$

$$\frac{dn}{dt} = \frac{1}{\tau_n}(\mathcal{R}(n)\Theta(E - E_n) - (1 - \Theta(E - E_n))n), \tag{4.19}$$

here τ_E, τ_n, E_s, E_h, E_n, and Re are all constants and $\Theta(x)$ is the Heaviside function. The equations are based on inclusion of descriptions of the dispersion relation (important for tissue), here given as $\mathcal{D}(n) = n^M$, and the restitution relation, chosen to be $\mathcal{R}(n) = (1 - (1 - e^{-Re})n)/(1 - e^{-Re})$. The voltage and the gating variable are plotted in figure 4.9.

The Karma model can also produce alternans in a cardiac ring. Increasing the values of M and/or Re tend to promote alternans and spiral-wave breakup in 2D.

4.3.4 The Beeler–Rueter model

The Beeler–Rueter (BR) model [2], published in 1977, is the first mathematical model explicitly for ventricular cells. The model has eight variables—voltage (V_m), three sodium variables (activation gates m and h and inactivation gate j), two calcium variables (activation gate d and inactivation gate f), one potassium gate (x_1), and the calcium concentration ($[Ca^{2+}]_i$). All six gating variables follow HH-type equations. The model includes four currents. i_{Na} is the fast sodium current responsible for the upstroke of the AP. It follows a similar form as the HH model with an additional inactivation gate j. i_{K1} is the time-independent outward potassium current and is an instantaneous function of V_m. i_{x1} is a time-activated potassium current that depends on the gating variable x_1. i_s is a slow inward calcium current that depends on the activation gate d and the inactivation gate f. The equations for the four currents and $[Ca^{2+}]_i$ are as follows:

$$i_{\text{ion}} = i_{Na} + i_{x1} + i_{K1} + i_s, \tag{4.20}$$

$$i_{\rm Na} = (\bar{g}_{\rm Na} m^3 hj + g_{\rm NaC})(V_m - E_{\rm Na}), \tag{4.21}$$

$$i_{x1} = \bar{i}_{x1} x_1, \tag{4.22}$$

$$\bar{i}_{x1} = 0.8(e^{0.04(V_m+77)} - 1.0)/e^{0.04(V_m+35)}, \tag{4.23}$$

$$\begin{aligned} i_{\rm K1} = {} & 0.35(4.0(e^{0.04(V_m+85)} - 1.0)/(e^{0.08(V_m+53)} + e^{0.04(V_m+53)}) \\ & + 0.2(V_m + 23)/(1.0 - e^{-0.04(V_m+23)})), \end{aligned} \tag{4.24}$$

$$i_s = \bar{g}_s df(V_m - E_s), \tag{4.25}$$

$$E_s = -82.3 - 13.0287 \ln([\mathrm{Ca}^{2+}]_i), \tag{4.26}$$

$$\frac{d[\mathrm{Ca}^{2+}]_i}{dt} = -10^{-7} i_s + 0.07(10^{-7} - [\mathrm{Ca}^{2+}]_i). \tag{4.27}$$

Here $\bar{g}_{\rm Na} = 4$ mS cm^{-2}, $g_{\rm NaC} = 0.003$ mS cm^{-2}, $E_{\rm Na} = 50$ mV, and $g_s = 0.09$ mS cm^{-2}. The membrane voltage, the calcium concentration, the six gating variables, and the four currents are plotted in figure 4.10. $[\mathrm{Ca}^{2+}]_i$ has units of 10^{-7} M L^{-1}; the gating variables are dimensionless; and the currents are divided by $C_m = 1$ μA cm^{-2}.

The BR model was the first ionic model simulated in 2D, by Courtemanche and Winfree in 1991 [9]. They showed for the first time in cardiac models spiral waves and spiral-wave breakup, one of the mechanisms for cardiac fibrillation. The breakup in the BR model can be inhibited by reducing the time constants of the f and d gates to reduce the variation in the wavelengths; a common variation often referred to as the modified Beeler–Reuter (MBR) model reduces both time constants by a factor of two [10]. The BR model also readily produces alternans in cells and tissue.

4.3.5 The Luo–Rudy model

The Luo–Rudy 1991 model (LR1) [38] is the first model for guinea pig ventricular myocytes. It is a modification of the BR model, so it has the same eight variables as the BR model. Figure 4.11 shows V_m, $[\mathrm{Ca}^{2+}]_i$, the gating variables, and the currents. The behavior of $[\mathrm{Ca}^{2+}]_i$ and the gating variables looks very similar to that of the BR model (see figure 4.10). The LR1 has a total of six currents: the fast sodium current, $i_{\rm Na}$; the time-independent potassium current, $i_{\rm K1}$; the time-dependent potassium current, $i_{\rm K}$; the plateau potassium current, $i_{\rm Kp}$; the background potassium current, $i_{\rm b}$; and the slow inward calcium current, $i_{\rm si}$. The formulations for $i_{\rm si}$ and $i_{\rm K}$ are identical to those for $i_{\rm s}$ and i_{x1} in the BR model. For $i_{\rm Na}$, the rate coefficients for the opening and closing of the sodium channels are modified to reproduce faster sodium current dynamics. $i_{\rm K1}$ is formulated completely differently than in the BR model. It is zero at high potentials and rises at low potentials. Two more currents were added to the LR1 model: $i_{\rm Kp}$ and $i_{\rm b}$. The sum of $i_{\rm K1}$, $i_{\rm Kp}$, and $i_{\rm b}$ are the total time-independent

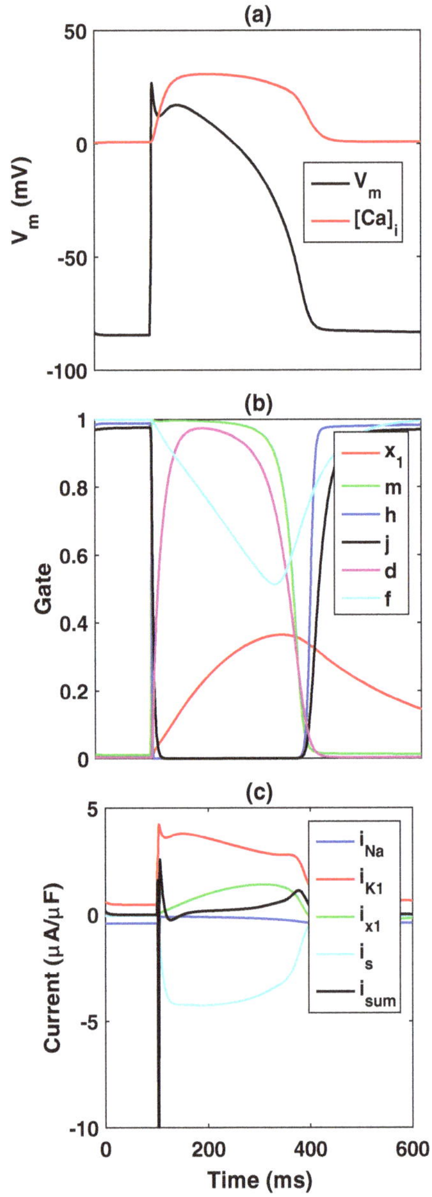

Figure 4.10. The AP, $[Ca^{2+}]_i$, gating variables and currents of the Beeler–Rueter model. $[Ca^{2+}]_i$ is in units of 2×10^{-7} M L^{-1}. i_{Na} and $i_{sum} = -i_{ion}$ fall outside the current axis limit in the plot.

potassium current in the LR1 model. This model can also produce spiral waves and spiral-wave breakup through the Doppler shift mechanism and its flat restitution curve [14].

An updated version of the LR1 model—the Luo–Rudy II (LRII)—was published by Luo and Rudy in 1994 [37]. Then any subsequent modification of this model has

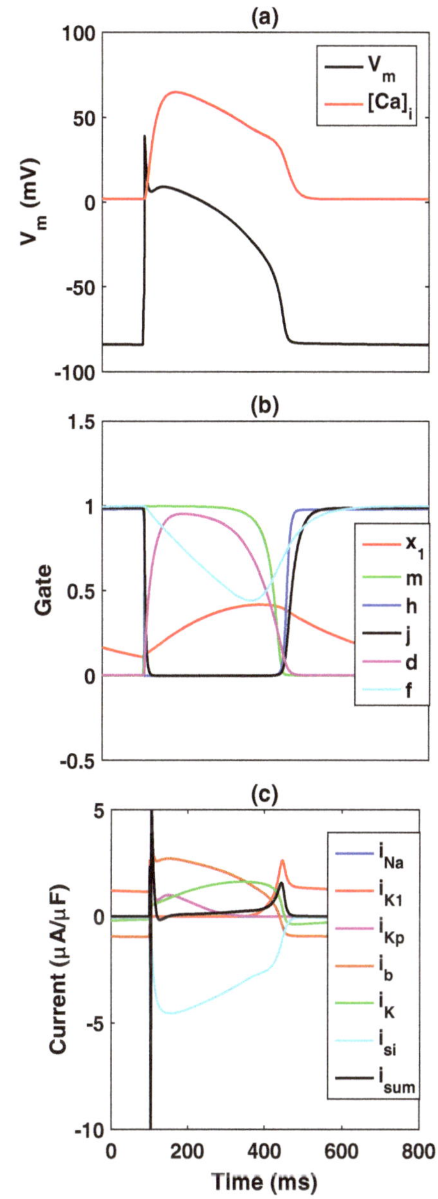

Figure 4.11. The AP, $[Ca^{2+}]_i$, gating variables, and currents of the Luo–Rudy model. $[Ca^{2+}]_i$ is in units of 10^{-7} M L^{-1}. Note that in (c), i_{Na} and $i_{sum} = -i_{ion}$ falls below the minimum of the y-axis. The LR1 model is similar to the BR model but includes two more currents.

been coined as LRD for Luo–Rudy-dynamic which have been incorporating dynamically varying sodium and potassium concentrations and more detailed intracellular calcium dynamics.

4.3.6 The Fenton–Karma model

The Fenton–Karma model (FK) is a phenomenological model with three variables, voltage (u), fast gating variable (v), and slow gating variable (w). It also includes three currents, fast inward (i_{fi}), slow outward (i_{so}), and slow inward (i_{si}), that represent the sodium, potassium and calcium currents, respectively. The equations of the FK model are as follows,

$$i_{\text{ion}} = i_{\text{fi}} + i_{\text{so}} + i_{\text{si}}, \tag{4.28}$$

$$i_{\text{fi}} = -\frac{v}{\tau_d}\Theta(u - u_c)(1 - u)(u - u_c), \tag{4.29}$$

$$i_{\text{so}} = \frac{u}{\tau_o}(1 - \Theta(u - u_c)) + \frac{1}{\tau_r}\Theta(u - u_c), \tag{4.30}$$

$$i_{\text{si}} = -\frac{w}{2\tau_{\text{si}}}(1 + \tanh(k(u - u_c^{\text{si}}))), \tag{4.31}$$

$$\frac{dv}{dt} = \frac{(1 - \Theta(u - u_c))(1 - v)}{\tau_v^-(u)} - \frac{\Theta(u - u_c)v}{\tau_v^+}, \tag{4.32}$$

$$\frac{dw}{dt} = \frac{(1 - \Theta(u - u_c))(1 - w)}{\tau_w^-} - \frac{\Theta(u - u_c)w}{\tau_w^+}, \tag{4.33}$$

$$\tau_v^-(u) = \Theta(u - u_v)\tau_{v1}^- + (1 - \Theta(u - u_v))\tau_{v2}^-. \tag{4.34}$$

This model has a total of 13 parameters: nine time constant parameters (τ_d, τ_o, τ_r, τ_{si}, τ_{v1}^-, τ_{v2}^-, τ_v^+, τ_w^-, and τ_w^+), three voltage threshold parameters (u_c, u_c^{si}, and u_v), and one additional parameter (k) describes the strength of the calcium current at high voltage values. One advantage of this model is that it can be used to reproduce data from experiments and other models. It is straightforward to adjust the threshold for excitation (u_c), the rate of rise of the upstroke (τ_d), the minimum diastolic interval (τ_{v1}^-), the minimum and maximum action potential duration, and properties of the APD/conduction velocity restitution curves. The original paper provided four parameter sets to reproduce the BR model, the modified BR model, the LR1 model, and guinea pig experimental data. Subsequent modifications of the FK model, include a reduction to a two variable model by Mitchell and Schaeffer [42] and a four-variable model by Bueno-Orovio *et al* [4] (also known as minimum model), that improved the action potential shape by introducing an additional variable s that contributes to the slow inward current (figure 4.12).

4.3.7 The ten Tusscher *et al* model

The ten Tusscher *et al* 2004 model [58] is one of the two first human ventricular models developed using human data (in the same year, Iyet *et al* also published a human ventricular model [27]; see section 4.2.1). The updated 2006 model [59] is

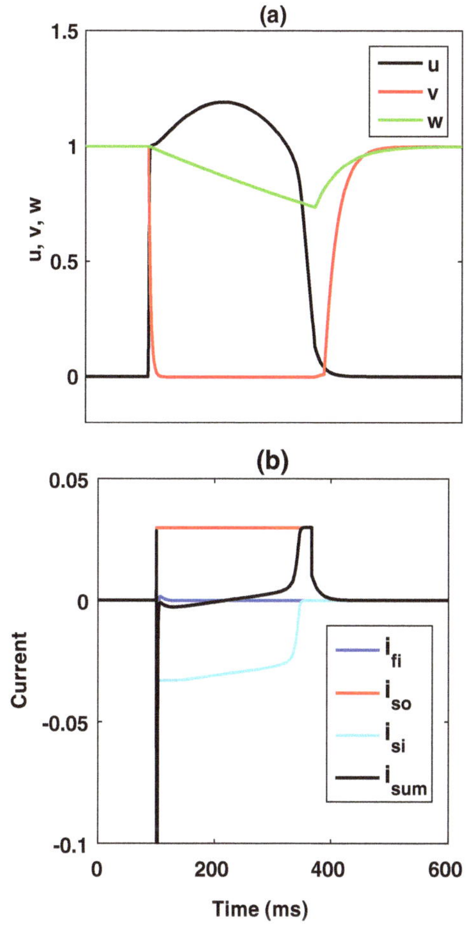

Figure 4.12. The AP (black line in (a)), the gating variables (green and red lines in (a)) and the ionic currents (b) of the Fenton–Karma model using the modified BR parameter set.

improved from the earlier version with a fuller description of the calcium dynamics using a Markov model for the calcium-induced–calcium-release (CICR) process. In this model, the cell is divided into three compartments: the bulk myoplasm, the sarcoplasmic reticulum (SR), and the SS. For computational efficiency, the Markov model was solved using quasi-steady-state assumptions.

The model has a total of 19 variables. Along with the transmembrane potential, it includes 12 gating variables: three sodium current gates (m, h, j), four calcium current gates (d, f, f_2, f_{Cass}), three time-dependent potassium current gates (x_{r1}, x_{r2}, x_s), and two transient outward potassium current gates (s, r). All twelve gating variables are of the HH type. There are also five dynamic ion concentration variables ($[Na^+]_i$, $[K]_i$, $[Ca^{2+}]_i$, $[Ca^{2+}]_{ss}$, and $[Ca^{2+}]_{SR}$) and one Markov variable (R) for the open probability of the RyR channels.

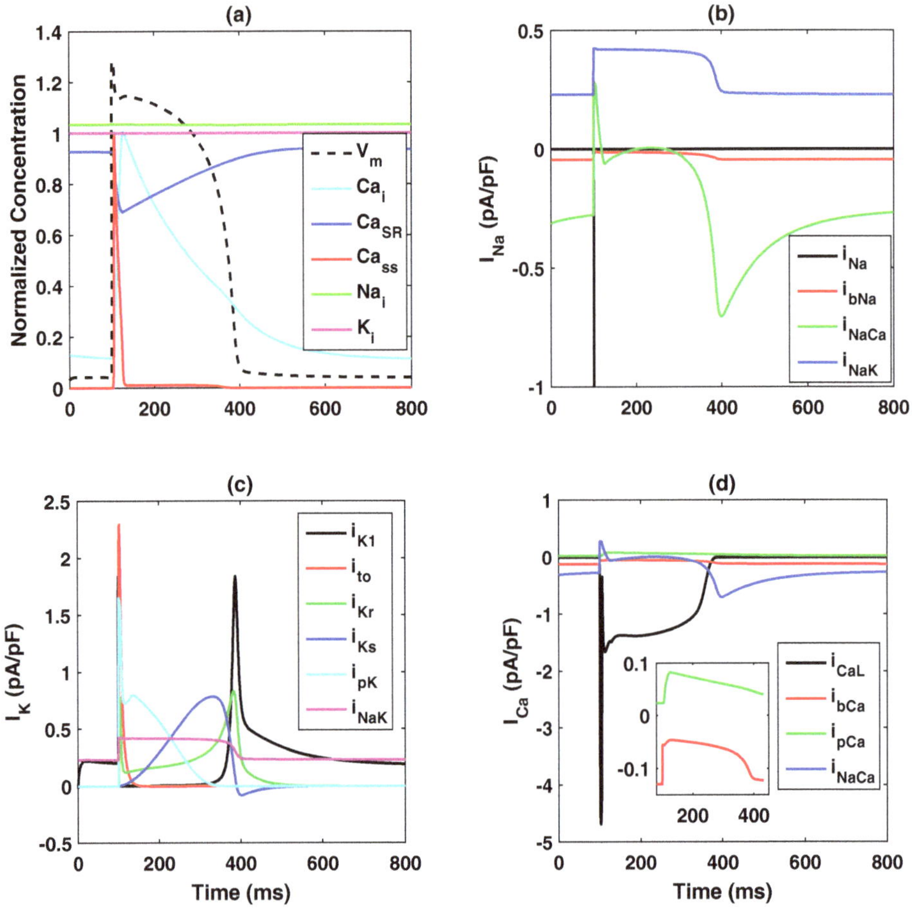

Figure 4.13. Ten Tusscher *et al* (2006) model with endocardial parameters. (a) Normalized transmembrane voltage and normalized ion concentrations. The normalization constants are $[Ca^{2+}]_i = 0.001$ mM L^{-1}, $[Ca^{2+}]_{SR} = 4$ mM L^{-1}, $[Ca^{2+}]_{ss} = 2$ mM L^{-1}, $[Na^{+}]_i = 10$ mM L^{-1}, $[K^{+}]_i = 138$ mM L^{-1}. (b) Four sodium-related membrane currents: i_{Na}, i_{bNa}, i_{NaCa}, and i_{NaK}. Note that i_{Na} falls outside the vertical range of the plot. (c) Six potassium-related membrane currents: i_{K1}, i_{to}, i_{Kr}, i_{Ks}, i_{pK}, and i_{NaK}. (d) Four calcium-related membrane currents: i_{CaL}, i_{bCa}, i_{pCa}, and i_{NaCa}.

The model includes 12 transmembrane currents: the fast Na^+ current (i_{Na}, depends on m, h, and j), the L-type Ca $^{2+}$ current (i_{CaL}, depends on d, f, f_2, and f_{Cass}), the transient outward K^+ current (i_{to}^+, depends on s and r), the slow/rapid delayed rectifier K^+ current (i_{Ks}/i_{Kr}, depends on x_s/x_{r1} and x_{r2}), the inward rectifier K^+ current (i_{K1}, function of V_m), the Na^+/Ca^{2+} exchanger current (i_{NaCa}, depends on $[Na^+]_i$ and $[Ca^{2+}]_i$), the Na^+/K^+ pump current (i_{NaK}, depends on $[Na^+]_i$), the background Na^+ and Ca $^{2+}$ currents (i_{bNa} and i_{bCa}), and the plateau currents (i_{pCa} and i_{pK}). The model can produce endocardial, epicardial, and midmyocardial action potentials with the conductances for i_{to} and I_{Ks} modified accordingly.

Figure 4.13 shows the normalized ion concentrations during an AP for the ten Tusscher *et al* (2006) model with the endocardial parameters. The normalization constants are $[Ca^{2+}]_i$ = 0.001 mM L^{-1}, $[Ca^{2+}]_{SR}$ = 4 mM L^{-1}, $[Ca^{2+}]_{ss}$ = 2 mM L^{-1}, [$[Na^+]_i$ = 10 mM L^{-1}, $[K^+]_i$ = 138 mM L^{-1}. During the upstroke of the AP, the L-type calcium channel is activated, resulting in an inward i_{CaL} and an increase in the subspace calcium concentration ($[Ca^{2+}]_{ss}$). $[Ca^{2+}]_{ss}$ activates the RyRs in the SR, leading to a large calcium release from the SR. This process is called calcium induced calcium release (CICR). $[Ca^{2+}]_{SR}$ decreases and $[Ca^{2+}]_i$ increases. Then the L-type channels close due to the positive V_m, resulting in a decrease in $[Ca^{2+}]_{ss}$, which no longer activates the RyR. Then the SERCA pump in the SR is activated and Ca is pumped back to the SR, resulting in the recovery of $[Ca^{2+}]_{SR}$ and the decrease of $[Ca^{2+}]_i$. In contrast to the calcium concentrations, $[Na^+]_i$ and $[K^+]_i$ change little during the AP.

Figure 4.13(b)–4.13(d) show the various Na^+, K^+, and Ca $^{2+}$ currents. The fast inward i_{Na} causes the upstroke of the AP, with fast outward i_{to} resulting in the immediate repolarization after the upstroke. The balance between the outward currents i_{pK} and i_{Ks} with the inward current i_{CaL} produces the plateau of the AP. Finally, the outward currents i_{K1} and i_{Kr} return the membrane to the resting potential.

4.3.8 The O'Hara *et al* model

The O'Hara *et al* model (often referred to as ORd) [51] is a widely used model for human ventricular myocytes. Its formulations for i_{CaL}, i_{Kr}, i_{NaCa}, and the inward rectifier currents were developed from new human cardiomyocyte data, so that compared to earlier human ventricular models it was based even more closely on species-specific data.

The model divides a cell into four compartments. Along with the bulk myoplasm and the subsarcolemmal space, which both appear in the ten Tusscher *et al* model, the O'Hara *et al* model divides the SR into two compartments: the JSR and NSR. The two compartments are connected by a calcium tranlocation flux from the NSR to the JSR defined as $J_{tr} = ([Ca^{2+}]_{NSR} - [Ca^{2+}]_{JSR})/\tau_{tr}$.

The model includes 16 transmembrane currents: the fast and late sodium currents (i_{Na} and i_{NaL}); the L-type calcium current including sodium and potassium currents through the L-type channels (i_{CaL}, i_{CaNa}, and i_{CaK}); the transient outward, rapid and slow delayed rectifier, and inward rectifier potassium currents (i_{to}, i_{Kr}, i_{Ks}, and i_{K1}); the myoplasmic and subspace components of the Na–Ca exchanger current (i_{NaCa_i} and $i_{NaCa_{ss}}$); the Na/K pump current (i_{NaK}); the background sodium, calcium, and potassium currents (i_{Nab}, i_{Cab}, and i_{Kb}); and the sarcolemmal calcium pump current (i_{pCa}).

A total of 41 variables are included: voltage; nine concentrations including sodium, potassium, and calcium in the myoplasm and the subspace along with calcium in the NSR and JSR, and CaMK (Ca/calmodulin-dependent protein kinase II); 31 gating variables including six i_{Na} gates, three i_{NaL} gates, six i_{to} gates, nine i_{CaL} gates, two i_{Kr} gates, two i_{Ks} gates, one i_{K1} gate, and two RyR channel gates. Other than one i_{CaL} gate (n), all the gating variables are of the HH type.

The ORd model can reproduce experimental results such as AP morphology, APD rate dependence, and restitution and rate dependence of $[Ca^{2+}]_i$ and $[Na^{+}]_i$. Figure 4.14 shows the various ion concentrations and transmembrane currents with endocardial parameters.

The ORd model has been used as a surrogate to study the effects of drugs on multiple cardiac ion channels. Li *et al* [35] incorporated a nine-state Markov model of the hERG (human Ether-a-go-go-related gene) channel into the ORd model, thereby replacing the original i_{Kr} current, and studied the channel-blocking effects of 12 pro-arrhythmic drugs. Through this process, they discovered a metric (q_{net}, the total charge for a few selective currents) that correlates well with pro-arrhythmia risk

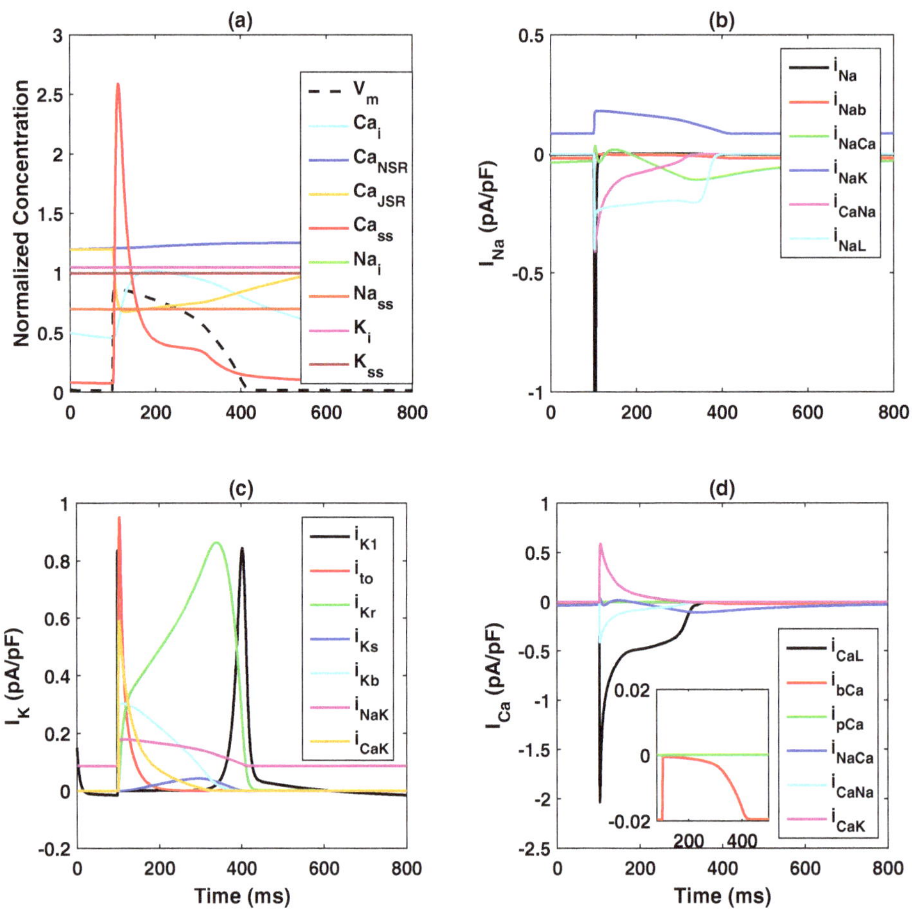

Figure 4.14. O'Hara *et al* (2011) model with endocardial parameters. (a) Normalized transmembrane voltage and eight normalized ion concentrations. The normalization constants are $[Ca^{2+}]_i$ = 0.000 2 mM L^{-1}, $[Ca^{2+}]_{NSR}$ = 1.0 mM L^{-1}, $[Ca^{2+}]_{JSR}$ = 1.0 mM L^{-1}, $[Ca^{2+}]_{ss}$ = 0.001 2 mM L^{-1}, $[Na^{+}]_i$ = 10.0 mM L^{-1}, $[Na^{+}]_{ss}$ = 10.0 mM L^{-1}, $[K^{+}]_i$ = 138.0 mM L^{-1}. (b) $[K^{+}]_{ss}$ = 145.0 mM L^{-1}. (b) Six sodium-related membrane currents: i_{Na}, i_{Nab}, i_{NaCa}, i_{NaK}, i_{CaNa}, and i_{NaL}. Note that i_{Na} falls outside the vertical range of the plot. (c) Seven potassium-related membrane currents: i_{K1}, i_{to}, i_{Kr}, i_{Ks}, i_{Kb}, i_{NaK}, and i_{CaK}. (d) Six calcium-related membrane currents: i_{CaL}, i_{bCa}, i_{pCa}, i_{NaCa}, i_{CaNa}, and i_{CaK}.

levels [12], and they quantified the uncertainties in risk assessment propagated from the uncertainties in patch-clamp data [5]. This study potentially initiated a new paradigm for assessing pharmacological effects using computer simulations.

4.3.9 Some dynamics of cardiac models in 2D

During normal sinus rhythm, the electrical signals emerging from the His–Purkinje system into the ventricles propagates an almost simultaneous smooth wave activation throughout the heart. All the models presented here can produce waves when cells are connected and simulated as a tissue in one, two and three dimensions, and can be used to study normal propagation and abnormal dynamics responsible for electrophysiological arrhythmias. Functional reentry, for example, happens when static or dynamically induced heterogeneties disrupt the smooth propagation creating a conduction block in one region of the heart, forming a spiral wave that continuously rotates upon itself. The frequency of the rotation is usually higher than that of sinus rhythm, thereby causing the heart to contract faster at this higher frequency and with less force. Reentry is one of the most common mechanisms for tachycardia and can be simulated with these models in two and three dimensions. It is widely believed that fibrillation results when the spiral wave further breaks into multiple waves and the tissue becomes highly unsynchronized, and these models can also show various mechanisms [14] that can show these multiple spiral wave states, some in their original parameters and others by modifications in parameters to mimic disease conditions or arrhythmic drug effects.

The dynamics of spiral waves is commonly characterized by the spiral wave tip trajectory. The tip of the spiral wave is often defined as the phase singularity where the wave front meets the wave back and can be calculated as the point with zero normal velocity [16] (other mathematical definitions exist). The type of tip trajectory is closely related to the excitability of the system. In cardiac models, the excitability can be adjusted by modifying the conductance of the sodium channel (g_{Na}). When excitability increases, the tip trajectory changes from a circular core to various shapes with petals including epicycloidal (inward petal), cycloidal (petals with translational motion) and hypercycloidal (outward petals), to hypermeandering and linear cores. Figure 4.15 shows the change of the tip trajectory of the FK model with the MBR parameter set by varying the conductance for the fast inward current (g_{fi}).

Figure 4.16 shows the behavior of spiral waves initiated using the cardiac models described here, except for the Noble model as it is for Purkinje fibers. There are multiple numerical methods for the integration of these models in two and three dimensions [48], in here we used finite differences [28] coded and run in WebGL using the Abubu.js library [30, 31]. The grid size of all the simulations was 512×512, and the domain size was $18 \times 18\ \mathrm{cm}^2$ except for the MFHN and Karma models (512×512 in dimensionless units). The time step was $\delta t = 0.1$ ms and the diffusion coefficient was $1.0 \times 10^{-3}\ \mathrm{cm}^2\ \mathrm{ms}^{-1}$ except for the MFHN and the Karma model (1.0). The initial conditions of all the state variables are either from the original papers or from the downloaded CellML files. At the beginning of the simulation, a

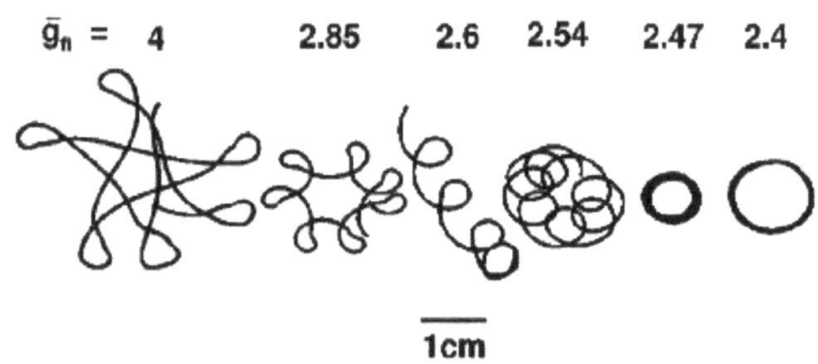

Figure 4.15. Change of the tip trajectory as the excitability is decreased in the Fenton–Karma model. Left to right: hypermeandering, hypocycloidal, cycloidal, epicycloidal, small circular core, large circular core. Figure reproduced from [16] with permission.

plane wave was initiated, propagating from left to right in the square domain. When the wave passed the mid-point of the domain, the voltage of the bottom half of the domain was set to be above the excitation threshold to break the symmetry. Then a spiral wave was initiated. The total simulation time ranged from 1 s to 5 s. See the attached video files. For the ORd model, midmyocardial parameters were used. For the ten Tusscher model, epicardial parameters were used and the membrane capacitance was set to 0.185 μF cm^{-2}. For the Morotti *et al* model, we used the epicardial parameters. For the FK model, the BR parameter sets were used. For the Karma model, the parameters from table 1 in [32] were used.

For both the MFHN and the ORd models, spiral waves have circular cores, and for the Morotti *et al* model the trajectory is hypocycloidal (outward petals) The Karma the BR, LRI, FK, and ten Tusscher *et al* models all show sustained wave break.

While the simulations shown here are in square domains with homogeneous dynamics. Variations in ion channel conductances can be included to account for action potential dispersion existing in the heart as well as fiber anisotropy to perform realistic 3D heart studies [34, 39] with all these cell models.

4.4 Discussion

In this chapter, we have given an overview of mathematical models of the electrophysiology of cardiac cells, including a classification scheme across four modeling dimensions. We also have described eight models in detail and provided a gallery of images showing spiral-wave behavior believed to be associated with many forms of tachycardiac and fibrillation.

Modeling is an exercise in developing a simplified representation of a system; as such, it requires the modeler to make choices of what features to include based on the intended use of the model. No model can be designed to reproduce all possible phenomena [60] and, as for example in the case of the Noble model, not all phenomena that can be reproduced arise in models from the proper mechanisms. Furthermore, differences in modeling choices as well as specific data used in developing models mean that not all models designed to represent the same system give rise to the same behavior [4, 6, 7], as seen by the difference in spiral-wave

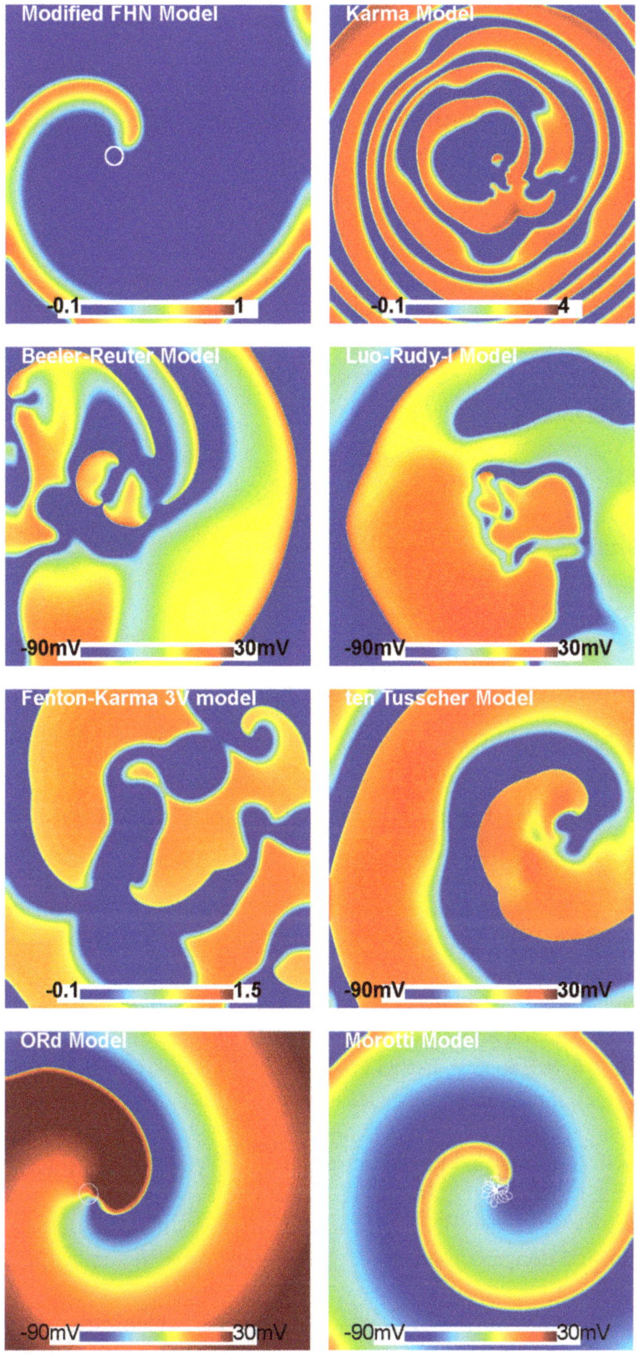

Figure 4.16. The behavior of spiral waves in 2D for eight cardiac models. Videos available at https://iopscience.iop.org/book/978-0-7503-2064-1.

dynamics for the ten Tusscher *et al* and O'Hara *et al* models of human ventricular cells.

For these reasons, care should be taken when using models to make predictions. The developing field of uncertainty quantification may provide new tools for interpreting predictions made from models.

List of video files

Please find below a list of video files of 2D spiral waves and spiral wave breakups of eight cardiac models that are available in the multimedia tab at https://iopscience.iop.org/book/978-0-7503-2064-1.

1. MFHN_2D.avi,
2. Karma_2D.avi,
3. FHN3V_2D.avi,
4. BR_2D.avi,
5. LRI_2D.avi,
6. tenTusscher_2D.avi,
7. ORd_2D.avi,
8. Morotti_2D.avi.

Acknowledgements

This work was supported in part by the National Science Foundation under grants CNS-1446312 and CNS-1446675 and the National Institutes of Health under grant 1R01HL143450-01. The authors thank Conner Herndon (Georgia Tech) for his valuable input.

References

[1] The cellml project–cellml

[2] Beeler G W and Reuter H 1962 A modification of the Hodgkin–Huxley equations applicable to Purkinje fibre action and pace-maker potentials *J. Physiol.* **268** 177–210

[3] Bondarenko V E, Szigeti G P, Bett G C, Kim S J and Rasmusson R L 2004 Computer model of action potential of mouse ventricular myocytes *Am. J. Physiol. Heart Circ. Physiol.* **287** H1378–403

[4] Bueno-Orovio A, Cherry E M and Fenton F H 2008 Minimal model for human ventricular action potentials in tissue *J. Theor. Biol.* **253** 544–60

[5] Chang K C *et al* 2017 Uncertainty quantification reveals the importance of data variability and experimental design considerations for in silico proarrhythmia risk assessment *Front. Physiol.* **8** 917

[6] Cherry E M and Evans S J 2008 Properties of two human atrial cell models in tissue: restitution, memory, propagation, and reentry *J. Theor. Biol.* **254** 674–90

[7] Cherry E M and Fenton F H 2007 A tale of two dogs: analyzing two models of canine ventricular electrophysiology *Am. J. Physiol.-Heart Circ. Physiol.* **292** H43–55

[8] Cherry E M and Fenton F H 2008 Visualization of spiral and scroll waves in simulated and experimental cardiac tissue *New J. Phys.* **10** 125016

[9] Courtemanche M and Winfree A T 1991 Re-entrant rotating waves in a Beeler-Reuter based model of two-dimensional cardiac electrical activity *Int. J. Bifurcation Chaos Appl. Sci. Eng* **1** 431–44

[10] Courtemanche M 1996 Complex spiral wave dynamics in a spatially distributed ionic model of cardiac electrical activity *Chaos* **6** 579–600

[11] DiFrancesco D and Noble D 1985 A model of cardiac electrical activity incorporating ionic pumps and concentration changes *Philos. Trans. R. Soc. Lond. B Biol. Sci.* **307** 353–98

[12] Dutta S, Chang K C, Beattie K A, Sheng J, Tran P N, Wu W, Wu M, Strauss D G, Colatsky T and Li Z 2017 Optimization of an in silico cardiac cell model for proarrhythmia risk assessment *Front. Physiol.* **8** 616

[13] Espinosa L 1998 L'Echange Na+ /Ca2+ dans l'Hypertrophie Ventriculaire D'Altitude chez le Rat: Etude Electrophysiologique et Utilisation du Modele 'Oxsoft Heart' https://models.cellml.org/exposure/3ead51a9899108b4f8c8177c0b5f2421/espinosa_1998_normal.cellml/view

[14] Fenton F H, Cherry E M, Hastings H M and Evans S J 2002 Multiple mechanisms of spiral wave breakup in a model of cardiac electrical activity *Chaos* **12** 852–92

[15] Fenton F H and Cherry E M 2008 Models of cardiac cell *Scholarpedia* **3** 1868

[16] Fenton F H and Karma A 1998 Vortex dynamics in three-dimensional continuous myocardium with fiber rotation: filament instability and fibrillation *Chaos* **8** 20–47

[17] Fenton F H, Cherry E M and Glass L 2008 Cardiac arrhythmia *Scholarpedia* **3** 1868

[18] Fenton F H, Cherry E M, Hastings H M and Evans S J 2002 Real-time computer simulations of excitable media: Java as a scientific language and as a wrapper for C and Fortran programs *Biosystems* **64** 73–96

[19] Filippi S, Gizzi A, Cherubini C, Luther S and Fenton F H 2014 Mechanistic insights into hypothermic ventricular fibrillation: the role of temperature and tissue size *Europace* **16** 424–34

[20] FitzHugh R 1961 Impulses and physiological states in theoretical models of nerve membrane *Biophys. J.* **1** 445–66

[21] Fleckenstein A 1983 History of calcium antagonists *Circ. Res.* **52** I3–16

[22] Goette A *et al* 2016 EHRA/HRS/APHRS/SOLAECE expert consensus on atrial cardiomyopathies: definition, characterization, and clinical implication *Europace* **18** 1455–90

[23] Herndon C and Fenton F H Long refractory period protects alligators from cardiac fibrillation across temperature changes *Manuscript under construction*

[24] Hilgemann D W and Noble D 1987 Excitation-contraction coupling and extracellular calcium transients in rabbit atrium: reconstruction of basic cellular mechanisms *Proc. R. Soc. Lond. B Biol. Sci.* 230 163–205

[25] Hodgkin A L and Huxley A F 1952 A quantitative description of membrane currents and its application to conduction and excitation in nerve *J. Physiol.* **117** 500–44

[26] Hund T J and Rudy Y 2004 Rate dependence and regulation of action potential and calcium transient in a canine cardiac ventricular cell model *Circulation* **110** 3168–74

[27] Iyer V, Mazhari R and Winslow R L 2004 A computational model of the human left-ventricular epicardial myocyte *Biophys. J.* **87** 1507–25

[28] Ji Y C and Fenton F H 2016 Numerical solutions of reaction-diffusion equations: Application to neural and cardiac models *Am. J. Phys.* **84** 626–38

[29] Ji Y C, Gray R A and Fenton F H 2015 Implementation of contraction to electrophysiological ventricular myocyte models, and their quantitative characterization via post-extrasystolic potentiation *PLoS One* **10** e0135699

[30] Kaboudian A, Cherry E M and Fenton F H 2019 Real-time interactive simulations of large-scale systems on personal computers and cell phones: toward patient-specific heart modeling and other applications *Sci. Adv.* **5** eaav6019

[31] Kaboudian A, Cherry E M and Fenton F H 2019 Large-scale interactive numerical experiments of chaos, solitons and fractals in real time via GPU in a web browser *Chaos Solitons Fract.* **121** 6–29

[32] Karma A 1994 Electrical alternans and spiral wave breakup in cardiac tissue *Chaos* **4** 461–72

[33] Kogan B J 2009 *Introduction to Computational Cardiology: Mathematical Modeling and Computer Simulation* (Berlin: Springer)

[34] Krishnamoorthi S *et al* 2014 Simulation methods and validation criteria for modeling cardiac ventricular electrophysiology *PLoS One 9* **9** e114494

[35] Li Z, Dutta S, Sheng J, Tran P N, Wu W, Chang K, Mdluli T, Strauss D G and Colatsky T 2017 Improving the in silico assessment of proarrhythmia risk by combining hERG (human ether-á-go-go-related gene) channel-drug binding kinetics and multichannel pharmacology *Circ. Arrhythm. Electrophysiol.* **10** 3351–71

[36] Lombardo D M, Fenton F H, Narayan S M and Rappel W-J 2016 Comparison of detailed and simplified models of human atrial myocytes to recapitulate patient specific properties *PLoS Comput. Biol.* **12** e1005060

[37] Luo C H and Rudy Y 1991 A dynamic model of the cardiac ventricular action potential. I. Simulations of ionic currents and concentration changes *Circ. Res.* **74** 1071–96

[38] Luo C H and Rudy Y 1991 A model of the ventricular cardiac action potential. Depolarization, repolarization, and their interaction *Circ. Res.* **68** 1501–26

[39] Lyon A, Bueno-Orovio A, Zacur E, Ariga R, Grau V, Neubauer S, Watkins H, Rodriguez B and Minchole A 2018 Electrocardiogram phenotypes in hypertrophic cardiomyopathy caused by distinct mechanisms: apico-basal repolarization gradients versus purkinje-myocardial coupling abnormalities *EP Europace* **20** iii102–12

[40] Mahajan A *et al* 2008 A rabbit ventricular action potential model replicating cardiac dynamics at rapid heart rate *Biophys. J.* **94** 392–410

[41] Matsuoka S, Sarai N, Kuratomi S, Ono K and Noma A 2003 Role of individual ionic current systems in ventricular cells hypothesized by a model study *Jpn. J. Physiol.* **53** 105–23

[42] Mitchell C C and Schaeffer D G 2003 A two-current model for the dynamics of cardiac membrane *Bull. Math. Biol.* **65** 767–93

[43] Morotti S, McCulloch A D, Bers D M, Edwards A G and Grandi E 2016 Atrial-selective targeting of arrhythmogenic phase-3 early afterdepolarizations in human myocytes *J. Mol. Cell. Cardiol.* **96** 63–71

[44] Nagumo J, Arimoto S and Yoshizawa S 1962 An active pulse transmission line simulating nerve axon *Proc. IRE* **50** 2061–70

[45] Nemtsas P, Wettwer E, Christ T, Weidinger G and Ravens U 2010 Adult zebrafish heart as a model for human heart? An electrophysiological study *J. Mol. Cell. Cardiol.* **48** 161–71

[46] Niederer S A, Fink M, Noble D and Smith N P 2009 A meta-analysis of cardiac electrophysiology computational models *Exp. Physiol.* **94** 486–95

[47] Niederer S A, Ter Keurs H E D J and Smith N P 2009 Modelling and measuring electromechanical coupling in the rat heart *Exp. Physiol.* **94** 529–40

[48] Niederer S A *et al* 2011 Verification of cardiac tissue electrophysiology simulators using an *n*-version benchmark *Philos. Trans. R. Soc. A: Math. Phys. Eng. Sci.* **369** 4331–51

[49] Noble A, Garny A and Noble P J 2012 How the Hodgkin–Huxley equations inspired the Cardiac Physiome Project *J. Physiol.* **59** 2613–28

[50] Noble D 1962 A modification of the Hodgkin–Huxley equations applicable to Purkinje fibre action and pace-maker potentials *J. Physiol.* **160** 317–52

[51] O'Hara T, Virag L, Varro A and Rudy Y 2011 Simulation of the undiseased human cardiac ventricular action potential: Model formulation and experimental validation *PLoS Comput. Biol.* **7** e1002061

[52] Puglisi J L and Bers D M 2001 LabHEART: an interactive computer model of rabbit ventricular myocyte ion channels and Ca transport *Am. J. Physiol. Cell. Physiol.* **281** C2049–60

[53] Pullan A J, Cheng L K and Buist M L 2005 *Mathematically modelling the electrical activity of the heart: from cell to body surface and back again* (Singapore: World Scientific)

[54] Rice J J, Winslow R L and Hunter W C 1999 Comparison of putative cooperative mechanisms in cardiac muscle: length dependence and dynamic responses *Am. J. Physiol.* **276** H1734–54

[55] Rogers J M and McCulloch A D 1994 A collocation-Galerkin finite element model of cardiac action potential propagation *IEEE Trans. Biomed. Eng.* **41** 743–57

[56] Sato D, Bers D M and Shiferaw Y 2013 Formation of spatially discordant alternans due to fluctuations and diffusion of calcium *PLoS One* **8** e85365

[57] Shannon T R, Wang F, Puglisi J, Weber C and Bers D M 2004 A mathematical treatment of integrated Ca dynamics within the ventricular myocyte *Biophys. J.* **87** 3351–71

[58] ten Tusscher K H W J, Noble D, Noble P J and Panfilov A V 2004 A model for human ventricular tissue *Am. J. Physiol. Heart Circ. Physiol.* **286** H1573–89

[59] ten Tusscher K H W and Panfilov A V 2006 Alternans and spiral breakup in a human ventricular tissue model *Am. J. Physiol. Heart Circ. Physiol.* **291** H1088–100

[60] Uzelac I, Ji Y C, Hornung D, Schröder-Scheteling J, Luther S, Gray R A, Cherry E M and Fenton F H 2017 Simultaneous quantification of spatially discordant alternans in voltage and intracellular calcium in Langendorff-perfused rabbit hearts and inconsistencies with models of cardiac action potentials and ca transients *Front. Physiol.* **8** 819

[61] Xu A and Guevara M R 1998 Two forms of spiral-wave reentry in an ionic model of ischemic ventricular myocardium *Chaos* **8** 157–74

IOP Publishing

Modeling and Simulating Cardiac Electrical Activity

Trine Krogh-Madsen and David J Christini

Chapter 5

Modeling cardiomyocyte signaling pathways

Daniel Gratz, Alexander Winkle, Amara Greer-Short and Thomas J Hund

Cardiac cells have evolved elaborate signaling networks to support the coordinated response of the cell, organ, and ultimately organism to environmental stimuli, including physical and biochemical signals. Importantly, cardiac cell signaling pathways are tuned to respond across temporal and spatial scales to specific stimuli. Furthermore, dysregulation of cell signaling pathways is commonly observed in cardiovascular disease and arrhythmia. Quantitative analysis of cardiac cell signaling faces a number of unique challenges. The sheer number of constituent elements may pose a barrier to detailed characterization of the system (e.g. local concentrations and/or phosphorylation status/activity of kinases, target proteins) *in vitro* let alone *in vivo*. Furthermore, signaling pathways are often imbued with a degree of pleiotropy where the same network gives rise to multiple functional outputs. Adding to the complexity, signaling pathways tend to be highly nonlinear, lacking a well-defined beginning, middle and end. Together, the complex aspects of signaling provide both obstacles and exciting opportunities for mathematical modeling to generate a cohesive understanding of cardiac cell function. Despite these difficulties, mathematical modeling has been applied extensively to codify complex cell signaling networks across systems, including the heart. This chapter provides an introduction to the general approach for modeling cell signaling pathways followed by discussion of specific examples from efforts to understand the elaborate network of protein kinases and downstream targets which support the relatively rapid cardiac response to stress.

5.1 Introduction to modeling of protein kinase signaling pathways

One of the fundamental challenges in modeling is deciding the appropriate scope, with the tradeoff that including more detail (e.g. ancillary components) tends to expand the predictive and expositive power of the model at the expense of tractability [33]. In some cases, the need for detailed predictions is the driving force behind the model, making the added complexity acceptable. Other times, however,

doi:10.1088/978-0-7503-2064-1ch5

the need for simplicity and intuitive understanding is more important. That is, it is important to recognize that each model is usually developed with a specific application in mind, and one must be careful to assess *a priori* whether a particular question resides within the model's intended scope, and, if not, whether the model can be modified or an entirely new model is needed. The tension between power and tractability is reflected in the differences between three general classes of models, which will be discussed in this chapter (figure 5.1):

1. **Phenomenological models**—'Blackbox' representations, which empirically map an input to an output for a signaling pathway. In these models, there is often only one input (e.g. concentration of a receptor agonist), which then causes a specific response in the cell (e.g. increase in sodium current).
2. **Biophysical models**—Built from chemical and biological first principles and fit with data from *in vitro* and *in vivo* experiments, these models usually focus on one (or a small number) of signaling pathways, similar to phenomenological models, but incorporate biophysical detail on the reactions underlying the pathway.
3. **'Systems' models**—These models focus on high level understanding of how a host of interacting components give rise to emergent behavior of the intact system, with less concern for detailed representation or understanding of each individual component. 'Systems' models are particularly useful in cases where the network of interest is expansive, phenotypes are complex, and data are abundant and from disparate sources.

This chapter will discuss all three types of models, beginning with the phenomenological and biophysical models and then moving on to the higher level 'systems' models. Regardless of the specific class, it is helpful to delineate steps that are typically taken when developing a model:

1. **Determine the scope**—At the outset of a project, it is important to define the scope of the question or pathway of interest, which will in turn inform the desired class and type of model. At this stage, it is useful to map out the relevant stimuli and cellular responses for the phenomenon of interest. A literature review may be performed to identify critical signaling pathways and other elements to consider in the model. A network diagram may be useful to visualize the pathway(s) with desired inputs and outputs and lay the foundation for defining the mathematics at a later step. In some cases, the determination may be made that a model already exists to answer the question. However, modification or combination of existing models is often required. It is important to keep in mind that the scope of the project will often change as the project progresses, and this and subsequent steps will need to be revisited in an iterative fashion.
2. **Identify data sources**—Once a framework has been defined for the project, it is critical to conduct a thorough literature review to assess the availability and quality of experimental data to constrain the model. Sometimes, new experiments will be required. As much as possible, it is desirable to match experimental conditions across datasets, although some degree of variability

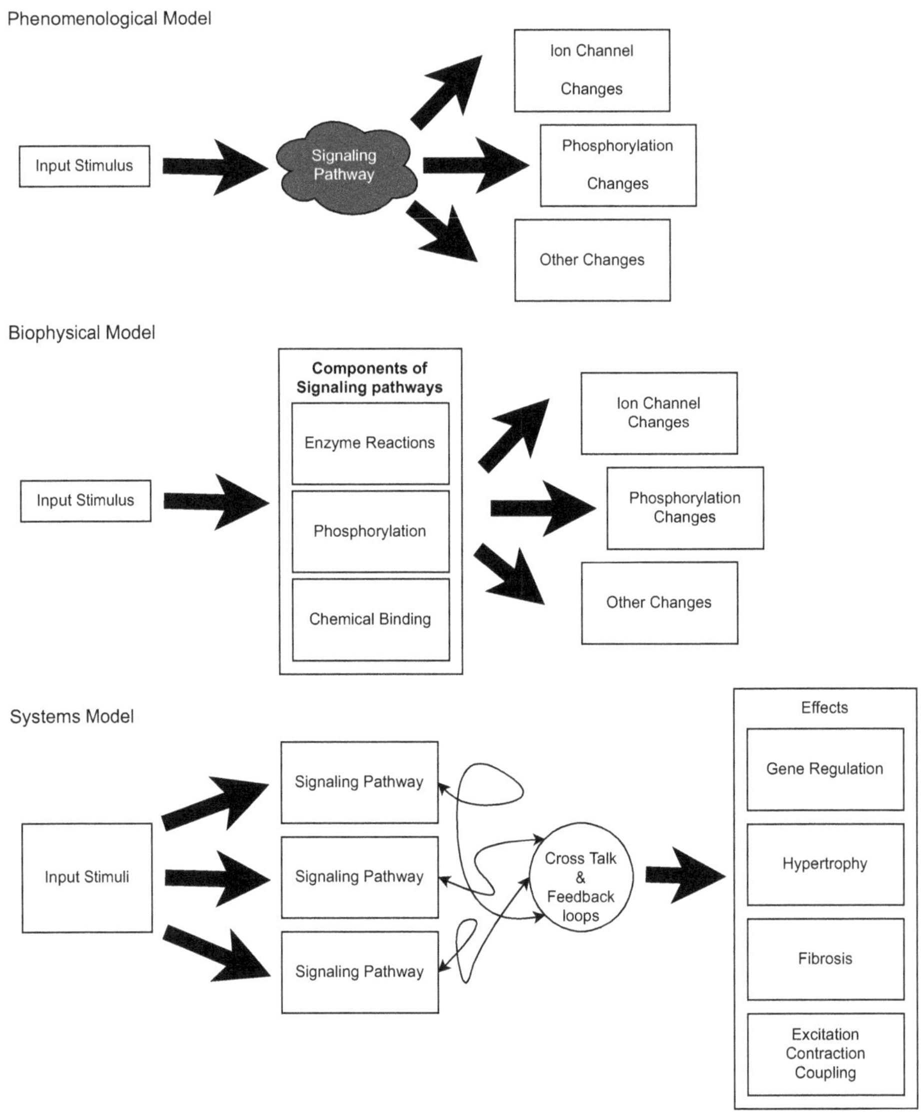

Figure 5.1. General classes of models used to study cardiac cell signaling pathways. (A) Phenomenological models take a 'black box' approach where critical system inputs (e.g. input stimulus) and outputs (e.g. ion channel or action potential changes) are considered but not specific biophysical details about the pathway itself. (B) Biophysical models explicitly represent molecular components of the pathway (e.g. biochemical reactions). (C) Systems models are best suited for higher level efforts focused, for example, on nonlinear interactions between multiple pathways giving rise to complex phenotypes (e.g. gene programs, hypertrophy, fibrosis).

between experiments/labs is unavoidable. It is helpful to take detailed notes on experimental conditions for each dataset as well as standardizing units. At this stage, it is useful to think about which data will be used to constrain the model and which will be used for validation (final step).

3. **Formulate the mathematics**—In parallel with data collection, it is possible to construct the skeleton of the model by converting the network diagram from Step 1 into formal equations. The nature of the equations will depend on the selected model class for the project. To the extent possible, it is advisable to map parameters from the equations to available experimental data.
4. **Estimate unknown model parameters**—Model parameters (e.g. enzyme concentrations, reaction rates) may often be taken directly from the literature or fit to available data. It is useful at the beginning of this step to define a coherent optimization strategy for how data fitting will be performed and 'goodness of fit' evaluated. Nevertheless, it is unlikely that a value will be known for every model parameter at the outset. *De novo* experiments may be needed to define a critical parameter. Sometimes, even if an exact value is not known, a 'ballpark' estimate for a parameter will be sufficient.
5. **Translate into computer code**—After the mathematics and key parameters have been defined (at least on a first pass), the task may begin of translating the model into a computer programming language with incorporation into existing models if applicable. The exact language will depend on the scope of the problem, user preference, and available resources. Initial parameter sensitivity analysis may be useful at this step to help determine the importance of precisely defining certain parameters.
6. **Validate the model**—The final step is to validate the model using data that has not been applied in the previous steps. Typically, data on the cellular outputs of interest (e.g. action potential recordings or Ca^{2+}/contractility measurements) are used for this purpose. In the event that model validation fails (typically the outcome in the early stages of a project), the researcher will need to revisit previous steps and reflect on possible sources of error, which may involve improper parameter fitting or incorrect model assumptions. Sometimes, failed validation may lead to an important discovery by identifying a critical component not accounted for previously in the model.

Before discussing specific examples of the different signaling pathway models, we will discuss a few relevant concepts from modeling biochemical reactions with a focus on the mathematics and corresponding data commonly used to build the model. The easiest and perhaps most utilized way to simulate cell signaling is through a phenomenological approach, which typically focuses on capturing the steady-state effects of activating (or inhibiting) a particular pathway on select downstream targets. In this case, model parameters are fit to experimental measurements on the response of key cell properties (e.g. major ion currents) to a defined stimulus. Depending on availability of data, the model may describe the response to a single concentration of a receptor agonist, for example, or perhaps a range of concentrations to be able to simulate dose response curves. Phenomenological models may also involve defining a set of simplified rules derived from experimental observations to map an input to an output. Such empirically derived 'maps' have been applied extensively to study dynamics of cardiac action potential generation and conduction (e.g. action potential restitution curve) [35].

While phenomenological models are advantageous for their relative simplicity, in some cases a biophysical model is necessary to provide the desired level of mechanistic insight. In general, using a biophysical model is more conducive to studying dynamic changes in cell function with acute activation of a signaling pathway, as opposed to just examining the new steady-state associated with a perturbation. Furthermore, biophysical models allow for the study of complex interactions, involving, for example, cross talk between signaling pathways, which is particularly important for Ca^{2+}-dependent signaling cascades in cardiac myocytes where intracellular Ca^{2+} changes by an order of magnitude on beat-to-beat basis.

5.1.1 Biophysical representation of chemical reactions

Biophysical models commonly apply mass action kinetics (the law of mass action) to formulate the mathematics describing a simple chemical reaction. For example, given two chemical species A and B that react to form a product AB, the law of mass action states that the rate of the corresponding reaction is proportional to the concentrations of A and B. Thus, for the reaction: $\mathrm{A} + \mathrm{B} \underset{k_r}{\overset{k_f}{\rightleftharpoons}} \mathrm{AB}$, the forward reaction rate is $k_f[A][B]$ and the reverse rate is $k_r[AB]$. At equilibrium, $k_f[A][B] = k_r[AB]$. It is straight forward to derive mathematical equations to describe this reaction and estimates for k_f and k_r may often be derived from published experimental data. Relevant to our discussion of protein kinase networks, the law of mass action may be applied to enzyme mediated reactions where a substrate S binds to an enzyme E and produces a product P, without affecting the enzyme.

$$\mathrm{E} + \mathrm{S} \underset{k_r}{\overset{k_f}{\rightleftharpoons}} \mathrm{ES} \overset{k_{\mathrm{cat}}}{\longrightarrow} \mathrm{E} + \mathrm{P}, \tag{5.1}$$

Applying the law of mass action to equation (5.1) with some key assumptions gives rise to the widely used Michaelis–Menten kinetics equation (equation (5.2)), which describes the formation rate for P as a function of the concentrations of S and E [3]:

$$\frac{d[P]}{dt} = V_{\mathrm{max}}\frac{[S]}{K + [S]} = k_{\mathrm{cat}}[E]_{\mathrm{tot}}\frac{[S]}{K + [S]}, \tag{5.2}$$

where V_{max} is the maximum rate of the reaction and K is the concentration of S at which $\frac{d[P]}{dt}$ is half of V_{max}. K, also referred to as the half-saturation coefficient, is a function of k_{cat} (the catalytic rate constant), k_f (the forward rate constant), and k_r (the reverse rate constant). One of two different assumptions must hold for the Michaelis–Menten equation to be applied. In the first case, commonly referred to as the quasi-equilibrium-assumption (QEA), $\mathrm{E} + \mathrm{S} \underset{k_r}{\overset{k_f}{\rightleftharpoons}} \mathrm{ES}$ is assumed to be at equilibrium, with the enzyme binding to the substrate at a much faster rate than production of product (k_{cat}/k_r is much smaller than 1) and $K = K_d = k_r/k_f$, where K_d is the dissociation constant. An alternative assumption is that $[ES]$ is a constant throughout the reaction, also known as the quasi-steady-state-assumption (QSSA).

In this case, $K = K_M = \frac{k_r + k_{\text{cat}}}{k_f}$ is called the Michaelis constant, but the overall equation remains unchanged. In both the QEA and QSSA cases, it is also assumed that formation of the product is irreversible. In biological systems, the assumption of irreversibility is often true when $[P]$ is very small, either because the reaction has not been running for very long, or because P is being used or transported elsewhere. When such conditions are not met, the Michaelis–Menten equation does not apply and other more complex models may be necessary.

The Hill–Langmuir equation (or Hill equation, equation (5.3)) is a hyperbolic curve derived from the law of mass action, similar to the Michaelis–Menten equation, to describe receptor binding to a ligand:

$$P = C \cdot \frac{L^n}{K^n + L^n}, \tag{5.3}$$

where P is the concentration of receptor with bound ligand, L is the unbound ligand concentration, C is the total receptor concentration, K is the ligand concentration resulting in P to be half its maximal value (analogous to half-saturation coefficient for Michaelis–Menten equation), and n is the Hill coefficient used to describe cooperativity of binding.

In the case of cell signaling networks, the Hill equation may be used to describe binding of a ligand to a receptor, while Michaelis–Menten kinetics may be assumed for phosphorylation of an ion channel by a protein kinase, for example. The Hill equation is sometimes interpreted to describe the physiological response to a ligand (rather than receptor concentration). In this approach, a model typically tracks two populations of a system component: for example, bound and unbound receptors or non-phosphorylated and phosphorylated channels.

5.1.2 Physical derivation of rate constants for biochemical reactions

Once the appropriate mathematics have been defined for a reaction, it is necessary to define associated parameters. In the case of biochemical reactions relevant to cell signaling, oftentimes it is important to be able to derive accurate values for chemical reaction rate constants. The Scatchard equation (equation (5.4)) is commonly used in experiments to derive K_d by calculating $\bar{n}$, the mean ligand bound per receptor

$$K_d = [L]\left(\frac{1}{\bar{n}} - 1\right). \tag{5.4}$$

In experiments, it is important to ensure the kinetic environment is as close to biological conditions as possible, as the value of K_d will vary with temperature and pH. For this reason, it is ideal to use *in vitro* tissue preparations for these studies [36]. The most widely-accepted method for the study of receptor and enzyme kinetics employs the use of a radio-tracer to monitor changes in bound and unbound enzyme or receptor.

Individual tissue preparations are supplanted with concentrations of ligand ranging several orders of magnitude, centered around the expected dissociation constant K_d [36]. It is important in these studies to ensure that the radioactive tag chosen does not have an effect either direct or indirect on the pathway being monitored. If non-specific binding (NSB) is expected, a competitive molecule must be chosen and introduced in parallel studies to determine its effect on the total binding. This is then subtracted from the total measured binding to give the specific binding of the desired complex, yielding saturation and K_d measures [21].

The tissue is incubated with a tagged ligand followed by a wash to remove unbound ligand. The incubation period is chosen to be long enough to allow the system to reach equilibrium, but not so long that the radioactivity of the tracer is diminished. The post-incubation wash is necessary so that only the bound complexes are included in the measurement. The presence of these bound complexes can then be quantified by radiographic techniques such as autoradiography or scintillation counting.

5.2 Mathematical modeling of electrical signaling

Having established some of the fundamental concepts related to mathematical modeling of cell signaling, we now turn to specific examples of how modeling has been applied to generate new insight into the function of signaling pathways with a particular focus on cardiac electrophysiology. Of course, mathematical modeling has a rich and storied history in the larger canon related to excitable cells, touched on briefly here but addressed in greater detail in chapter 1 of this book and elsewhere [13, 38]. In fact, our fundamental understanding of action potential generation is built upon mathematical modeling efforts from almost 70 years ago. Leveraging the advent of the voltage clamp technique, experimental tractability of the giant squid axon, and a set of prescient assumptions, Hodgkin and Huxley formulated a set of first order ordinary differential equations that reproduced key features of the action potential [16]. Briefly, their Nobel prize-winning effort began with the concept that flow of ions (mainly Na^+ and K^+) across the cell membrane through independent voltage- and time-dependent pathways (ionic independence) gave rise to the action potential. Importantly, the assumption of ionic independence allowed them to design elegant experiments to isolate and characterize the Na^+ and K^+ currents, which were then integrated back into a mathematical model of the intact system. The success of their approach spawned countless action potential models corresponding to a wide variety of cell types, including nerve, muscle and pancreatic cells [5, 12, 31]. However, it is perhaps the cardiac field where the paradigm and associated model established by Hodgkin and Huxley have been most widely and successfully applied. After publication of the Hodgkin and Huxley equations in 1952, the British biologist Denis Noble modified their formulation to produce a more cardiac-like action potential with a much longer plateau phase compared to the neuron [32]. The Noble group and others would refine their models over the ensuing decades to produce more detailed and accurate representations of action potentials from different species and regions of the heart made possible by a host of technological advances. Discovery of the patch clamp technique provided some of the first concrete experimental evidence for existence of ion channels, thereby confirming a central assumption of the Hodgkin–Huxley paradigm some 25 years later. Patch clamp also led to an explosion in available electrophysiology data to constrain and validate physiological action potential models, as did subsequent advances in molecular biology and biochemistry.

An important conceptual breakthrough that led not only to a leap in the utility but also complexity of action potential models came with the recognition that ion flux across the cell membrane would not only change the transmembrane potential but also affect intracellular ion concentrations potentially on a beat-to-beat basis. This paradigm shift began with efforts to incorporate dynamic changes in the intracellular concentration of Ca^{2+} (Ca^{2+} transient) into a model of the ventricular action potential [29]. Eventually, a host of so-called 'second generation' models would emerge, which incorporated a detailed accounting of dynamic changes in Na^{+}, K^{+} and Ca^{2+} over the course of the action potential [8, 9, 37]. Among the most successful and widely utilized second generation models is the Luo–Rudy dynamic model of mammalian ventricular cell [26], notable for its detailed characterization of a relatively large number of ion channels, careful attention to dynamic ion concentration changes including key pumps and exchangers as well as buffers critical for normal ion homeostasis, and multiple compartments for Ca^{2+} cycling (network and junctional sarcoplasmic reticulum (SR) spaces in addition to cytoplasm). Action potential models have continued to evolve to allow for simulation of genetic variants linked to arrhythmia, drug interactions, and other complex phenomena (addressed elsewhere in this book) in a wide range of species, including human.

5.3 Mathematical modeling of protein kinase signaling pathways

Intracellular Ca^{2+} serves as a critical second messenger in cardiac myocytes, acting as a signal for a large number of essential cellular functions including contraction, hypertrophy, and apoptosis [25]. The cellular effects of changes in Ca^{2+} are mediated in part by a host of Ca^{2+}-dependent kinases and phosphatases, which directly target ion channels, pumps, myofilament proteins and other elements of the cell machinery to acutely regulate their function. Cyclic AMP is another important second messenger with well characterized effects on cell function due in large part to its ability to activate protein kinase A (PKA) and downstream targets. A similar paradigm holds for other biochemical signals, including reactive oxygen species (ROS), glycans, fatty acids, and lipids. Thus, protein kinase signaling networks reside at the hub of a highly efficient communication network to link a wide range of stress stimuli to acute changes in cell function. Importantly, dysregulation of cell signaling pathways is a hallmark of cardiovascular disease and arrhythmia [19, 44]. Early attempts to simulate protein kinase signaling employed a phenomenological approach whereby steady-state effects of kinase activation were implemented by scaling model parameters associated with target proteins (e.g. ion channel conductances). An early example of this approach comes from a partner paper to the Luo-Rudy dynamic model [27], in which key model parameters (L-type Ca^{2+} channel conductance and rate of Ca^{2+} release from the sarcoplasmic reticulum) were scaled to simulate β-adrenergic receptor (βAR) activation and arrhythmia in the setting of Ca^{2+} overload. In this case, no new equations were needed to simulate

βAR activation as existing model parameters were adjusted to fit steady-state experimental measurements.

A more recent example comes from Grandi, Bers and colleagues who incorporated effects of the βAR agonist isoproterenol into a model of a human atrial myocyte [14]. Similar to the approach used in the previous example, model parameters were scaled to fit experimental data on cell response to isoproterenol. However, our understanding of the β-adrenergic pathway had developed considerably since 1994 allowing for Grandi and colleagues to more precisely model the effects known to be caused by PKA and cAMP supported by a large amount of experimental data from the literature. The model was shown to have triggered activity in the presence of isoproterenol and a partial block of the ultra-rapid delayed rectifier K^+ current (IKur), whereas the effect of isoproterenol alone did not immediately cause inappropriate depolarizations, suggesting a protective role for IKur in opposing the effects of increased depolarizing current from the L-type calcium channel.

The phenomenological approach depicted in the examples from the Rudy and Bers groups remains in wide use to study kinase signaling networks and is attractive in that it eliminates the need to model complex dynamics over a range of agonist concentrations while still having the ability to make interesting predictions. However, with the turn of the century and millennium, a new generation of action potential models emerged which employed a more detailed biophysical approach that accounted for dynamic kinase activity and effects on key substrates [18, 41].

5.3.1 Dynamic models of CaMKII signaling

Ca^{2+}/calmodulin-dependent protein kinase II (CaMKII) is a ubiquitous and multifunctional kinase with diverse roles in cardiac pathophysiology (figure 5.2) [4, 17, 44]. The α and β CaMKII isoforms are primarily present in neuronal tissue, while γ and δ are expressed throughout the body, including the heart. The kinase forms as a holoenzyme composed of 12 monomers which are each comprised of an association, regulatory, and catalytic domain. Under basal conditions, the regulatory domain is bound to the catalytic domain to inhibit kinase activity (autoinhibitoin). Ca^{2+}/CaM binding to the CaMKII regulatory domain disrupts binding with the catalytic domain and allows for subsequent phosphorylation of target proteins, including neighboring CaMKII subunits (autophosphorylation). Autophosphorylation and/or oxidation of the regulatory domain prevents association between the regulatory and catalytic domains even in the absence of Ca^{2+}/CaM, thereby conferring a Ca^{2+}-independent activation state (autonomy).

Once activated, CaMKII phosphorylates serine/threonine residues on a large host of intracellular targets, including the voltage-gated Na^+ channel ($Na_v1.5$), L-type Ca^{2+} channel (Cav1.2), SR ryanodine receptor Ca^{2+} release channel (RyR2), and phospholamban (PLB) (figure 5.2). Not surprisingly, based on its targets, acute CaMKII hyperactivity is linked with altered Na^+ and Ca^{2+} homeostasis, membrane excitability, and arrhythmias [28]. Chronic CaMKII hyperactivity, as demonstrated in CaMKII overexpression models, results in heart failure and promotes

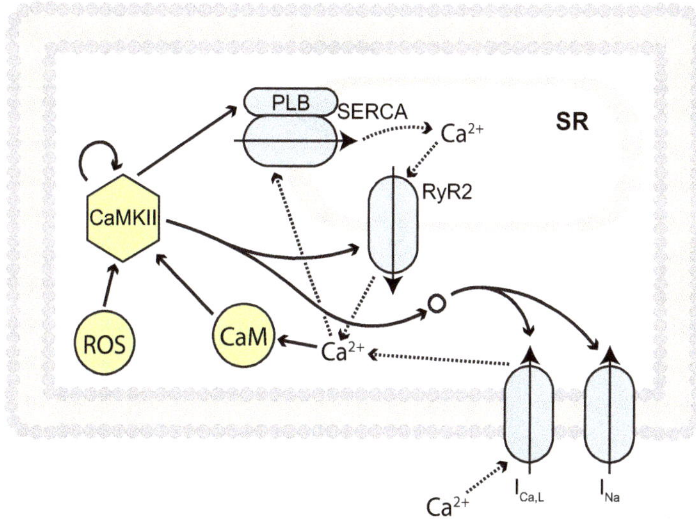

Figure 5.2. Schematic of the CaMKII signaling pathway integrated in a model of the cardiac action potential. Ca^{2+}/calmodulin-dependent protein kinase II (CaMKII) may be activated by Ca^{2+}/calmodulin (CaM) or reactive oxygen species (ROS). Activated CaMKII in turn phosphorylates downstream targets to alter function of the L-Type Ca^{2+} current (ICa,L), voltage-gated Na^+ current (INa), sarcoplasmic reticulum (SR) ryanodine receptor Ca^{2+} release channel (RYR2) and phospholamban (PLB). CaMKII targets are depicted in Hund–Rudy dynamic model [18] and Soltis–Saucerman model [43].

Ca^{2+}-mediated arrhythmias. Furthermore, studies of heart failure have shown that CaMKII regulates structural remodeling, such as hypertrophy, apoptosis, and fibrosis [7, 24, 45].

Due to its dependence on intracellular Ca^{2+} and regulation of ion channels involved in generation of the action potential, CaMKII makes for an attractive target for a more biophysical modeling approach. The Hund–Rudy dynamic (HRd) canine ventricular model [18] was an early effort to incorporate a dynamic representation of CaMKII signaling into an integrated model of the action potential and Ca^{2+} transient [19]. In the HRd model, CaMKII is activated by Ca^{2+} in a restricted space, where Ca^{2+} reaches much higher concentrations than in the bulk cytoplasm. CaMKII is modeled as having three possible states: inactive ($CaMKII_{inactive}$), Ca^{2+}/calmodulin bound active state ($CaMKII_{bound}$), and autonomous active state ($CaMKII_{trap}$) [15] (figure 5.3). The model tracks the fraction of CaMKII subunits in the active state rather than the exact concentration. CaMKII is capable of phosphorylating downstream targets in either the $CaMKII_{bound}$ or $CaMKII_{trap}$ state. Calmodulin is assumed to be saturated in the model and therefore not the limiting reactant. Key model parameters were assumed to be instantaneous functions of CaMKII activity and fit to data from the literature. This model is an example of a hybrid approach using biophysical detail to track kinase activity with a phenomenological approach to link kinase activity to targets.

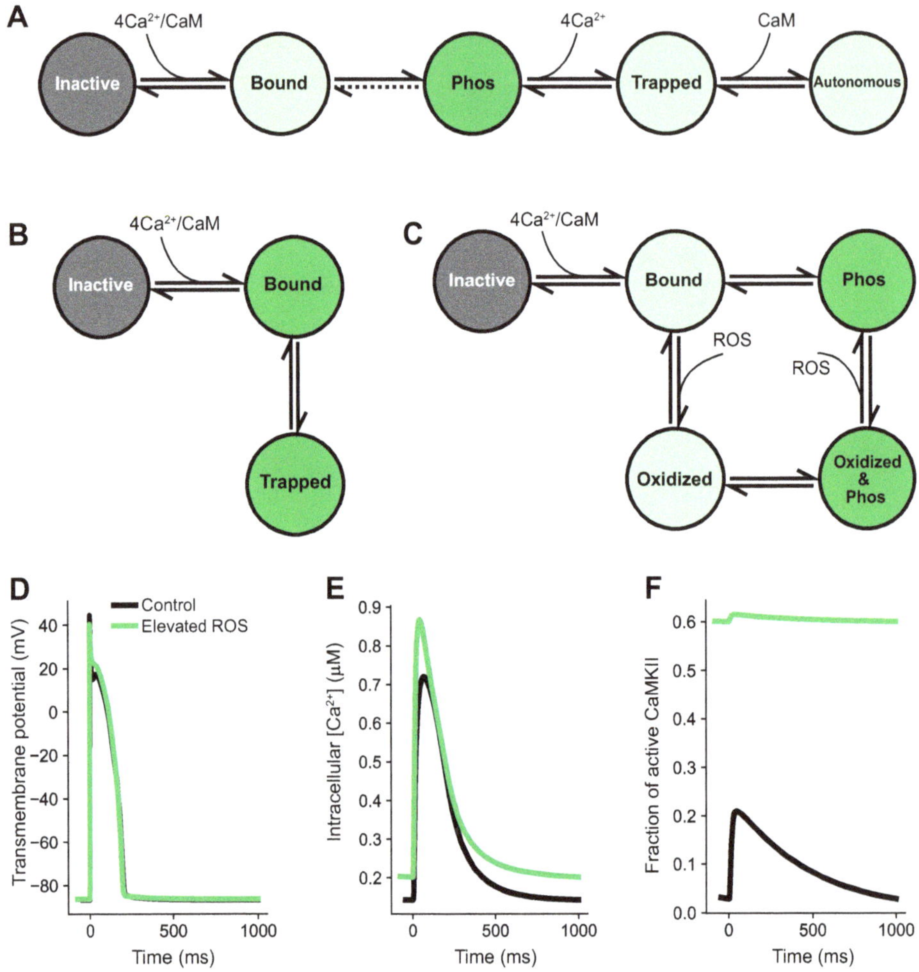

Figure 5.3. Mathematical models of CaMKII activation. (A) Depiction of inactive (grey) and active (green with dark green indicating maximal activation) CaMKII states in the model of Dupont *et al* [10]. Briefly, binding of Ca^{2+}/calmodulin ($4Ca^{2+}$/CaM) induces a transition from an inactive state to a bound, active state. Subsequent autophosphorylation induces a maximally active state (Phos), after which unbinding of Ca^{2+} and then CaM gives rise to trapped and autonomous active states, respectively. (B) Simplified model implemented in Hund–Rudy dynamic model [18]. (C) Later versions of the HRd model incorporated oxidized active states [6]. (D), (E) Simulated action potentials, Ca^{2+} transients and CaMKII activity in the HRd model for normal conditions (black lines) and in the presence of 5 μM ROS (green lines).

The core model of CaMKII activity is made up of two equations: one to calculate the fraction $\text{CaMKII}_{\text{bound}}$ (equation (5.5)) and another equation to track the change in $\text{CaMKII}_{\text{trap}}$ (equation (5.6)).

$$\mathrm{CaMKII}_{\mathrm{bound}} = \mathrm{CaMKII}_o \cdot (\mathrm{CaMKII}_{\mathrm{inactive}} + \mathrm{CaMKII}_{\mathrm{bound}}) \cdot \frac{[\mathrm{Ca}]_{\mathrm{ss}}}{[\mathrm{Ca}]_{\mathrm{ss}} + K_{m,\mathrm{CaMKII}}}. \tag{5.5}$$

In equation (5.5), CaMKII_o is the fraction of CaMKII sites bound at equilibrium, which is the maximum possible value for $\mathrm{CaMKII}_{\mathrm{bound}}$. CaMKII_o is then multiplied by the fraction of CaMKII which is available to be bound: $\mathrm{CaMKII}_{\mathrm{inactive}} + \mathrm{CaMKII}_{\mathrm{bound}}$, reflecting that inactive CaMKII can be bound, and bound CaMKII may remain bound. The resulting term, $\mathrm{CaMKII}_o \cdot (\mathrm{CaMKII}_{\mathrm{inactive}} + \mathrm{CaMKII}_{\mathrm{bound}})$ is analogous to C, the total receptor concentration, from the Hill–Langmuir equation (equation (5.5)). Multiplying this term by $\frac{[\mathrm{Ca}]_{\mathrm{ss}}}{[\mathrm{Ca}]_{\mathrm{ss}} + K_{m,\mathrm{CaMKII}}}$, where $[\mathrm{Ca}]_{\mathrm{ss}}$ is the concentration of calcium in the restricted space and $K_{m,\mathrm{CaMKII}}$ is the half-saturation coefficient of CaMKII, provides the fraction of CaMKII binding sites which are occupied by Ca^{2+}.

$$\frac{d(\mathrm{CaMKII}_{\mathrm{trap}})}{dt} = \alpha \cdot \mathrm{CaMKII}_{\mathrm{bound}}(\mathrm{CaMKII}_{\mathrm{bound}} + \mathrm{CaMKII}_{\mathrm{trap}}) - \beta \cdot \mathrm{CaMKII}_{\mathrm{trap}}. \tag{5.6}$$

Equation (5.6), describing the kinetics of CaMKII trapping, is adopted from a model of neuronal CaMKII developed by Hanson and Schulman [15], with constants α and β the on and off rates, respectively, for autophosphorylation of CaMKII to produce the trapped state, $\mathrm{CaMKII}_{\mathrm{trap}}$. $\mathrm{CaMKII}_{\mathrm{bound}}(\mathrm{CaMKII}_{\mathrm{bound}} + \mathrm{CaMKII}_{\mathrm{trap}})$ is the CaMKII available to be trapped. Notice that $\mathrm{CaMKII}_{\mathrm{trap}}$ appears in both terms, reflecting cooperativity of the trapping process. Equations (5.5) and (5.6) may be used together, to determine the fraction of CaMKII in each of the possible states, most importantly the fraction of active CaMKII ($\mathrm{CaMKII}_{\mathrm{trap}} + \mathrm{CaMKII}_{\mathrm{bound}} = \mathrm{CaMKII}_{\mathrm{active}}$), which is then used, to determine the effect of CaMKII on its downstream substrates.

In a subsequent version of the HRd model, the CaMKII module was revised to allow for not only regulation of activity by autophosphorylation but also oxidation [6]. The addition of oxidation was motivated by the fact that elevated reactive oxygen species are commonly observed in cardiovascular disease. Rate constants from the literature were mostly used for the reaction model with validation from *in vitro* measurements of CaMKII activity as a function of reactive oxygen species (H_2O_2) and Ca^{2+}/calmodulin concentrations (figure 5.3). The model predicted that increased CaMKII activation, due in part to increased oxidative stress, promoted pro-arrhythmia changes in action potential generation and conduction in tissue adjacent to a myocardial infarct (border zone).

Today, there are a variety of cardiac action potential models available that include representation of the CaMKII pathway (reviewed in detail in Onal *et al* [34]). Models have evolved to include novel aspects of CaMKII signaling, including additional kinase targets, multiple compartments for kinase activation, and interaction with other signaling networks.

5.3.2 Dynamic models of the β1-adrenergic signaling pathway

Another early example of using a biophysical based approach to simulate dynamic kinase activity comes from the work of Saucerman, McCulloch and colleagues who successfully incorporated β-adrenergic signaling into a model of the rat action potential (figure 5.4) [41, 42]. This sophisticated model relies heavily on Michaelis–Menten kinetics under the QEA assumption to model kinase activation and encompasses the signaling pathway from the membrane receptor to downstream effects on PLB, RyR2 and L-type Ca^{2+} channel. The detailed model divides the pathway into six modules: four to model proteins and reactions, and two to model effects on ion currents/fluxes. Each module was designed and developed based on data from the literature regarding biochemical interactions in the pathway.

At the center of the model is the β-adrenergic receptor module, with three primary species: the β1 receptor, the G_s protein and ligand (in this case, isoproterenol). The β1-adrenergic receptor binds the extracellular ligand to allow for subsequent binding to the receptor of the intracellular G_s protein. Binding of both the ligand and the G_s protein catalyzes an enzymatic reaction whereby G_s is split into $G_{s\,\alpha}$ and $G_{\beta\gamma}$, which

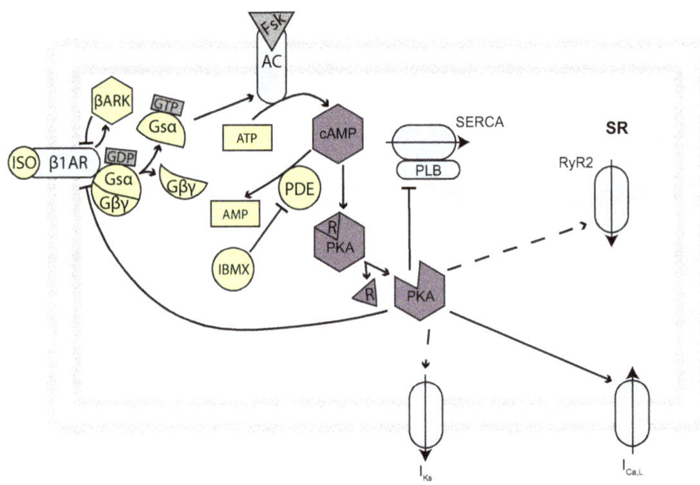

Figure 5.4. Schematic of β-adrenergic signaling pathway.The β-adrenergic signaling pathway, depicting isoproterenol (ISO) binding to the β-adrenergic receptor (β1AR). This causes the G protein subunit α (Gsα) to exchange a guanosine diphosphate (GDP) for a guanosine triphosphate (GTP) making it active and releasing it from the complex. The Gsα subunit then activates adenylyl cyclase (AC), which catalyzes the transformation of ATP into cyclic adenosine monophosphate (cAMP). cAMP then binds to Protein Kinase A (PKA) on its regulatory subunit (R) allowing the catalytic subunit to be released which then phosphorylates numerous targets. The targets included in the original Saucerman model [41] are shown with solid lines, while the targets which were added in the Soltis–Saucerman model are shown as dashed lines. PKA inhibits β1AR and Phospholamban (PLB). The inhibition of PLB reduces its ability to inhibit the SR Ca^{2+}-ATPase (SERCA) allowing it to pump more Ca^{2+} into the sarcoplasmic reticulum (SR). PKA also alters the activity sarcoplasmic reticulum (SR) ryanodine receptor Ca^{2+} release channel (RYR2) the L-Type Ca^{2+} current (ICa,L) and the slow delayed K^+ current (IKs). Other abbreviations used in the figure are β-adrenergic-receptor-kinase (βARK), G protein $\beta\gamma$subunits (G$\beta\gamma$), Forskolin (Fsk), Phosphodiesterase (PDE), 3-isobutyl-1-methylxanthine (IBMX), and adenosine monophosphate (AMP).

are used by the downstream module. Briefly, $G_{s\,\alpha}$ leads to the production of cAMP, which in turn activates PKA to phosphorylate its ion channel targets. Ion channel targets are assumed to reside in either a non-phosphorylated or phosphorylated state with distinct current properties (figure 5.4).

Mass action kinetics and Michaelis–Menten kinetics are used to transform the modules into a sets of differential and algebraic equations. Once the governing equations for a set of reactions have been defined, reasonable values for the associated parameters (e.g. rate constants between states, and concentrations) must be defined, using data from the literature mostly from *in vitro* enzyme assays. An illustrative example of the application of Michaelis–Menten kinetics can be found in the fifth module of the model, which ties PKA activity to the phosphorylation of the PLB. In a default unphosphorylated state, phospholamban (PLB) inhibits the sarcoplasmic reticulum Ca^{2+} ATPase pump responsible for for transportation of Ca^{2+} into the sarcoplasmic reticulum. By phosphorylating PLB, PKA inhibits PLB, reducing its inhibitory effect on the ATPase Pump. The net effect is that PKA phosphorylation allows the ATPase Pump to transport more Ca^{2+} into the sarcoplasmic reticulum. Acting in reverse, PP1 (protein phosphatase-1) dephosphorylates PLB, slowing the ATPase Pump and opposing the effect of PKA. Both PKA and PP1 act as enzymes which fulfil the assumptions of of Michaelis–Menten kinetics.

$$[\mathrm{PKA}] + [\mathrm{PLB}] \rightleftharpoons [\mathrm{PKA}][\mathrm{PLB}] \xrightarrow{\mathrm{k_{PKAPLB}}} [\mathrm{PKA}] + [\mathrm{PLB_p}]. \tag{5.7}$$

PKA and PLB are assumed to be in quasi-equilibrium, and the reverse reaction from PLB_p to PLB is not significant.

$$[\mathrm{PP1}] + [\mathrm{PLB}_p] \rightleftharpoons [\mathrm{PP1}][\mathrm{PLB_p}] \xrightarrow{\mathrm{k_{PP1PLB}}} [\mathrm{PP1}] + [\mathrm{PLB}]. \tag{5.8}$$

The QEA assumption is again used and the reverse reaction is negligible. Transforming (equation (5.7)) and (equation (5.8)) into Michaelis–Menten kinetic equations and combining them results in the rate of PLB phosphorylation:

$$\frac{d[\mathrm{PLB}_p]}{dt} = k_{\mathrm{PKAPLB}}[\mathrm{PKA}]\frac{[\mathrm{PLB}]}{K_{m,\mathrm{PKAPLB}} + [\mathrm{PLB}]} - k_{\mathrm{PP1PLB}}[\mathrm{PP1}] \frac{[\mathrm{PLB}_p]}{K_{m,\mathrm{PP1PLB}} + [\mathrm{PLB}_p]}. \tag{5.9}$$

Where the PKA reaction yields the rate of PLB phosphorylation and the PP1 reaction yields the rate of dephosphorylation.

It is worth noting that the task of constraining parameters is not trivial even in a small module with relatively few components. For example, if some of the constants are unknown, or their values vary greatly in the literature, finding the optimal or correct value may be difficult. To add to the difficulty, precise measurements of each module independent of the other modules and separate from the data used to calculate the constants are often not available. As a result, the individual modules were not validated independently, but rather were validated at various stages along the pathway. Finally, overall effects on ion currents and transmembrane potential were compared qualitatively to experiments from myocytes exposed to a range of ISO concentrations.

5.3.3 Mathematical modeling of cross talk between PKA and CaMKII signaling pathways

One of the key advantages of biophysical models over phenomenological modeling, is that they allow for the study of interactions (crosstalk) between different signaling pathways. Soltis and Saucerman developed an action potential model that for the first time included both β-adrenergic and CaMKII signaling pathways [43]. The β-adrenergic signaling pathway discussed in the previous section was integrated with a model of CaMKII activation [40] into a model of the rabbit ventricular action potential and Ca^{2+} cycling. The model incorporated various reformulations of ion channel kinetics (L-type Ca^{2+} channel, $Na_v1.5$ and multiple K^+ channels) as well as changes to PLB due to phosphorylation. Similar to the previous effort, the signaling pathways were divided into modules with validation through comparison of features related to Ca^{2+} cycling to experimental values, at baseline and in a CaMKII overexpression model. The model allowed for investigation of the effects of β-adrenergic stimulation on the CaMKII pathway with the finding that β-adrenergic stimulation acted synergistically to increase CaMKII activity and arrhythmogenesis measured by occurrence of inappropriate afterdepolarizations following pacing.

5.4 Systems biology: the next frontier for mathematical modeling of cell signaling

Systems biology is a highly interdisciplinary approach used to study phenomenon in diverse fields from economics to particle physics and provides an exciting alternative to the approaches already discussed to understand the relationship between perturbations in cell signaling and cardiac function. There are many different definitions of what distinguishes a systems approach from the traditional 'reductionist' approach. In general, the reductionist approach is defined as breaking down (or reducing) a system into its smallest components, which are then analyzed to gain insight into a systems level phenomenon. In the biology of signaling pathways, the reductionist approach would dissect the pathway to its chemical components and reactions. By studying components in isolation, this approach allows for a high degree of certainty, as the inputs and environment can be tightly controlled. Of course, system components do not operate in isolation and a system may give rise to emergent behaviors that are difficult or impossible to understand by studying isolated components. To understand the function of each, it is critical to understand its role in the pathway, which necessarily involves understanding the pathway as a whole. In other words, systems analysis focuses on synergy, the way the whole is more than simply the sum of its parts.

Systems may be viewed as a network of *stocks* and *flows*, which operate together to perform a *function* [30]. We will define each of these terms separately as each is a key component of a systems model.

1. **Stocks** are the quantities of components in the system, which can be anything from the concentration of a molecule to the number of cells of a specific type.
2. **Flows** are the interactions between stocks or between a stock and the environment, for example, the reaction that turns one concentration into another, or the signaling that causes inflammatory cells to rush into an injury site. Flows may be uni- or bi-directional depending on reversibility of the process. While flows may vary in complexity, they are often more difficult to quantify than stocks. Nevertheless, flows imbue the system with its character and define how the system evolves over time.
3. **Function** is the operational output of the intact system. The function may be a physiological response at any scale from the molecule (e.g. concentration of second messenger signaling to another system, gene changes) to the cell (e.g. cell excitability, cell growth), organ (arrhythmia, hypertrophy) or even population (e.g. incidence of arrhythmia across individuals) level.

A network, or graph, is a common way to represent a system by representing stocks as nodes and flows as directed edges connecting nodes. Constructing a network in this manner is conducive to advanced mathematical analysis and a deeper understanding of the system (figure 5.5).

To demonstrate application of a systems approach to cell signaling, we will use the MAPK/ERK signaling pathway, implicated in cell growth and healing from injury in heart and other tissues. Broadly, epidermal growth factor (EGF) binds to an EGF receptor, which acts as an enzyme to allow Ras to swap a GDP for a GTP and triggers a cascade of MAP3K, MAP2K, and MAPK activation leading ultimately to changes in gene regulation through recruitment of various transcription factors. In this example, the levels of EGF, phosphorylated MAPK and active transcription factors may all be thought of as stocks with flows as directed connections between these elements and function as the ultimate effect on cell growth and division.

In a systems approach, it is important to consider the tradeoff between sophistication and tractability of the model as discussed previously. Determining the appropriate scope for the model will depend heavily on the question of interest. Considering the MAPK/ERK example, a researcher may want to know the following which would help define the scope: what are the critical factors that regulate EGF levels? How do changes in the gene expression profile manifest in functional outputs? What are the critical intermediary steps between a change in EGF and cell function? At the same time, before expanding a model, it may be useful to consider the tradeoff in terms of complexity. For example, will adding additional producers of EGF increase my ability to answer a specific question? Do changes in gene expression or other parameters vary on a timescale that is relevant to the question of interest? It may be determined that certain system features,

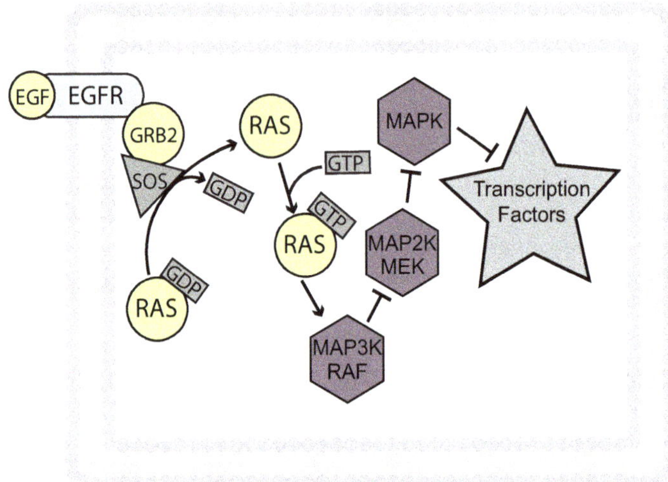

Figure 5.5. Schematic of the MAPK/ERK signaling pathway. The MAPK/ERK signaling pathway begins with epidermal growth factor (EGF) binding to the epidermal growth factor receptor (EGFR). This activates EGFR allowing growth factor receptor-bound protein 2 (GRB2) to dock on an intracellular site. Son of Sevenless (SOS) then may bind to GRB2 to induce its activation. Once active, SOS induces unbinding of guanosine diphosphate (GDP) from a member of the Ras GTPase subfamily (RAS). RAS then binds guanosine triphosphate (GDP), which allows RAS to activate proto-oncogene c-RAF (RAF), the first kinase in the kinase cascade. For this reason, in the MAPK/ERK pathway RAF can technically be called Mitogen-activated protein kinase kinase kinase (MAP3K). RAF then phosphorylates MEK (MAP2K) which finally activates Mitogen-activated protein kinase (MAPK). MAPK then phosphorylates numerous transcription factors to induce nuclear translocation and altered gene expression.

although undoubtedly important for physiology, may not require the same level of detail or attention as others that are more directly relevant to the primary question.

The types of questions appropriate for a systems model are typically different than for more traditional approaches. Instead of addressing specific questions about molecular interactions or structures, systems approaches facilitate operational questions about the pathway as a whole [2]. How does the pathway process its input signals? How is the information stored and transmitted through the pathway? Can the pathway be tuned to achieve a desired output? Does the system have a linear response or a more complicated response? In the MAPK/ERK example, a systems approach might address how the input EGF is processed to generate gene regulation changes (functional output). The MAPK/ERK pathway is an example of how synergy can be necessary for answering such questions. In the pathway, MAPK activity is dynamic but regular periodic (sinusoidal wave). When the concentration of EGF is increased, the frequency of MAPK activation increases, leading to higher activity of gene regulations. The discovery of the dynamic changes in MAPK was made by monitoring its activity levels over time [1]. By measuring the activity of MAPK over time the experimenters were able to characterize the whole pathway, without needing to break it up into its specific components, and discovered a result that would have been difficult to predict from the pathway's individual parts.

While a systems approach may be applied at any temporal or spatial scale, it offers unique advantages for higher level models, involving multiple pathways, with complex interactions. In such models, data is typically derived from disparate sources and experiments (e.g. various -omics based strategies). One example comes from the Saucerman lab, which built a systems model of cardiac hypertrophy [39] (figure 5.6). In this model, the inputs are the change in expression of a node, and the nodes are components of signaling pathways such as isoproterenol, EGF or CaMKII. The outputs are gene changes and physiological changes, such as cell area or SERCA expression. The model was built using information from the literature to determine the essential components (nodes) for each pathway. A network component list was generated including information about interactions with other components. Normalized Hill equations were used to model activation of each node [23]. The equations were then parameterized with standard values, rather than by fitting specific values to data, which would be difficult for the model with 106 nodes (stocks) and 192 edges (flows). To validate the model, 114 input–output relationships from experimental measurements in rat neonatal ventricular myocytes were used. The relationships were then converted into qualitative data by treating each output as a ternary value: upregulated, downregulated or unchanged. Overall,

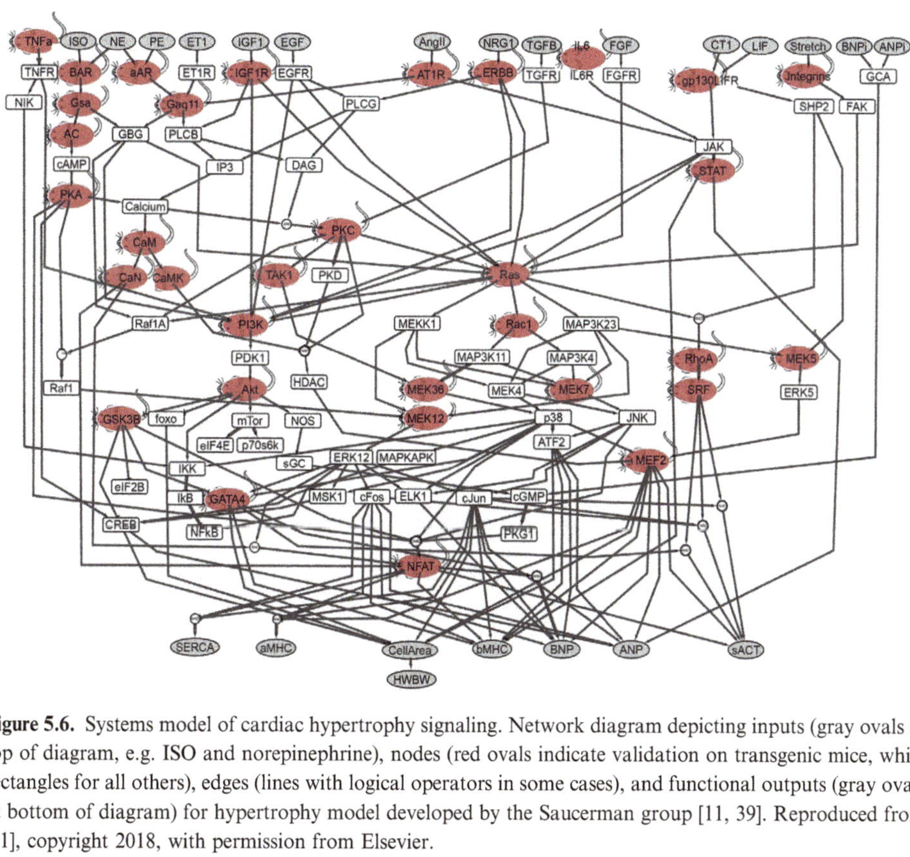

Figure 5.6. Systems model of cardiac hypertrophy signaling. Network diagram depicting inputs (gray ovals at top of diagram, e.g. ISO and norepinephrine), nodes (red ovals indicate validation on transgenic mice, white rectangles for all others), edges (lines with logical operators in some cases), and functional outputs (gray ovals at bottom of diagram) for hypertrophy model developed by the Saucerman group [11, 39]. Reproduced from [11], copyright 2018, with permission from Elsevier.

the model correctly predicted 109 of the relationships, which is impressive in light of the fact that standard values were used for equation parameters.

More recently, the model was updated to predict changes in hypertrophy in mice [11]. Additional flows were added to the network based on the literature, and the model was tested against data from 52 transgenic mice. In each transgenic mouse, a gene was up-regulated or down-regulated, and some measurements of the effects this had on the mouse were recorded. The gene corresponded to some protein or gene (node) in the mathematical model. Changing the regulation of this gene in the model would then provide predations of changes in both outputs and intermediary nodes. To compare the model predictions to the actual observed results, each change was converted to ternary logic (up-regulated, down-regulated or no change.) The model was able to predict the changes in output nodes correctly for 78% of cases, while for internal nodes it achieved an accuracy of 69%. As the exact degree of regulation is often difficult to measure in signaling pathways, i.e. for intermediate results the concentrations and fluctuations in molecules may not be well known. This also limits the accuracy of predicting interactions between pathways. However, future work characterizing signaling pathways will improve accuracy and scope [20] making models such as this one increasingly useful tools.

With the development of various -omics (transciptomics, proteomics and metabolomics) experiments, the data for building large scale pathway models has exploded in quantity and quality. In one example, a model of gene regulation in coronary atherosclerosis was constructed [22]. Samples from 22 human patients with atherosclerosis were used together with gene regulatory changes from microarray experiments. After the genes differentially regulated in atherosclerosis were analyzed, multiple techniques were used to define relevant pathways. First, documented signaling pathways were referenced to determine whether target genes were differentially expressed compared to a non-diseased patient to construct a model of pathway regulation. A second approach was also implemented whereby published abstracts were mined for associations between different genes/proteins. These associations were then compiled into a network of genes and their relationships. From this large network they created a subnetwork of relationships only between the genes of interest from their data. Both approaches were successful in generating large models of the pathways involved in atherosclerosis.

5.5 Conclusions and future directions

Mathematical modeling has been applied successfully to gain insight into the function of cell signaling pathways in cardiovascular physiology. In particular, modeling studies have been very useful in understanding acute signaling phenomenon, on a time scale of milliseconds to hours. Of course, perturbation to cell signaling may induce changes on much longer time scales. For example, CaMKII activation induces acute changes in ion channel function as discussed above but under pathological conditions promotes maladaptive remodeling in cardiac structure and function, which likely contributes to long-term progression of disease. Going forward, there is an opportunity for mathematical modeling to bridge the disparate time scales relevant for cardiac pathophysiology. At the same time, with increasing availability of large datasets from high throughput '-omics' studies, even from human patients, it is expected that mathematical modeling will only become more important for extracting mechanistic and predictive information. In particular, systems models, which have been gaining traction in biology with advancement in technology, stand to benefit the most from the increasing availability of data. That said, considerable effort will be required to construct systems models with predictive ability for complex diseases like heart failure. In the end, it is exciting to consider a potential role for hybrid models, combining features of systems approaches with mechanistic fidelity of biophysical models. Such models may be able to strike the appropriate balance between detail and tractability, while spanning spatial and temporal scales to generate valuable insight into the heart's function in health and disease.

References

[1] Albeck J G, Mills G B and Brugge J S 2013 Frequency-modulated pulses of ERK activity transmit quantitative proliferation signals *Mol. Cell.* **49** 249–61

[2] Antebi Y E, Nandagopal N and Elowitz M B 2017 An operational view of intercellular signaling pathways *Curr. Opin. Syst. Biol.* **1** 16–24

[3] Berg J M, Tymoczko J L and Stryer L 2002 *Biochemistry* 5th edn (New York: W H Freeman)

[4] Braun A P and Schulman H 1995 The multifunctional calcium/calmodulin-dependent protein kinase: from form to function *Annu. Rev. Physiol.* **57** 417–45

[5] Bursztyn L, Eytan O, Jaffa A J and Elad D 2007 Modeling myometrial smooth muscle contraction *Ann. N. Y. Acad. Sci.* **1101** 110–38

[6] Christensen M D, Dun W, Boyden P A, Anderson M E, Mohler P J and Hund T J 2009 Oxidized calmodulin kinase II regulates conduction following myocardial infarction: a computational analysis *PLoS Comput. Biol.* **5** 1000583

[7] Colomer J M, Mao L, Rockman H A and Means A R 2003 Pressure overload selectively up-regulates Ca^{2+}/calmodulin-dependent protein kinase II *in vivo Mol. Endocrinol.* **17** 183–92

[8] Dibb K, Trafford A, Zhang H and Eisner D 2015 A model model: a commentary on DiFrancesco and Noble (1985) 'A model of cardiac electrical activity incorporating ionic pumps and concentration changes' *Philos. Trans. R. Soc. Lond. Ser. B, Biol. Sci.* **370**

[9] DiFrancesco D and Noble D 1985 A model of cardiac electrical activity incorporating ionic pumps and concentration changes *Philos. Trans. R. Soc. Lond. Ser. B Biol. Sci.* **307** 353–98

[10] Dupont G, Houart G and De Koninck P 2003 Sensitivity of CaM kinase II to the frequency of Ca^{2+} oscillations: a simple model *Cell Calcium* **34** 485–97

[11] Frank D U, Sutcliffe M D and Saucerman J J 2018 Network-based predictions of *in vivo* cardiac hypertrophy *J. Mol. Cell. Cardiol.* **121** 180–9

[12] Fridlyand L E, Jacobson D A and Philipson L H 2013 Ion channels and regulation of insulin secretion in human β-cells: a computational systems analysis *Islets* **5** 1–15

[13] Glynn P, Unudurthi S D and Hund T J 2014 Mathematical modeling of physiological systems: an essential tool for discovery *Life Sci.* **111** 1–5

[14] Grandi E, Pandit S V, Voigt N, Workman A J, Dobrev D, Jalife J and Bers D M 2011 Human atrial action potential and Ca^{2+} model: sinus rhythm and chronic atrial fibrillation *Circ. Res.* **109** 1055–66

[15] Hanson P I, Meyer T, Stryer L and Schulman H 1994 Dual role of calmodulin in autophosphorylation of multifunctional CaM kinase may underlie decoding of calcium signals *Neuron* **12** 943–56

[16] Hodgkin A L and Huxley A F 1952 A quantitative description of membrane current and its application to conduction and excitation in nerve *J. Physiol.* **117** 500–44

[17] Hund T J and Mohler P J 2015 Role of CaMKII in cardiac arrhythmias *Trends Cardiovasc. Med.* **25** 392–7

[18] Hund T J and Rudy Y 2004 Rate dependence and regulation of action potential and calcium transient in a canine cardiac ventricular cell model *Circulation* **110** 3168–74

[19] Hunter J J and Chien K R 1999 Signaling pathways for cardiac hypertrophy and failure *New Engl. J. Med.* **341** 1276–83

[20] Kang J H, Lee H-S, Kang Y-W and Cho K-H 2016 Systems biological approaches to the cardiac signaling network *Brief. Bioinform.* **17** 419–28

[21] Keen M 1999 Definition of receptors in radioligand-binding experiments *Methods Mol. Biol.* **106** 187–96

[22] King J Y *et al* 2005 Pathway analysis of coronary atherosclerosis *Physiol. Genomics* **23** 103–18

[23] Kraeutler M J, Soltis A R and Saucerman J J 2010 Modeling cardiac β-adrenergic signaling with normalized-Hill differential equations: comparison with a biochemical model *BMC Syst. Biol.* **4** 157

[24] Kreusser M M *et al* 2009 The isoform of CaM kinase II is required for pathological cardiac hypertrophy and remodeling after pressure overload *Proc. Natl. Acad. Sci.* 106 2342–7

[25] Landstrom A P, Dobrev D and Wehrens X H T 2017 Calcium signaling and cardiac arrhythmias *Circ. Res.* **120** 1969–93

[26] Luo C H and Rudy Y 1994 A dynamic model of the cardiac ventricular action potential. I. Simulations of ionic currents and concentration changes *Circ. Res.* **74** 1071–96

[27] Luo C H and Rudy Y 1994 A dynamic model of the cardiac ventricular action potential. II. Afterdepolarizations, triggered activity, and potentiation *Circ. Res.* **74** 1097–113

[28] Maier L S, Zhang T, Chen L, DeSantiago J, Brown J H and Bers D M 2003 Transgenic CaMKIIδ C overexpression uniquely alters cardiac myocyte Ca^{2+} handling *Circ. Res.* **92** 904–11

[29] McAllister R E, Noble D and Tsien R W 1975 Reconstruction of the electrical activity of cardiac Purkinje fibres *J. Physiol.* **251** 1–59

[30] Meadows D H 2008 *Thinking in Systems: A Primer* 1st edn (Chelsea: New York)

[31] Mueller J K and Tyler W J 2014 A quantitative overview of biophysical forces impinging on neural function *Phys. Biol.* **11** 051001

[32] Noble D 1960 Cardiac action and pacemaker potentials based on the Hodgkin–Huxley equations *Nature* **188** 495–7

[33] Noble D 2002 Modelling the heart: insights, failures and progress *BioEssays* **24** 1155–63

[34] Onal B, Unudurthi S D and Hund T J 2014 Modeling CaMKII in cardiac physiology: from molecule to tissue *Front. Pharmacol.* **5** 9

[35] Qu Z and Weiss J N 2015 Mechanisms of ventricular arrhythmias: from molecular fluctuations to electrical turbulence *Annu. Rev. Physiol.* **77** 29–55

[36] Qume M 1999 Overview of ligand-receptor binding techniques *Methods Mol. Biol.* **106** 3–23

[37] Rasmusson R L, Clark J W, Giles W R, Shibata E F and Campbell D L 1990 A mathematical model of a bullfrog cardiac pacemaker cell *Am. J. Physiol.* **259** H352–69

[38] Rudy Y and Silva J R 2006 Computational biology in the study of cardiac ion channels and cell electrophysiology *Q. Rev. Biophys.* **39** 57–116

[39] Ryall K A, Holland D O, Delaney K A, Kraeutler M J, Parker A J and Saucerman J J 2012 Network reconstruction and systems analysis of cardiac myocyte hypertrophy signaling *J. Biol. Chem.* **287** 42259–68

[40] Saucerman J J and Bers D M 2008 Calmodulin mediates differential sensitivity of CaMKII and calcineurin to local Ca^{2+} in cardiac myocytes *Biophys. J.* **95** 4597–612

[41] Saucerman J J, Brunton L L, Michailova A P and McCulloch A D 2003 Modeling β-adrenergic control of cardiac myocyte contractility in silico *J. Biol. Chem.* **278** 47997–8003

[42] Saucerman J J and McCulloch A D 2004 Mechanistic systems models of cell signaling networks: A case study of myocyte adrenergic regulation *Prog. Biophys. Mol. Biol.* **85** 261–78

[43] Soltis A R and Saucerman J J 2010 Synergy between CaMKII substrates and β-adrenergic signaling in regulation of cardiac myocyte Ca^{2+} handling *Biophys. J.* **99** 2038–47

[44] Swaminathan P D, Purohit A, Hund T J and Anderson M E 2012 Calmodulin-dependent protein kinase II: Linking heart failure and arrhythmias *Circ. Res.* **110** 1661–77

[45] Zhang T, Maier L S, Dalton N D, Miyamoto S, Ross J, Bers D M and Brown J H 2003 The δc isoform of CaMKII is activated in cardiac hypertrophy and induces dilated cardiomyopathy and heart failure *Circ. Res.* **92** 912–9

Chapter 6

Modelling cardiomyocyte energetics

Edmund J Crampin, Michael Pan, Kenneth Tran, Shourya Ghosh and Vijay Rajagopal

The heart is a pump. Like all engines, the heart requires energy, from fuel, in order to generate the mechanical work of pumping blood around the circulation. Additionally, like all cells but unlike most engineered machines, heart cells also require a continuous flux of energy in order to maintain their structural integrity, to maintain normal homeostatic processes, and to maintain far-from equilibrium conditions which enable the generation of the action potential and calcium signalling. The manner in which energy is converted from fuel molecules and used to do work is very different to most engineered systems. In heart cells the energy conversion process is slow, involves a very large number of separate steps, and is highly regulated. These topics, and specifically how mathematical models can be developed to study bioenergetics of cardiomyocytes, are the focus of this chapter.

6.1 Introduction to cardiomyocyte energetics

The normal functioning of a cardiomyocyte depends critically on a stable and constant supply of energy. The hydrolysis of the molecule adenosine triphosphate (ATP), generated in the mitochondria, provides energy to sub-cellular processes (ATPases) which consume ATP to maintain electro-chemical gradients away from thermodynamic equilibrium, and directly to acto-myosin cross bridges to generate force (Ingwall 2002). The primary energy-consuming processes relevant for normal cardiomyocyte function are the sodium pump (Na^+/K^+ ATPase), SERCA (the sarcoplasmic endoplasmic reticulum calcium ATPase) and the force-generating actin–myosin ATPase cross-bridges, shown in figure 6.1. Under normal physiological conditions the relative contributions of each of these processes to cellular energy consumption are: Na^+/K^+ ATPase 10%; SERCA 15%; cross-bridges 75% (Schramm *et al* 1994).

A second, key aspect is that regulation of energy demand and supply in the myocardium is crucial for maintaining normal cardiac function. Critically, the

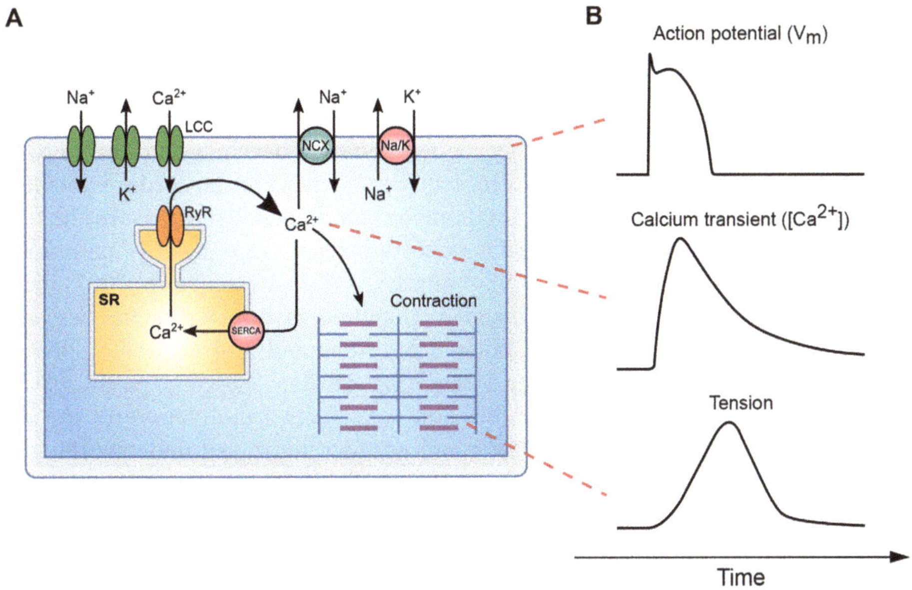

Figure 6.1. Cardiac cell physiology. (A) Schematic of the cardiac cell; (B) The components of cardiac physiology give rise to a wide range of measurable signals, including the membrane potential that arises from ionic currents (top), the intracellular Ca^{2+} transient arising from the movement of Ca^{2+} between compartments (middle), and the tension arising from the activation of contraction by Ca^{2+} (bottom). Abbreviations: SR, sarcoplasmic reticulum; LCC, L-type Ca^{2+} channel; Na/K, Na^+/K^+ ATPase; SERCA, sarcoplasmic/endoplasmic Ca^{2+} ATPase; RyR, ryanodine receptor; NCX, sodium–calcium exchanger; V_m, membrane potential.

supply of ATP from the mitochondria must match demand under varying workloads. Although a number of mechanisms have been proposed by which mitochondrial generation of ATP from adenosine diphosphate (ADP) and inorganic phosphate (Pi) can vary to match changing workload in the myocardium, identifying the underlying regulatory pathways remains controversial. A remarkable observation of cardiac bioenergetics is that these energy-metabolic compounds (ATP, ADP, Pi) remain at relatively constant concentrations in the cytoplasm over a wide range of cardiac workload (often reported as oxygen consumption rate), suggesting the presence of a very precise and robust control system. Balaban and co-workers (Balaban *et al* 1986, Katz *et al* 1989) report that up to 10-fold increase of the rate of ATP consumption leads to only minor changes in the concentrations of cellular ATP, ADP, Pi and CrP (phospho-creatine, a key ATP buffer). This led these authors to conclude that the substrates for oxidative phosphorylation (ADP and Pi) are not responsible for the regulation of mitochondrial ATP production, on the basis that they do not change significantly with rate of work (Balaban 1990), and that there must therefore exist other feedback mechanisms responsible for this metabolic stability. However, identifying the actual mechanisms and pathways that regulate the coupling between the energy demand and supply processes remains controversial (Saks *et al* 2006).

Recent insights into the nature of this control mechanism have come from data-informed mathematical modelling (Beard 2006, Tran *et al* 2015), discussed below. The development of biophysically-based and thermodynamically-consistent mathematical models of the key energy producing and consuming processes in the cell provides an important methodology with which to develop and test different hypotheses as to how ATP supply is matched to demand in the cardiomyocyte.

Furthermore, understanding the fundamental mechanisms that govern the transduction of energy at the myocyte level is important because a number of pathologies, including cardiac ischaemia (Katz 1977) and heart failure (Neubauer 2007) including diabetic cardiomyopathy are linked to impaired myocardial energetics, stemming from a mismatch between cellular energy demand and energy supply. The recent popularity of modulating and targeting cardiac energetic pathways as a therapy for heart failure (Neubauer 2007) provides further motivation for the value of quantitative mathematical modelling as a tool for improved understanding and providing insights into potential therapeutic approaches.

6.2 Mathematical modelling of biochemical energetics

In analogy with electrical circuits, biochemical reactions proceed due to a thermodynamic 'force'—the equivalent of the electrical potential, or voltage—which results in a reaction flux—the equivalent of electrical current flow. First we review the general theory of modelling biochemical reactions, and then we see how a reformulation of this approach to include basic thermodynamic considerations enables consideration of energetics.

6.2.1 Modelling biochemical systems

Biochemical networks are a collection of chemical species and the reactions the species are involved in. In many cases, the rates of each reaction follow the law of mass action, where the rate of reaction is proportional to the amount of each substrate. More specifically, if $\{S_i\}$ represents the set of (indexed) species of a network, for a reversible reaction j,

$$\sum_{i\in E} n_{ij}^{+} S_i \underset{k_j^-}{\overset{k_j^+}{\rightleftharpoons}} \sum_{i\in P} n_{ij}^{-} S_i, \tag{6.1}$$

with reactants $\{S_i\colon i \in E\}$, products $\{S_i\colon i \in P\}$ and stoichiometric coefficients n_{ij}^{+} and n_{ij}^{-}, the rate of the forward reaction is

$$v_j^+ = k_j^+ \prod_{i\in E} x_i^{n_{ij}^+}, \tag{6.2}$$

and the rate of the reverse reaction is

$$v_j^- = k_j^- \prod_{i\in P} x_i^{n_{ij}^-}, \tag{6.3}$$

where x_i [mol] is the amount of S_i (Keener and Sneyd 2009, Ederer and Gilles 2007). The stoichiometric coefficients are non-negative integers which prescribe the number of molecules of each reactant species that are involved in the forward or reverse reaction (hence zero for those molecular species which are not involved in a particular reaction). The total rate of reaction v_j [mol s^{-1}] is then

$$v_j = v_j^+ - v_j^- = k_j^+ \prod_{i \in E} x_i^{n_{ij}^+} - k_j^- \prod_{i \in P} x_i^{n_{ij}^-}. \tag{6.4}$$

While irreversible reactions can be modelled by excluding either the forward or reverse reaction, the principle of microscopic reversibility states that any physical process can in principle be reversed. Thus, in reality, all reactions are reversible (Polettini and Esposito 2014). This is an important consideration for bioenergetic modelling: irreversible reactions cannot be considered within a thermodynamically formulated model.

General biochemical systems can be modelled by summing together contributions of the reactions to each of the species. A convenient form for expressing this relationship is

$$\frac{d\mathbf{X}}{dt} = \mathbf{N} \cdot \mathbf{v}(\mathbf{X}), \tag{6.5}$$

where $\mathbf{X}$ is a vector containing the molar amount of each species, $\mathbf{v}(\mathbf{X})$ is a vector containing the reaction rates described by equation (6.4), and $\mathbf{N}$ is the stoichiometric matrix with entries $\mathbf{N}_{ij} = n_{ij}^- - n_{ij}^+$ that relates the fluxes to the rates of change of each species (Palsson 2006).

Two example biochemical reactions

Figure 6.2 shows two examples of biochemical reaction cycles. The reaction scheme in figure 6.2(A), corresponding to the reaction scheme

$$A \underset{k_1^-}{\overset{k_1^+}{\rightleftharpoons}} B \underset{k_2^-}{\overset{k_2^+}{\rightleftharpoons}} C \underset{k_3^-}{\overset{k_3^+}{\rightleftharpoons}} A, \tag{6.6}$$

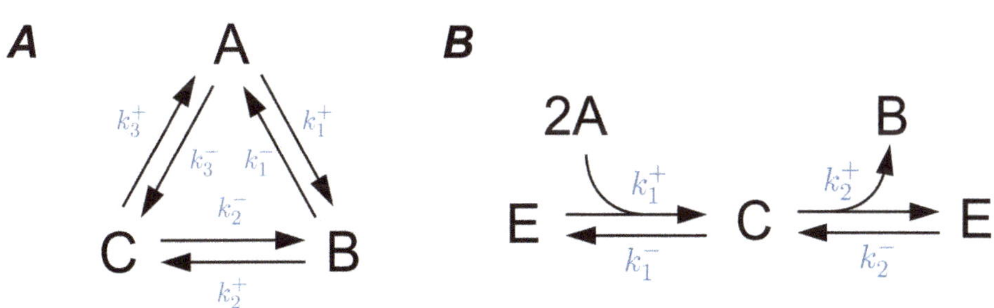

Figure 6.2. Two example biochemical reactions.

can be represented using

$$\mathbf{X} = \begin{bmatrix} x_A \\ x_B \\ x_C \end{bmatrix}, \quad \mathbf{N} = \begin{bmatrix} -1 & 0 & 1 \\ 1 & -1 & 0 \\ 0 & 1 & -1 \end{bmatrix}, \quad \mathbf{v} = \begin{bmatrix} v_1 \\ v_2 \\ v_3 \end{bmatrix} = \begin{bmatrix} k_1^+ x_A - k_1^- x_B \\ k_2^+ x_B - k_2^- x_C \\ k_3^+ x_C - k_3^- x_A \end{bmatrix}, \tag{6.7}$$

where k_1^+, k_1^-, k_2^+, k_2^-, k_3^+ and k_3^- are the kinetic parameters associated with the reactions. Similarly, the reaction scheme illustrated in figure 6.2(B) can be written as

$$E + 2A \underset{k_1^-}{\overset{k_1^+}{\rightleftharpoons}} C \underset{k_2^-}{\overset{k_2^+}{\rightleftharpoons}} E + B, \tag{6.8}$$

and can be represented using

$$\mathbf{X} = \begin{bmatrix} x_A \\ x_B \\ x_E \\ x_C \end{bmatrix}, \quad \mathbf{N} = \begin{bmatrix} -2 & 0 \\ 0 & 1 \\ -1 & 1 \\ 1 & -1 \end{bmatrix}, \quad \mathbf{v} = \begin{bmatrix} v_1 \\ v_2 \end{bmatrix} = \begin{bmatrix} k_1^+ x_A^2 x_E - k_1^- x_C \\ k_2^+ x_C - k_2^- x_B x_E \end{bmatrix}, \tag{6.9}$$

Stoichiometry and conserved quantities

The stoichiometric representation in equation (6.5) is useful because it enables systems-level analyses using established matrix methods. For example, if we define **G** as the left nullspace matrix of **N**, which contains a basis for the kernel of $\mathbf{N}^T$ in its rows (i.e. $\mathbf{G} \cdot \mathbf{N} = 0$), we find that

$$\mathbf{G} \cdot \dot{\mathbf{X}} = \mathbf{G} \cdot \mathbf{N}\mathbf{v} = 0. \tag{6.10}$$

Thus, the entries of $\mathbf{G} \cdot \mathbf{X}$ are pools of species that do not change over time and are therefore known as conserved moieties (‘moieties’ are recurring molecular species) (Palsson 2006, Klipp *et al* 2009). Conservation of adenosine (ATP + ADP + AMP) is an example of a conserved moiety that is often observed in metabolic systems (Haraldsdóttir and Fleming 2019, Schuster and Hilgetag 1995). In the example described in equations (6.8)–(6.9), a possible left nullspace matrix is

$$\mathbf{G} = \begin{bmatrix} 0 & 0 & 1 & 1 \\ 1 & 2 & 0 & 2 \end{bmatrix}, \tag{6.11}$$

The first row of **G** describes conservation of enzyme ($x_E + x_C =$ constant), while the second row describes conservation of substrate ($x_A + 2x_B + 2x_C =$ constant).

Similarly, the right nullspace matrix **H** can be defined as a matrix containing a basis for the kernel of **N** in its columns (i.e. $\mathbf{N} \cdot \mathbf{H} = 0$). Then if $\mathbf{v} = \mathbf{Hc}$ for some column vector **c**,

$$\dot{\mathbf{X}} = \mathbf{N} \cdot \mathbf{Hc} = 0. \tag{6.12}$$

Hence **H** defines the space of reaction fluxes that do not affect the concentrations of species, which are known as cycles (Palsson 2006, Klipp *et al* 2009). Note that in an isolated biochemical system, all steady-state cycles must be zero because reactions with finite flux dissipate energy, and systems with non-zero steady-state fluxes would generate heat without a source of energy, violating the laws of thermodynamics. Therefore, the right nullspace matrix also defines a set of constraints (known as detailed balance, or the Wegscheider conditions) that the kinetic parameters must satisfy to be thermodynamically consistent (Ederer and Gilles 2007, Beard *et al* 2004, Liebermeister *et al* 2010). A shortcoming of this purely kinetic representation is that parameters are not independent: they cannot be adjusted independently without the possibility of violating this constraint. For example, the parameters in the reaction scheme in equation (6.6) must be chosen such as to satisfy the constraint

$$\frac{k_1^+ k_2^+ k_3^+}{k_1^- k_2^- k_3^-} = 1, \tag{6.13}$$

in order to be thermodynamically consistent. In other words, while at first glance there appear to be six kinetic parameters to be specified in this model, in reality there are only five independent parameters. As will be discussed below in subsection 6.2.2, an alternative energetic formulation avoids this issue.

In the context of cardiac modelling, metabolite concentrations vary at rates which are slow relative to the kinetics of enzyme cycles. Transport processes such as SERCA and Na^+/K^+ ATPase can therefore be considered as thermodynamic cycles operating close to steady state (but not equilibrium). Thermodynamic constraints analogous to equation (6.13) are applicable to these cycles, with additional terms to account for the energy entering and leaving the cycle via the metabolites.

6.2.2 Energy-based modelling of biochemical processes

Here we extend the familiar kinetic approach to incorporate thermodynamic considerations, which we will illustrate using these two simple biochemical examples contrasted with thermodynamically consistent models.

Chemical potential

Each chemical species stores energy—analogous to a capacitor in electrical circuits. The propensity for this energy storage to drive a chemical reaction is determined by the chemical potential μ [J mol^{-1}], which depends on the species concentration, according to the relationship

$$\mu = RT \ln(Kx), \tag{6.14}$$

where $R = 8.314$ J mol^{-1} K^{-1} is the universal gas constant, T [K] is the absolute temperature of the system, x [mol] is the molar amount, and K [mol^{-1}] is the thermodynamic constant for species x. An exactly equivalent form of this expression is

$$\mu = \mu^0 + RT \ln(x) \tag{6.15}$$

$$\mu^0 = RT \ln(K), \tag{6.16}$$

where μ^0 is the standard chemical potential for species x. This provides the relationship between the species thermodynamic parameter K and the standard chemical potential μ^0.

Reaction rates
Chemical potentials drive reactions, generating molar flow rates v [mol s^{-1}] (by analogy, electrical potential drives electron flow producing an electric current). The reaction rate is determined by the affinity, A, which is a sum of chemical potentials for the forward and reverse reactions, according to the constitutive relationship

$$v = \kappa(e^{A^f/RT} - e^{A^r/RT}), \tag{6.17}$$

where κ [mol s^{-1}] is a reaction-specific rate constant, A^f [J mol^{-1}] is the sum of chemical potentials in the reactants, or the forward affinity, and A^r [J mol^{-1}] is the sum of chemical potentials in the products, or the reverse affinity.

Gibbs free energy change and reaction reversibility
Because reactions will only proceed in the direction of decreasing potential (by analogy, electric current flows in the direction of decreasing voltage), the overall affinity of reaction $A = A^f - A^r$ determines the direction of the flux v, which will be positive only if $A > 0$. The Gibbs free energy of reaction ΔG is commonly used to quantify the direction of reaction; this is related to the affinity by the relationship $\Delta G = -A$. Thus the reaction proceeds when $\Delta G < 0$.

Note that affinity A, and hence ΔG, depend on thermodynamic parameters K, but not reaction rate parameters κ. Thus, whether a reaction may proceed in the 'forward' direction or not depends on K, not on κ. If $\Delta G < 0$ then the reaction may proceed, however the *rate* at which the reaction proceeds depends on the reaction rate parameters κ, not the thermodynamic parameters.

Two example biochemical reactions—revisited
As we will see below, using the same examples as above, combining equations (6.16) and (6.17) results in the familiar mass action kinetics (Gawthrop and Crampin 2014). These examples indicate how the thermodynamic parameters (the species parameters K and the reaction-specific parameter κ) relate to the mass action parameters. The chemical potentials for each species in the reaction network in equation (6.6), figure 6.2(A), are

$$\mu_A = RT\ln(K_A x_A) \tag{6.18a}$$

$$\mu_B = RT\ln(K_B x_B) \tag{6.18b}$$

$$\mu_C = RT\ln(K_C x_C), \tag{6.18c}$$

and the reaction rates are given by the equations

$$v_1 = \kappa_1(e^{\mu_A/RT} - e^{\mu_B/RT}) = \kappa_1 K_A x_A - \kappa_1 K_B x_B \tag{6.19a}$$

$$v_2 = \kappa_2(e^{\mu_B/RT} - e^{\mu_C/RT}) = \kappa_2 K_B x_B - \kappa_2 K_C x_C \tag{6.19b}$$

$$v_3 = \kappa_3(e^{\mu_C/RT} - e^{\mu_A/RT}) = \kappa_3 K_C x_C - \kappa_3 K_A x_A, \tag{6.19c}$$

where the second equalities were derived by substituting equation (6.18). It follows straightforwardly that the rates of change of each species are

$$\frac{dx_A}{dt} = v_3 - v_1 \tag{6.20a}$$

$$\frac{dx_B}{dt} = v_1 - v_2 \tag{6.20b}$$

$$\frac{dx_C}{dt} = v_2 - v_3, \tag{6.20c}$$

Therefore equation (6.20) gives rise to the same mass-action dynamics as described by the kinetic scheme in equations (6.5) and (6.7), with

$$k_1^+ = \kappa_1 K_A \quad k_1^- = \kappa_1 K_B \tag{6.21a}$$

$$k_2^+ = \kappa_2 K_B \quad k_2^- = \kappa_2 K_C \tag{6.21b}$$

$$k_3^+ = \kappa_3 K_C \quad k_3^- = \kappa_3 K_A. \tag{6.21c}$$

An important observation is that whereas it is possible to specify kinetic parameters that give rise to thermodynamically infeasible systems, i.e. to choose kinetic parameter values inconsistent with the detailed balance constraint equation (6.13), it is impossible do so when specifying thermodynamic parameters. Noting that

$$\frac{k_1^+ k_2^+ k_3^+}{k_1^- k_2^- k_3^-} = \frac{\kappa_1 K_A \kappa_2 K_B \kappa_3 K_C}{\kappa_1 K_B \kappa_2 K_C \kappa_3 K_A} = 1, \tag{6.22}$$

therefore, all such sets of thermodynamic parameters are by definition thermodynamically feasible as they automatically satisfy the detailed balance constraint. Thus, in practice, whereas modellers using kinetic parameters must adhere to constraints such as equation (6.13) to satisfy thermodynamic consistency, these constraints are automatically satisfied if the thermodynamic approach is used. It should be noted here that there are six fully independent thermodynamic parameters, in contrast to the five independent kinetic parameters: the extra degree of freedom is used to determine the chemical potential of each individual species.

The chemical potentials for the species in the reaction given by equation (6.8), figure 6.2(B), are

$$\mu_A = RT \ln(K_A x_A) \tag{6.23a}$$

$$\mu_B = RT \ln(K_B x_B) \tag{6.23b}$$

$$\mu_E = RT\ln(K_E x_E) \tag{6.23c}$$

$$\mu_C = RT\ln(K_C x_C), \tag{6.23d}$$

and the reaction rates are

$$v_1 = \kappa_1(e^{A_1^f/RT} - e^{\mu_C/RT}) \tag{6.24a}$$

$$v_2 = \kappa_2(e^{\mu_C/RT} - e^{A_2^r/RT}). \tag{6.24b}$$

In contrast to the previous example, the presence of multi-substrate reactions requires chemical potentials on the forward and reverse sides of each reaction to be summed according to the reaction stoichiometry:

$$A_1^f = \mu_E + 2\mu_A \tag{6.25a}$$

$$A_2^r = \mu_E + \mu_B. \tag{6.25b}$$

By substituting into equation (6.24), the reaction rates are therefore

$$v_1 = \kappa_1 K_E K_A^2 x_E x_A^2 - \kappa_1 K_C x_C \tag{6.26a}$$

$$v_2 = \kappa_2 K_C x_C - \kappa_2 K_E K_B x_E x_B, \tag{6.26b}$$

and hence the rates of change for each species are described by the equations

$$\frac{dx_A}{dt} = -2v_1 \tag{6.27a}$$

$$\frac{dx_B}{dt} = v_2 \tag{6.27b}$$

$$\frac{dx_E}{dt} = v_2 - v_1 \tag{6.27c}$$

$$\frac{dx_C}{dt} = v_1 - v_2, \tag{6.27d}$$

which is equivalent to the kinetic representation in equation (6.9) with

$$k_1^+ = \kappa_1 K_E K_A^2 \quad k_1^- = \kappa_1 K_C \tag{6.28a}$$

$$k_2^+ = \kappa_2 K_C \quad k_2^- = \kappa_2 K_E K_B. \tag{6.28b}$$

As above, the thermodynamic approach ensures that detailed balance is satisfied.

This approach can easily be generalised to an arbitrarily large reaction network. Gawthrop *et al* (2015) show that a general biochemical network with forward stoichiometric matrix $\mathbf{N}^f$ with entries $\mathbf{N}_{ij}^f = n_{ij}^+$, reverse stoichiometric matrix $\mathbf{N}^r$ with entries $\mathbf{N}^r_{ij} = n^-_{ij}$, and stoichiometric matrix $\mathbf{N} = \mathbf{N}^r - \mathbf{N}^f$, can be represented using the equations

$$\frac{d\mathbf{X}}{dt} = \mathbf{N}\mathbf{v}(\mathbf{X}) \tag{6.29a}$$

$$\mathbf{v}(\mathbf{X}) = \boldsymbol{\kappa}\big(\mathbf{Exp}[\mathbf{N}^f\,\mathbf{Ln}(\mathbf{K}\cdot\mathbf{X})] - \mathbf{Exp}[\mathbf{N}^r\,\mathbf{Ln}(\mathbf{K}\cdot\mathbf{X})]\big), \tag{6.29b}$$

where **Exp** and **Ln** are the element-wise exponential and logarithm respectively, $\boldsymbol{\kappa}$ is a diagonal matrix containing reaction rate parameters κ and $\mathbf{K}$ is a diagonal matrix containing species parameters K.

6.2.3 Thermodynamics of ion movement

Generation of the cardiac action potential and the intracellular calcium transient which couples electrical excitation to mechanical contraction is due to movement of ions along and against electrochemical gradients between inter- and intra-cellular compartments. The thermodynamic approach outlined above readily extends to encompass movement of charged species across electrical gradients as, for example, when ions move across the cell membrane.

To extend the above thermodynamic analysis to incorporate movement of charge in an electrical potential, we consider transport of a charged species X^{z+} with charge z across a membrane with membrane potential $V_m = V_i - V_e$. For movement of charged species, if there is a potential difference across the membrane then the free energy change is extended to account for the electric potential, and is expressed as:

$$\Delta G_X = RT\ln\frac{[X^{z+}]_i}{[X^{z+}]_e} + zFV_m, \tag{6.30}$$

where $F = 96\,485$ C mol^{-1} is Faraday's constant. Here, the first term relates to the chemical potential, as before, and the second term to the membrane potential. At any point in time, we can determine the direction of X^{z+} movement across the membrane if we know the values for the transmembrane X^{z+} concentration gradient and the membrane potential. The membrane potential at which $\Delta G_X = 0$ is given by:

$$V_X = \frac{RT}{zF}\ln\frac{[X^{z+}]_e}{[X^{z+}]_i}, \tag{6.31}$$

which is the equilibrium potential or Nernst potential for X^{z+}, i.e. the membrane potential at which there will be no ion flux through an open membrane ion channel. This thermodynamic description of ion movement is used below to develop an energy-based model of the cardiac action potential. A convenient formalism for developing such energy-based models is provided by the bond graph approach, which is described next.

6.2.4 Bond graph formalism for biochemical energetics

A set of biochemical reactions forms a biochemical or metabolic network, represented in the stoichiometric matrix for the system. This network represents conservation of mass, indicating how one species or moiety (recurring molecular part) is converted into another. From a bioenergetics perspective, this network also represents the various pathways through which energy moves through the system. Energy flow can naturally be represented as a bond graph. This is a formalism widely used in systems engineering contexts to unify modelling across multiple physical domains by considering energy flow, using the concepts of potential (known in bond graph terminology as *effort*) and *flow* variables, whose product is power (Gawthrop and Smith 1996). In the biochemical context, chemical (or electro-chemical) potential μ (with units J mol^{-1}) drives reaction flux v (with units mol s^{-1}), with product power (J s^{-1}). Analogously, in the electrical domain electrical potential, voltage V (with units of V, or J C^{-1}) drives electrical flow, current I (with units of A or C s^{-1}), which have product $P = VI$ with units of J s^{-1}.

A bond graph represents the interactions between energy storage components and energy dissipation components within a system. The bond graph C component is a generic energy *storage* component corresponding to a *capacitor* in the electrical domain and the amount (or concentration) of a chemical species in the biomolecular domain. If x is the time integral of the flow variable, the electrical capacitor generates voltage V according to the *linear* relationship: $V = x/C$ where C is the capacitance. In contrast, the chemical species generates chemical potential μ according to the *nonlinear* relationship $\mu = RT \ln Kx$ (equation (6.16)). The bond graph R component is a generic energy *dissipation* component corresponding to a resistor in the electrical domain. In the electrical case, it generates current i according to the linear relationship: $i = \kappa \Delta V$ where κ is the conductance and ΔV is the net voltage across the resistor (i.e. the constitutive relationship between current and voltage for an electrical resistance). In the biochemical domain the R component corresponds to a chemical reaction. Assuming mass-action kinetics the R component generates molar flow v according to the *nonlinear* relationship (equation (6.17)): $v = \kappa(\exp A^f/RT - \exp A^r/RT)$ where A^f (the forward affinity) and A^r (the reverse affinity) are the total chemical potentials of the reactants and products of the reaction, respectively; κ is the reaction rate constant. Given the unique nature of this component, we give it a special name in biochemical systems: the Re component. In this context, we see that the mass action expression for the reaction rate is a constitutive relationship which relates the reaction rate (molar flow) to the affinities.

The biochemical network structure is represented using the concept of *bonds*. Bond graphs represent the flow of energy between components by the bond symbol '$\rightharpoondown$' where the direction of the harpoon defines the direction of positive energy flow (sign convention). Bonds connect components via *junctions* which transmit, but do not store or dissipate, energy. This allows the structure of the biochemical reaction network to be described. Bonds are connected using two types of junctions: the 0 junction where all impinging bonds have the same *potential*, and the 1 junction where all impinging bonds have the same *flow*. The expression for the flows associated with the 0 junction, and the potentials associated with the 1 junction, are determined by the energy conservation requirement.

Two example biochemical reactions—bond graph formalism

These ideas can be used to represent the two example reactions from figure 6.2 as bond graphs, shown in figure 6.3 (right-hand panels), which unambiguously represent the flow of energy through the biochemical network. The standard bond graph notation for components takes the form *<component type >:<name>*. For example, C:**A** is the C component corresponding to species A, and Re:**1** is the Re component corresponding to reaction 1. Thus each species has been mapped to a C component and each reaction to an Re component. Furthermore, the stoichiometry of the system is represented via connections to 0 and 1 junctions. The 0 junction is used to connect a species to multiple reactions, encoding conservation of mass laws such as equation (6.20) for figure 6.3(A) and equations (6.27)(c), (d) for figure 6.3(B). Similarly, the 1 junction connects one side of a reaction to multiple species, encoding conservation of potential laws such as equation (6.25) for figure 6.3(B). Therefore, this formulation incorporates all of the thermodynamic concepts outlined above in an unambiguous manner, while retaining explicitly the network structure and

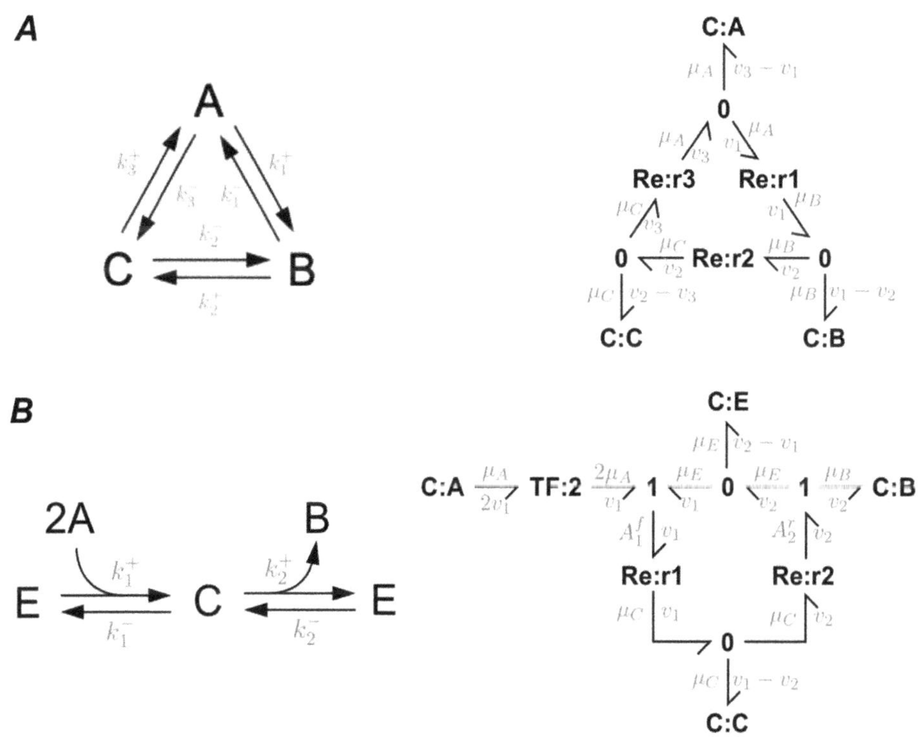

Figure 6.3. Two example biochemical reactions, showing annotated bond graph representations (right-hand panels).

stoichiometry of the biochemical system, and distinguishing between thermodynamic and purely kinetic parameters.

A major advantage of the bond graph formulation for modelling cardiac cell physiology is that the formalism naturally incorporates chemical, electrical and mechanical aspects of cellular function (Gawthrop *et al* 2017): the approach was originally developed in order to provide a common modelling framework for multi-domain systems engineering problems. Application of bond graphs to modelling biochemical systems was originally proposed in pioneering work from Oster *et al* (1971, 1973), and more recently in the context of contemporary systems biology by Gawthrop and Crampin (2014), Gawthrop *et al* (2015), Gawthrop and Crampin (2016). A key feature is that if a model can be represented as a bond graph, then it is in principle a *physically plausible* model (Gawthrop 2018) (i.e. consistent with physical laws, although it may not necessarily represent the true biological mechanism). Additional benefits for modelling of cell physiology are that the bond graph formulation promotes modular (Gawthrop and Crampin 2016) and hierarchical (Gawthrop *et al* 2015) approaches to model building, which have been identified as important for large-scale biochemical systems modelling and for assembling comprehensive models of cell function (Terkildsen *et al* 2008, Cooling *et al* 2008, Siekmann *et al* 2012).

In many cellular processes and biological systems, electrical and chemical processes are intimately linked. Examples include the movement of charges across the cell membrane (ion currents) generating the action potential; and redox (reduction–oxidation) reactions, which are central to many biochemical processes including the mitochondrial electron transport chain, described below. Below we briefly describe the use of bond graphs in modelling the cardiac cell action potential, showing that the formulation identifies issues to do with charge conservation which are often overlooked, in addition to the energy-based approach allowing a direct calculation of the energetic cost of action potential generation to the cell.

Charge conservation: drift in cardiac action potential models

As described above, energy-based modelling ensures that conserved quantities are appropriately represented, and furthermore provides useful tools for identification of non-obvious conserved moieties. Bond graphs allow modelling of electrical and chemical systems, and as a consequence conservation of charge is also represented. This can include more subtle conservation relationships too, in particular in larger systems some unanticipated mass-conserved moieties, and charge conservation, with important consequences for model behaviour and analysis.

The cardiac action potential is generated by currents from ion channels that interact through the membrane potential. In contrast to the above examples that involve chemical potential gradients (figure 6.3), ion channels involve electrochemical gradients (see subsection 6.2.3). Being a generic multi-physics representation, bond graphs are well suited for representing ion channels. The membrane potential is analogous to chemical species in that both can be modelled using C components. However, while chemical species are represented using a nonlinear relationship (equation (6.14)), the membrane potential is modelled as a linear electrical capacitor with the constitutive relationship $V = q/C$. Since Faraday's constant F provides the factor required to convert from electrical to chemical energy ($\mu = zFV$), summing the contributions of electrical and chemical species gives rise to equation (6.30) as shown earlier. Due to the analogy between electrical capacitors and chemical species, bond graph representations of ion channels are structurally similar to biochemical systems. However, while the thermodynamic constraints in biochemical systems apply to ion channels, ion channels have distinct kinetic relationships such as the linear and Goldman–Hodgkin–Katz current–voltage relations (Keener and Sneyd 2009); these can be modelled under the bond graph formalism by using dissipative R components with appropriate constitutive relationships. A further consideration is that ion channels are often regulated by gating processes that open and close the channel; these are described by additional chemical reactions that are linked to the ion transport process (Pan *et al* 2018a).

A bond graph model of cardiac electrophysiology was developed by Pan *et al* (2018a), based on the Luo and Rudy (1994) model updated with thermodynamically consistent representations of the NCX and Na^+/K^+ ATPase. A cell schematic diagram of the model is shown in figure 6.4. Because this model was built upon the bond graph framework, it automatically satisfies the laws of thermodynamics. As a result, it is possible to assess the energy consumption of an action potential. After pacing the model using a stimulus current, the energy cost of an action potential could be calculated by summing the integrated power dissipated by each ion channel and transporter (figure 6.5). For this particular model, the action potential consumed 305 nJ cm^{-2} of energy. Interestingly, this figure is 76% higher than a similar calculation made for the neuronal action potential (Gawthrop *et al* 2017).

Models of cardiac electrophysiology have increased in complexity over the past few decades, now accounting for Ca^{2+} transport, varying ion concentrations and slower membrane transport processes that maintain the ionic gradients required to drive the action potential. However, coinciding with this increase in complexity is the occurence of a phenomemon known as 'drift', where a model fails to reach a steady state limit cycle with prolonged simulation (Varghese and Sell 1997). This abnormal and unphysiological behaviour may cause the model to fail over long simulations, which may be important in examining longer term physiological

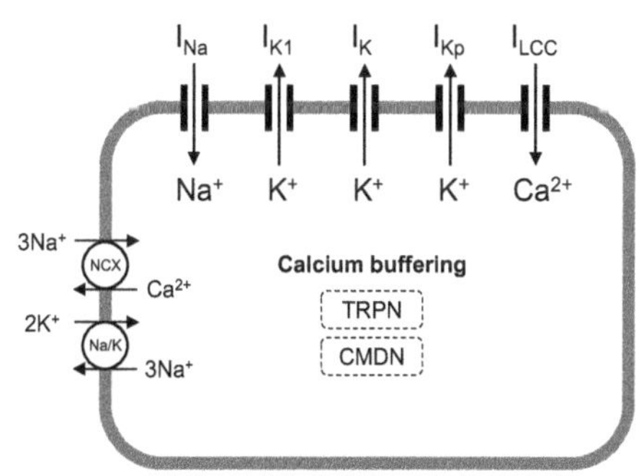

Figure 6.4. Cell schematic showing key components of a simplified bond graph model of cardiac electrophysiology. From Pan *et al* 2018a with permission of the Royal Society.

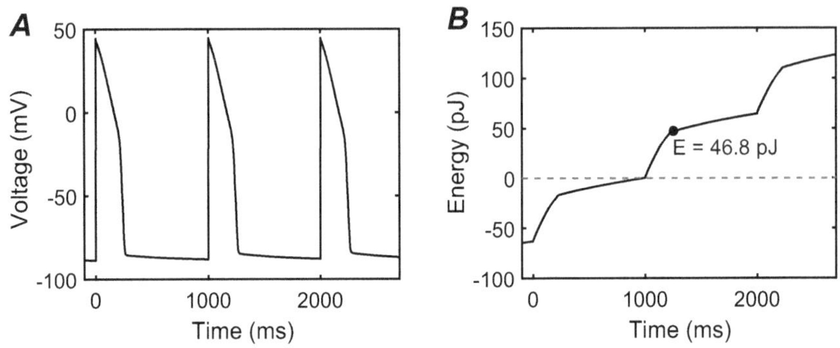

Figure 6.5. Simulations of a bond graph model of cardiac electrophysiology over three stimuli. (A) Membrane potential; (B) energy consumption. The black dot represents the energy consumption of the second action potential. Adapted from Pan *et al* 2018a with permission of the Royal Society.

changes such as ion accumulation in heart disease (Terkildsen *et al* 2007). Drift was investigated in more detail in some later studies (Hund *et al* 2001, Livshitz and Rudy 2009), which found that in certain models (Faber and Rudy 2000, Hund and Rudy 2004), stimulus currents that violated conservation of charge caused models to drift, and accounting for the movement of charge using K^+ ions helped these models to avoid drift. However, identifying equations linked to conservation of charge (if they exist) often requires domain-specific expertise and as a result, drift remains an unresolved issue in many models of cardiac electrophysiology.

The study by Pan *et al* (2018a) finds a more systematic method of detecting and resolving drift, observing that conservation of charge is an example of the more general concept of conserved moieties in biochemical networks (subsection 6.2.1). Since the membrane is represented as a C component, bond graphs enable analyses from biochemistry to be ported to electrophysiology. This systematic approach to conservation laws provides two advantages. Firstly, it is possible determine all the conservation laws of a system relevant for long-term behaviour. Secondly, an arbitrary stimulus current can be tested against an arbitrary model to check for drift without the need for simulation. To examine the relationship between conservation laws and drift, the model of cardiac electrophysiology (Figure 6.4) was simulated over long durations with both a nonconservative stimulus current using nonspecific charge and a conservative stimulus current with charge accounted for using K^+ ions (figure 6.6). Three variants of the model were simulated, with different ion concentrations fixed to constant values to alter the conserved moieties. While none of the models exhibits drift in response to the conservative current (blue lines), as expected, two of the variant models drift in response to the nonconservative current (figures 6.6(A), (B); red lines). By identifying the conserved moieties of each variant, charge was found to be conserved for the first two variants, but not the third. The presence of charge as a conserved moiety coincided with a model's susceptibility to drift; the two variants in figures 6.6(A), (B) drifted in response to a nonconservative stimuli because the stimulus currents caused irreversible changes in the total charge within the system (figure 6.6, bottom panels), whereas the third

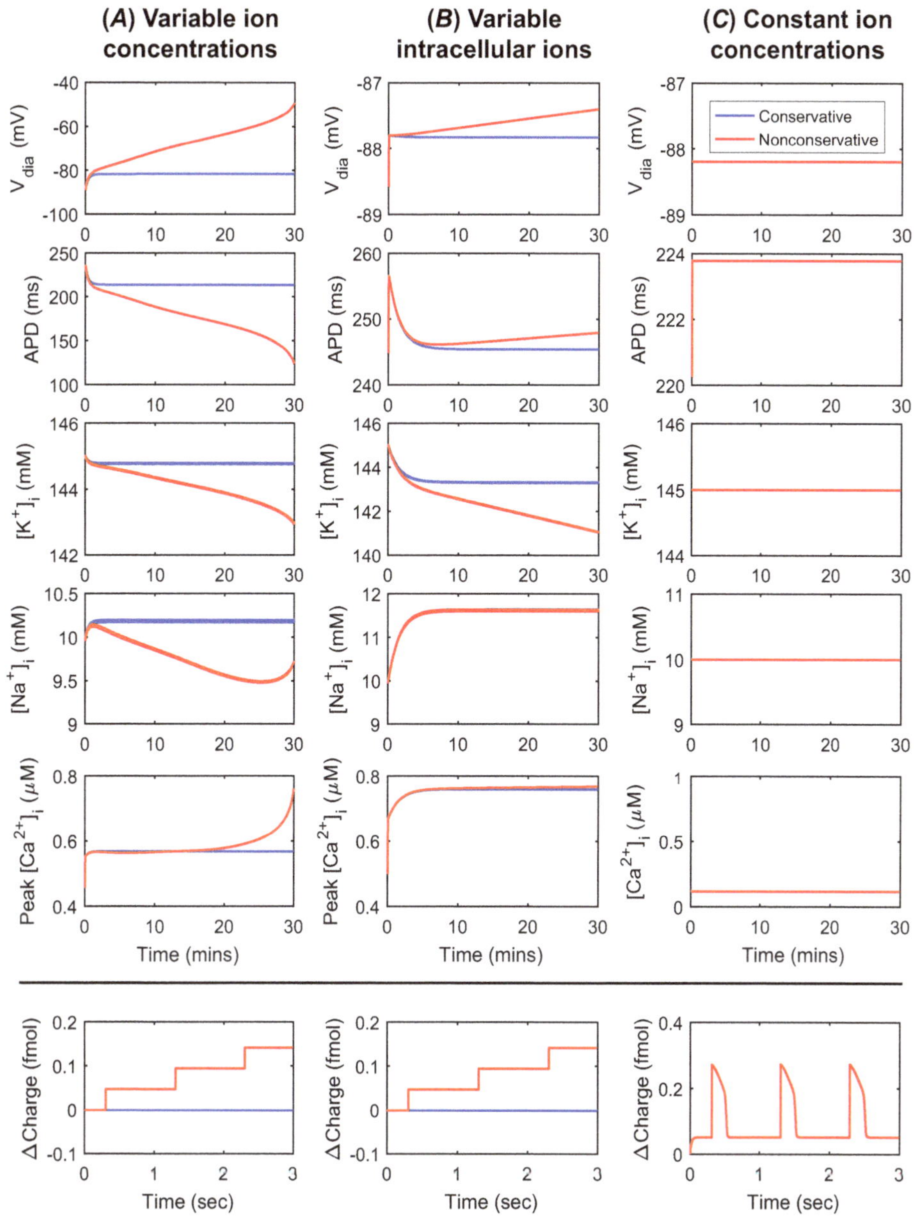

Figure 6.6. Long-term simulations of model of cardiac electrophysiology. The model was simulated under (A) dynamic ion concentrations; (B) fixed extracellular ion concentrations and dynamic intracellular ion concentrations; and (C) fixed ion concentrations. The overall charge of the system is plotted on the bottom panel. Note that charge is a conserved moiety for variants A and B, but not for variant C. From Pan *et al* (2018a) with permission of the Royal Society.

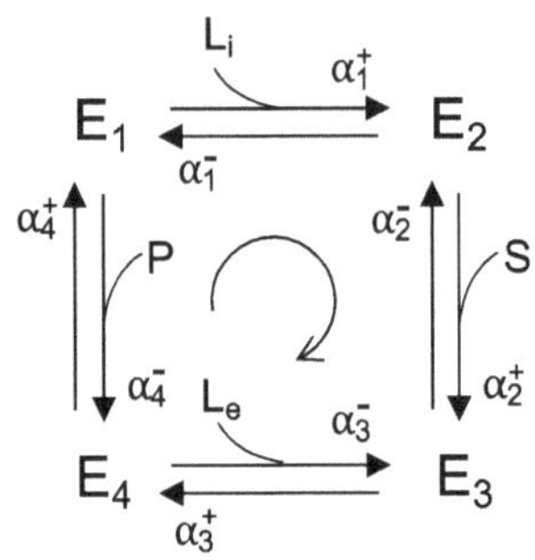

Figure 6.7. Schematic of an enzyme cycle, $E_1, \ldots, E_4$, where the transport of a ligand from an intracellular (L_i) to an extracellular (L_e) compartment is coupled to the conversion of a substrate (S) to a product (P). The α's denote the rate constants of the cycle.

variant in figure 6.6(C) was resistant to drift because charge was not a conserved quantity and therefore free to return to its original value.

The results in figure 6.6 imply that stimulus currents that contribute to any of the conserved moieties cause the model to drift. Therefore, if **G** is the left nullspace of the stoichiometric matrix, any stimulus current v_{stim} that does not satisfy the equation

$$\mathbf{G}v_{\text{stim}} = 0, \tag{6.32}$$

will cause the model to drift. Provided that models are described in sufficient biophysical detail, equation (6.32) provides a simple and general method for checking for drift in an arbitrary model of cardiac electrophysiology, and is more comprehensive than existing methods as it accounts for all conservation laws and not only charge.

6.3 Coupled reactions: thermodynamics of ATPases

Biochemical processes for which $\Delta G > 0$ will not proceed unless energy is input from another source. This can be achieved by coupling to a process with $\Delta G < 0$ for which the free energy change is sufficiently negative, such that the overall free energy change is negative. One scheme by which such coupling is achieved is in the context of ATPase enzyme cycles, described below.

6.3.1 Thermodynamic modelling of coupled enzyme cycles

Figure 6.7 illustrates an enzyme cycle in which transport of a Ligand L is coupled to the conversion of a substrate S to a product P. We can represent this coupled transport with a basic enzymatic cycle. Figure 6.7 shows a general four-state enzyme cycle which couples free energy from the conversion of a substrate S to form a product P, to the transport of a ligand L across an impermeable membrane from an intracellular (i) to an extracellular (e) compartment. Our convention will be that the *forward* reaction proceeds in the clockwise direction (ligand transport from extracellular e to intracellular i compartments, in this case).

Denoting P_i as the proportion (or equivalently the probability for a single protein) of enzyme complexes in state i, where i = E_1, E_2, E_3 and E_4, the chemical potential of a complex in state i is given by:

$$\mu_i = \mu_i^0 + RT \ln P_i, \tag{6.33}$$

where μ_i^0 is the standard chemical potential for state P_i (i.e. when $P_i = 1$). Similarly, denoting $[S]$ and $[P]$ as the concentrations of the substrate and product, respectively and $[L]_i$ and $[L]_e$ as the concentrations of ligand in each compartment, their chemical potentials can be expressed as:

$$\mu_S = \mu_S^0 + RT \ln[S] \tag{6.34}$$

$$\mu_P = \mu_P^0 + RT \ln[P] \tag{6.35}$$

$$\mu_{Li} = \mu_L^0 + RT \ln[L]_i \tag{6.36}$$

$$\mu_{Le} = \mu_L^0 + RT \ln[L]_e, \tag{6.37}$$

Starting at state E_1 and proceeding around the cycle in the clockwise direction, we can determine the net free energy change of the cycle by summing up the contributions from the partial reactions (Hill 1989):

$$\Delta G_{\text{cycle}} = \Delta G_S^0 + RT \ln \frac{[P]}{[S]} + RT \ln \frac{[L]_e}{[L]_i} \tag{6.38}$$

$$= \Delta G_S + \Delta G_L, \tag{6.39}$$

where ΔG_{cycle} is the net free energy change of the cycle, $\Delta G_S = \Delta G_S^0 + RT \ln([P]/[S])$ is the free energy change of substrate conversion to product with the standard free energy $\Delta G_S^0 = \mu_P^0 - \mu_S^0$, and $\Delta G_L = RT \ln([L]_e/[L]_i)$ is the free energy required to translocate the ligand across the membrane. Note that the chemical potential contributions of the enzyme states cancel as the end point of the cycle is the same as the starting point (E_1 in this case).

The direction of transport is determined by the sign of ΔG_{cycle}. If $\Delta G_{\text{cycle}} < 0$, the cycle will proceed in the forward direction and transport the ligand across the membrane at the expense of substrate. If $\Delta G_{\text{cycle}} > 0$, the cycle will reverse and generate substrate from the potential energy in the ligand concentration gradient across the membrane. An equilibrium is established when $\Delta G_{\text{cycle}} = 0$. Note that ΔG_{cycle} determines only the direction of cycling; the reaction kinetics, i.e. the *rate* of cycling, is determined by the rate constants in the cycle.

For modelling ATPase ion transporters, as will be considered below, $[L_e] > [L_i]$, i.e. L is moving *against* a concentration gradient, and so $\Delta G_L > 0$. Therefore, coupling to a process with sufficiently negative ΔG_S is required in order to drive the cycle forwards. These thermodynamic constraints are imposed by the enzymatic cycle and hold regardless of the specific enzyme under consideration.

6.3.2 ATP hydrolysis reaction: driving energetically unfavourable reactions

There are several ways in which coupling such as illustrated in figure 6.7 can be achieved. One involves the cotransport of several species across a membrane, wherein the energetically favourable transport of one species (down a concentration gradient, for example) provides sufficient free energy to transport another species against a transmembrane concentration gradient. In cardiomyocytes, examples include the sodium–calcium exchanger (NCX) which uses the large gradient in sodium concentration across the cell membrane to transport calcium out of the cell, and the various cotransporters and exchangers which control acid influx and efflux to regulate cellular pH (Crampin *et al* 2006, Crampin and Smith 2006, Roberts and Christini 2011).

Here, however, we focus on ATPases, for which transport is coupled to the ATP hydrolysis reaction, which has a large negative free energy change. The hydrolysis of ATP is assumed to follow the reaction:

$$\mathrm{MgATP} + \mathrm{H_2O} \rightleftharpoons \mathrm{MgADP} + \mathrm{Pi} + \mathrm{H^+}, \tag{6.40}$$

where the dominant form of ATP is magnesium-bound (MgATP) and other ATP-bound species have been ignored (Gupta *et al* 1978). Under normal physiological conditions the ATP hydrolysis reaction is maintained at approximately ten orders of magnitude away from equilibrium (Nicholls and Ferguson 2002). Free energy change is therefore large and negative, and the reaction can be coupled to drive reactions such as those mentioned above. In analogy to the example in figure 6.7, MgATP is the 'substrate' and its hydrolysis products MgADP and Pi are the 'products'.

The free energy change of ATP hydrolysis, $\Delta G_{\mathrm{MgATP}}$, is determined by the ratio of the concentrations of MgATP to its hydrolysis products:

$$\Delta G_{\mathrm{MgATP}} = \Delta G^0_{\mathrm{MgATP}} + RT \ln \frac{[\mathrm{MgADP}][\mathrm{Pi}]}{[\mathrm{MgATP}]}, \tag{6.41}$$

with R and T as previously defined and $\Delta G^0_{\mathrm{MgATP}} = -29.6$ KJ mol^{-1} is the free energy change under standard conditions: 1 M MgATP, MgADP and Pi; pH 7, and 37 °C (Guynn and Veech 1973). Under normal physiological conditions ([MgATP] = 5 mM, [MgADP] = 36 μM, [Pi] = 1 mM, pH 7), $\Delta G_{\mathrm{MgATP}} \approx -60$ KJmol^{-1}. Note that this calculation does not account for changes in pH, which is assumed to remain at pH 7, as defined in $\Delta G^0_{\mathrm{MgATP}}$. The approach can readily be modified to situations in which pH varies in a straightforward manner (Tran 2011).

This general coupled enzymatic cycle model can be used to describe the enzyme cycle kinetics and thermodynamics of the three major energy-consuming ATPases in the cardiomyocyte, where ΔG_S is replaced by $\Delta G_{\mathrm{MgATP}}$ and ΔG_L is replaced by either $\Delta G_{\mathrm{SERCA}}$, ΔG_{NaK} or ΔG_{XB} to represent SERCA, the $\mathrm{Na^+/K^+}$ ATPase, and cross-bridge cycling, as described in the following sections.

6.4 Modelling cardiomyocyte bioenergetics

The three key ATPases in the myocyte, in order of energy demand, are the acto-myosin ATPase (cross-bridge cycling), the SERCA pump and the Na^+/K^+ ATPase. SERCA and Na^+/K^+ ATPase are responsible for maintaining trans-membrane ionic gradients, while the acto-myosin ATPase is responsible for force production. For the purpose of deriving mathematical models, each of these ATPases is assumed to operate in a cyclic manner where the cycle of the enzyme is coupled to the free energy of ATP hydrolysis. Each of these processes (ion transport against an electrochemical gradient and force production) is associated with a positive free energy change, and therefore must be coupled to ATP hydrolysis. Although there are other active transporters, such as the sarcolemmal Ca^+ pump, their contribution to overall ATP consumption is relatively minor and they are not considered further here.

The cell can drive such energetically unfavourable reactions by coupling to an accessory process which has strong negative free energy change. Proteins which are ATPases, such as those considered here, achieve this through coupling to the hydrolysis reaction for ATP.

6.4.1 SERCA

Calcium is released from the sarcoplasmic reticulum (SR) intracellular calcium store with each beat, during the process of calcium-induced calcium release, which couples electrical excitation to contraction (Bers 2001). When ryanodine receptors (RyRs) open in response to Ca^+ entry to the cell on sarcolemmal L-type calcium channels (LCC), calcium is released from the SR flowing down a concentration gradient. The calcium released from the SR through opening of RyRs must be returned to the SR in order to prepare for the subsequent release. SR calcium concentration is maintained at a high level relative to cytoplasmic Ca^+ by moving calcium across the SR membrane, against the concentration gradient. This is achieved by the SERCA calcium pump, an ATPase in the SR membrane which couples ATP hydrolysis to Ca^+ transport, according to:

$$2Ca_i^{2+} + MgATP + H_2O \rightleftharpoons 2Ca_{sr}^{2+} + MgADP + Pi + H^+, \tag{6.42}$$

This enzymatic cycle can be modelled using the schema in figure 6.7, where the transported species *L* is replaced by Ca^+ which is transported from intracellular i to SR compartments sr, and substrate conversion to product is replaced by ATP hydrolysis (equation (6.40)). The stoichiometry is such that 2 Ca^+ ions are transported for each ATP molecule. This transporter was studied in detail by Tran *et al* (2009), in which modelling of available kinetic data was able to determine the key regulatory steps in the thermodynamic cycle, predicting a fully cooperative Ca^+-binding mechanism, as well as providing a thermodynamically plausible explanation for observed reversal of the Ca^+ transport under conditions where the concentrations of MgADP and Pi are high and MgATP is kept low, resulting in ATP synthesis (Makinose and Hasselbach 1971). As there is no SR membrane potential, the charge on Ca^+ is not relevant for SERCA. A bond graph formulation of this model of SERCA has subsequently been developed by Pan *et al* (2018b).

6.4.2 Na^+/K^+ ATPase

Rapid Na^+ influx through voltage gated Na^+ channels into the cardiomyocyte following a depolarising stimulus is due to the large electrochemical gradient for Na^+ across the membrane. This gradient must be maintained through ATP-consuming pumping of Na^+ out of the cell. In this case, the stoichiometry of the ion transport is that 3 Na^+ are transported against 2 K^+ in the opposite direction, hence there is a net translocation of charge. Therefore it is necessary to extend the scheme in figure 6.7 to accommodate the thermodynamic dependence on the membrane potential.

Denoting the intracellular concentration of Na^+ as $[Na^+]_i$ and the extracellular concentration as $[Na^+]_e$, let us consider first a membrane with no electric potential gradient. The chemical potential of Na^+ in the intracellular compartment can be written as:

$$\mu_{Na_i} = \mu^0_{Na} + RT \ln [Na^+]_i, \tag{6.43}$$

where μ^0_{Na} is the standard free energy of Na^+ (or, equivalently, the species thermodynamic parameter for Na^+, see equation (6.14)) which takes the same value in each compartment; R is the ideal gas constant, and T is the absolute temperature. Similarly, the chemical potential for Na^+ in the extracellular compartment is:

$$\mu_{Na_e} = \mu^0_{Na} + RT \ln [Na^+]_e. \tag{6.44}$$

It is useful to express these quantities in terms of chemical potentials because it allows us to determine the direction of a reaction, or in this case, the direction of Na^+ movement across the sarcolemma when Na^+ channels open. As the standard free energies are the same in each compartment, the change in chemical potential, or the free energy change of Na^+ translocation across the sarcolemma (from outside to inside) is given by:

$$\begin{aligned} \Delta G_{Na} &= \mu_{Na_i} - \mu_{Na_e} \\ &= RT \ln \frac{[Na^+]_i}{[Na^+]_e}, \end{aligned} \tag{6.45}$$

Reactions with negative free energy change may occur spontaneously. If $[Na^+]_e > [Na^+]_i$ this gives $\Delta G_{Na} < 0$, and the difference in chemical potential will favour the influx of Na^+ into the cell. If $\Delta G_{Na} > 0$, an efflux of Na^+ will be favoured. An equilibrium occurs when $\Delta G_{Na} = 0$, in which case, there is no net movement of Na^+ because the intracellular and extracellular concentrations are identical.

For movement of charged species, if there is a potential difference across the membrane then the free energy change is extended to account for the electric potential, and is expressed as:

$$\Delta G_{\mathrm{Na}} = RT \ln \frac{[\mathrm{Na}^+]_\mathrm{i}}{[\mathrm{Na}^+]_\mathrm{e}} + z_{\mathrm{Na}} F V_m, \tag{6.46}$$

where z_{Na} is the charge of the ion (in this case, for sodium, +1), F is Faraday's constant and V is the potential difference across the sarcolemmal membrane. At any point in time, we can determine the direction of Na^+ movement across the membrane if we know the values for the Na^+ gradient and the membrane potential. At equilibrium, when $\Delta G_{\mathrm{Na}} = 0$, the equilibrium potential across the membrane is given by:

$$V_{\mathrm{Na}} = \frac{RT}{F} \ln \frac{[\mathrm{Na}^+]_\mathrm{e}}{[\mathrm{Na}^+]_\mathrm{i}}, \tag{6.47}$$

which is the Nernst potential for Na^+, i.e. the membrane potential at which there will be no sodium ion flux. Under physiological concentrations ($[\mathrm{Na}^+]_\mathrm{i}$ = 12 mM, $[\mathrm{Na}^+]_\mathrm{e}$ = 140 mM), V_{Na} = +66 mV at 37 °C. Since the resting membrane potential is approximately −85 mV, the opening of voltage-gated Na^+ channels therefore strongly favours an influx of Na^+, which depolarises the membrane potential and initiates the action potential. The above derivations can be applied to any ionic species in the cell. The free energy of K^+ translocation is similarly given by:

$$\Delta G_{\mathrm{K}} = RT \ln \frac{[\mathrm{K}^+]_\mathrm{i}}{[\mathrm{K}^+]_\mathrm{e}} + F V_m, \tag{6.48}$$

while the Nernst potential for K^+ is given by:

$$V_{\mathrm{K}} = \frac{RT}{F} \ln \frac{[\mathrm{K}^+]_\mathrm{e}}{[\mathrm{K}^+]_\mathrm{i}}. \tag{6.49}$$

Under physiological conditions ($[\mathrm{K}^+]_\mathrm{i}$ = 130 mM, $[\mathrm{K}^+]_\mathrm{e}$ = 4mM), V_{K} = −93 mV at 37 °C, which is close to the sarcolemmal resting membrane potential. During the depolarisation phase of the action potential, the potential difference is greater, so potassium efflux is favoured more.

Na^+ is subsequently removed from the cytoplasm by the Na^+/K^+ ATPase. The stoichiometry of the Na^+/K^+ ATPase is such that it transports 3 Na^+ *out* of the cell to 2 K^+ transported *in*:

$$3\mathrm{Na}_\mathrm{i}^+ + 2\mathrm{K}_\mathrm{e}^+ \rightleftharpoons 3\mathrm{Na}_\mathrm{e}^+ + 2\mathrm{K}_\mathrm{i}^+. \tag{6.50}$$

The net free energy change for ion translocation for the pump, following the above notation, is therefore given by

$$\Delta G = 2\Delta G_{\mathrm{K}} - 3\Delta G_{\mathrm{Na}} \approx 45 \text{ kJ mol}^{-1}, \tag{6.51}$$

which is large and positive, i.e. not spontaneously occurring under normal physiological conditions, but is within the ≈60 kJ mol^{-1} supplied by ATP hydrolysis. Coupling of ion translocation to ATP hydrolysis is achieved by the Na^+/K^+ ATPase. The enzymatic cycle schema in figure 6.7 can be used to describe the overall reaction

$$3Na_i^+ + 2K_e^+ + MgATP + H_2O \rightleftharpoons 3Na_e^+ + 2K_i^+ + MgADP + Pi + H^+. \quad (6.52)$$

A thermodynamically consistent model for this transport process based on these ideas was proposed by Smith and Crampin (2004), who also considered formal mathematical approaches to simplifying thermodynamic cycle models of this type. This study found a set of parameters consistent with published data, and noted that with this set of parameters the pump reversal current was negligible at positive free energies arising from varying ADP concentration, consistent with earlier experimental observations (Jansen *et al* 2003), who showed that by decreasing the free energy available from ATP hydrolysis, Na^+ started to accumulate in the cell. The model predicts however that by decreasing MgATP and increasing the magnitude of the naturally occurring Na^+ and K^+ gradients, the pump could be made to run in reverse at an appreciable rate, transporting Na^+ into the cell, generating a net inward current and synthesising ATP, which is consistent with earlier experimental observations (Garrahan and Glynn 1967, Lew *et al* 1970).

The model was subsequently used by Terkildsen *et al* (2007) in a study which considered extracellular K^+ accumulation, hyperkalaemia, during ischemic disease in particular showing that a balance between inactivation and activation of the Na^+/K^+ ATPase during ischemia, due to changing metabolic and ionic conditions as ischemia progresses, underlies the experimentally observed triphasic accumulation of extracellular K^+. The Na^+/K^+ ATPase model has more recently been updated by Pan *et al* (2018b) using bond graph formalism to ensure thermodynamic consistency.

6.4.3 Force generation

Cyclical force development, shortening and relaxation of the cardiomyocyte triggered by the intracellular Ca^+ transient is achieved by the expenditure of metabolic energy from ATP hydrolysis. The sliding filament theory of muscle (Huxley 1957) proposes that at the molecular level force is generated through cyclic formation, shortening (force development) and release of a population of cross-bridges formed between interweaving actin and myosin filaments. Force generation through interactions between actin and myosin filaments has received considerable attention both in terms of direct biophysical observation and experimental interrogation, and in mathematical modelling.

The end products of ATP hydrolysis are inhibitory to the cross-bridge cycle. This inhibition becomes particularly important during periods of hypoxia or ischemia where reduction in ATP available and accumulation of Pi, ADP and protons (H^+) can interfere with the interactions between actin monomers and myosin heads on the thick filaments. In order to explore the influence of these energy metabolites on force production under physiological and disease conditions, a thermodynamically-consistent and metabolite-sensitive model of cardiac cross-bridge cycling is needed.

The force generated in a myofibril is due to interactions of many actin–myosin filaments, each of which consists of formation and dynamics of a number of actin–myosin cross bridges with nontrivial interactions. The most detailed models of active force development therefore require consideration of population-level, stochastic

(Monte Carlo) and potentially spatial interactions (Smith 1998). Sliding-filament models are based on strain-dependent distributions of cross-bridge populations, and have been extended to include metabolite interactions (Pate and Cooke 1989), including the recent model of Tewari *et al* (2015, 2016) which is consistent with a broad range of experimental data on force development and metabolic data. However, the explicit representation of a spatial dimension (cross-bridge strain) requires the solution of partial differential equations (PDEs), which renders them computationally expensive and not ideal for integrating into cell or tissue models. Mean-field approximation models have been proposed that average over the microscopic details of cross-bridge population distributions and instead describe the behaviour of the entire cross-bridge population to an average cross-bridge cycle, while retaining sufficient mechanistic detail in order to couple force development to ATP hydrolysis in an appropriate manner.

The Rice *et al* (2008) model, based on earlier work by Razumova *et al* (1999), deals specifically with representing a comprehensive set of experimental characterizations including steady state force-length, force-calcium, length-calcium and force-velocity relationships; isometric twitches (force developed at fixed length after Ca^+ activation); force development after a sudden length change; cell shortening twitches, and length effects on the intracellular Ca^+ transient, using an approximate, mean-field cross-bridge cycle model. This model provides a foundation onto which a biochemical scheme of ATP hydrolysis is integrated to develop a thermodynamically consistent and metabolite-sensitive model of cross-bridge cycling kinetics. Tran *et al* (2010) presented an extended version of the Rice *et al* (2008) model which updates the cross-bridge cycle by thermodynamically coupling to ATP hydrolysis, introducing new dependencies on metabolite concentrations. While free energy transduction for the cross-bridge (XB) cycle (ΔG_{XB}) is somewhat different from the two pumps described above, as the free energy is used to generate force (to do work) rather than coupled to another transport process, from a modelling perspective a similar approach can be taken:

$$XB_{bound} + MgATP + H_2O \rightleftharpoons XB_{unbound} + MgADP + Pi. \tag{6.53}$$

In the model by Tran *et al* (2010), three new parameters are used to describe the thermodynamic cycle and metabolite dependencies, which were fitted using available data and validated against a range of metabolic and sinusoidal length change protocols. They used the model to investigate the effects of changing metabolite concentrations, observed with the onset of ischemia, on the development of active force.

One important addition to the model is a modification to make the cycle in principle reversible. Typically it is assumed in such models that the cross-bridge cycle is irreversible. In fitting the model to data, the value of the introduced reverse rate constant is calculated to be very small. This small reverse rate constant restricts the cross-bridge from cycling backward to any appreciable extent, however, simulation of muscle stretch under non-physiological conditions where the overall free energy was artificially set to be positive produced only a very small reversal flux.

This is consistent with earlier experimental studies that failed to demonstrate generation of ATP through the application of external work to myocytes (Infante *et al* 1964, Curtin *et al* 1969, Curtin and Davies 1973, Linari *et al* 2003). Further discussion on non-reversibility of the cross-bridge cycle was provided by Loiselle *et al* (2010).

6.4.4 Mitochondrial ATP generation

Mitochondria make use of reduction–oxidation (redox) reactions in order to generate ATP. As first described by Mitchell (1961), mitochondria use a chain of redox reactions to move protons (H^+) across the mitochondrial inner membrane, thereby generating across the membrane an electrochemical gradient (the so-called proton-motive force, PMF). The PMF is subsequently coupled to ATP synthesis from ADP and Pi as protons pass back across the membrane through the F_1F_0-ATP-synthase. A key component of this system is the mitochondrial electron transport chain (ETC). Electrons are transported by, and gain energy from, a series of redox reactions, and this energy is partially transduced by pumping protons across the inner membrane, primarily in three protein complexes (CI, CIII and CIV) in the mitochondrial inner membrane, as shown in figure 6.8. The overall chemical reaction of the electron transport chain is:

$$2\text{NADH} + \text{O}_2 + 18\text{H}_x^+ \rightleftharpoons 2\text{NAD}^+ + 2\text{H}_2\text{O} + 16\text{H}_\text{i}^+, \tag{6.54}$$

Because mitochondria transduce energy, a thermodynamically-consistent approach to modelling is desirable. Beard and colleagues have developed the most comprehensive such models to date (Beard 2005, Wu *et al* 2007, Bazil *et al* 2016). A bond graph model of mitochondrial oxidative phosphorylation based on modelling

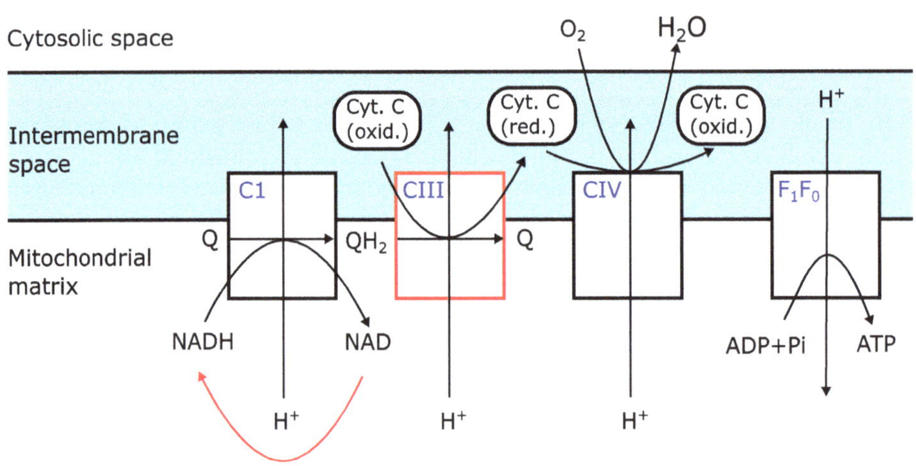

Figure 6.8. Schematic of the major components of the electron transport system in the mitochondrial matrix. The red arrow (dehydrogenase flux) and red square (Complex III) denote the reactions that have explicit Pi regulation.

the redox reactions associated with complexes CI, CIII and CIV of the mitochondrial electron transport chain has been given by Gawthrop (2017).

The Beard (2005) mitochondrial model consists of three separate compartments: the matrix space, the intermembrane space (IM) and the cytosolic space (figure 6.8). The external substrates are NADH and O_2 and the external products are NAD^+ and H_2O. The TCA cycle and other NADH-producing reactions are not explicitly modelled. Instead, a phenomenological driving force is used to simulate the phosphate-dependent rate of reduction of NAD^+ to NADH (dehydrogenase flux) via the reaction:

$$NAD^+ \rightleftharpoons NADH + H^+. \tag{6.55}$$

The phenomenological driving force is regulated by the concentration of Pi in the mitochondrial matrix (red arrow in figure 6.8), which acts as a feedback mechanism to stimulate mitochondrial respiration when energy demand increases.

Ubiquone in reduced form QH2 and oxidised form Q (sometimes denoted coenzyme Q or CoQ) is recycled around CI and CIII; cytochrome c in reduced form Fe^{2+} and oxidised form Fe^{3+} is recycled around CIII and CIV and the two cycles intersect. The reaction at Complex I is described by

$$H^+ + NADH + Q \rightleftharpoons NAD^+ + QH_2 + 4\Delta H^+, \tag{6.56}$$

where the notation ΔH^+ is used to indicate the H^+ that are transferred from the matrix to the cytosol, against the electrochemical gradient. Q and QH_2 exist in the mitochondrial matrix. Complex II is not considered as it does not contribute to the proton gradient. FAD and $FADH_2$ are not included in the model. Complex III is

$$QH_2 + 2\text{cytC(ox)}^{3+} \rightleftharpoons Q + 2\text{cytC(red)}^{2+} + 2H^+ + 4\Delta H^+. \tag{6.57}$$

The oxidised and reduced forms of cytochrome c are present in the IM space. Although four protons are pumped out of the matrix, the total number of charges transferred is only two because the redox transfer from QH_2 to cytochrome c generates two matrix H^+ for each turnover of the reaction. Pi regulation is introduced here and shown to be necessary to fit Pi-dependent mitochondria data (Beard 2005). This reaction is highlighted by the red square in figure 6.8. The flux through complex III (J_{C3}) is sensitive to the matrix Pi concentration. Complex IV is

$$2H^+ + 2\text{cytC(red)}^{2+} + \frac{1}{2}O_2 \rightleftharpoons 2\text{cytC(ox)}^{3+} + H_2O + 2\Delta H^+. \tag{6.58}$$

In this final respiratory chain reaction, oxygen is reduced to water with a transfer of two protons and four charges across the inner membrane. ATP synthesis is then

$$H^+ + ADP + Pi \rightleftharpoons n_A\Delta H^+ + ATP, \tag{6.59}$$

where $n_A = 3$ is the number of protons transported each time this reaction occurs. The synthesis of ATP from the phosphorylation of ADP is catalysed by the F_1F_0-ATP-synthase. ATP is transported into the IM space by the ANT antiporter which simultaneously brings in ADP, and is coupled to mitochondrial membrane potential as it involves net charge translocation. Pi transport into the matrix (to synthesise ATP) is coupled to the proton gradient by the PiHt co-transporter. In the IM space, the transport of ADP, AMP, ATP and Pi across the outer membrane and into the cytosol is modelled as passive diffusion.

Phosphates are transferred between ATP, ADP and AMP in the IM and cytosolic spaces via the adenylate kinase reaction:

$$2\mathrm{ADP} \rightleftharpoons \mathrm{ATP} + \mathrm{AMP}. \tag{6.60}$$

In the cytosolic space, the phosphate from ATP is transferred to creatine via the creatine kinase reaction:

$$\mathrm{ATP} + \mathrm{Cr} \rightleftharpoons \mathrm{CrP} + \mathrm{ADP} + \mathrm{H}^+, \tag{6.61}$$

generating creatine phosphate (CrP). This reaction acts as a buffer to maintain the ATP concentration at near constant levels in the face of high work demand.

Subsequent modelling efforts from Beard and colleagues have included the TCA cycle and other components to provide an increasingly comprehensive model of mitochondrial biophysics. Gawthrop *et al* (2019) have developed a simplified, thermodynamically-consistent model for the ETC using the bond graph approach to choose suitable constitutive expressions to describe overall reaction steps in the ETC. Using this approach they were able to derive an explicit formula relating the the flux through the electron transport chain to the PMF generated across the mitochondrial membrane, which was shown to capture the behaviour of complex mechanistic models as well as explain experimental data. The advantage of this simplified approach is that it enables development of simplified yet thermodynamically consistent (hence physically plausible) models which can be incorporated into large-scale biochemical models, without requiring full mechanistic description of the underlying processes.

6.4.5 Coupling ATP supply and demand in the cardiomyocyte

Tran *et al* (2015) coupled the Beard (2005) mitochondrial supply model to a cardiomyocyte electrophysiology, calcium signalling and mechanics model (Crampin and Smith 2006) updated with metabolite-sensitive thermodynamic representations of SERCA (Tran *et al* 2009), Na^+/K^+ ATPase (Terkildsen *et al* 2007) and cross-bridges (Tran *et al* 2010). The model is illustrated in figure 6.9. Coupling of ATP generation and consumption in this model was achieved by quantifying the movements of the ions and metabolites that are in common to both models: more specifically MgATP, MgADP and Pi. These metabolites regulate both the direction and rate of the active processes in the cytosol as well as the rate of ATP generation by the mitochondria. Mitochondrial ATP generation is regulated by the dependence of the dehydrogenase and complex III reactions on inorganic phosphate, as well as by the 'mass-action' effect of changing concentrations of reactants and products of metabolic reactions. As discussed further below, a key focus of the model was to investigate the effects of dynamic metabolite concentrations on ATP consumption in a working cardiomyocyte.

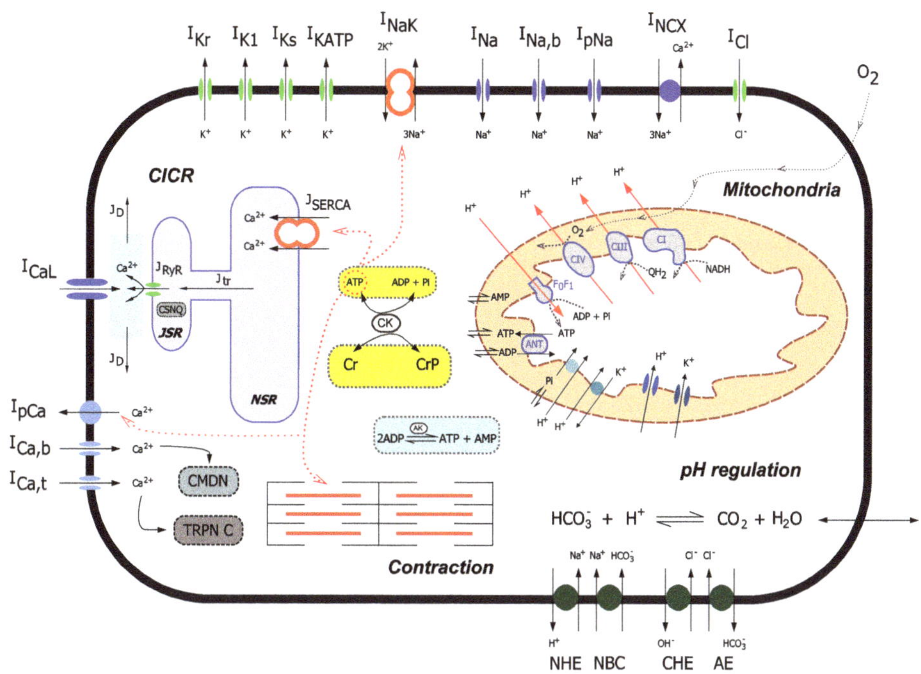

Figure 6.9. Schematic representation of the myocyte model including the Beard (2005) model of mitochondrial oxidative phosphorylation. Reprinted from Tran *et al* 2015 with permission of the Physiological Society.

Other mechanisms incorporated into the model by which mitochondrial and cytoplasmic processes are strongly coupled include proton concentration (i.e. pH) which plays a crucial role in regulating both mitochondria and myocyte function. Mitochondria use the proton gradient across the mitochondrial inner membrane to generate ATP in oxidative phosphorylation. In the cytosol, protons regulate many key processes including EC coupling. Potassium also plays an important role coupling mitochondrial to electrophysiological behaviour. K^+ is transported into the intermembrane space via the mitochondrial H^+/K^+ anti-porter and passively diffuses back out. In the myocyte, K^+ plays an important role in repolarising the sarcolemmal potential after excitation.

Model validation and comparisons to data

The steady state ATP consumption rates reached in the model at 2 Hz stimulation frequency are compared in table 6.1 to experimental data from Schramm *et al* (1994) measured from guinea pig trabeculae. Quantitatively, the total ATP consumption rates from the integrated model compare well to the experimental observations. Comparison of measured oxygen consumption rates to those predicted by the model over varying myocyte workloads presents an important test of the model. Figure 6.10 shows predictions of steady metabolite concentrations as a function of ATP consumption rate (varying cardiomyocyte workload) from the model. The top panel of figure 6.10 shows the rate of oxygen consumption (MVO_2) calculated from

Table 6.1. Comparing average ATPase consumption rates (in units of mM ms^{-1}) from the fully integrated Tran *et al* (2015) myocyte model to data from guinea pig trabeculae from Schramm *et al* (1994). The relative contribution (%) of each of the ATPases is given in parentheses. Model and experimental data values were obtained at 2 Hz stimulation frequency.

(mM ms^{-1})	Na^+/K^+ ATPase	SERCA	Cross-bridge	Total
Tran *et al model*	2.84×10^{-5} (5%)	5.60×10^{-5} (10%)	5.01×10^{-4} (85%)	5.85×10^{-4}
Schramm *et al data*	4.96×10^{-5} (9%)	8.27×10^{-5} (15%)	4.20×10^{-4} (76%)	5.51×10^{-4}

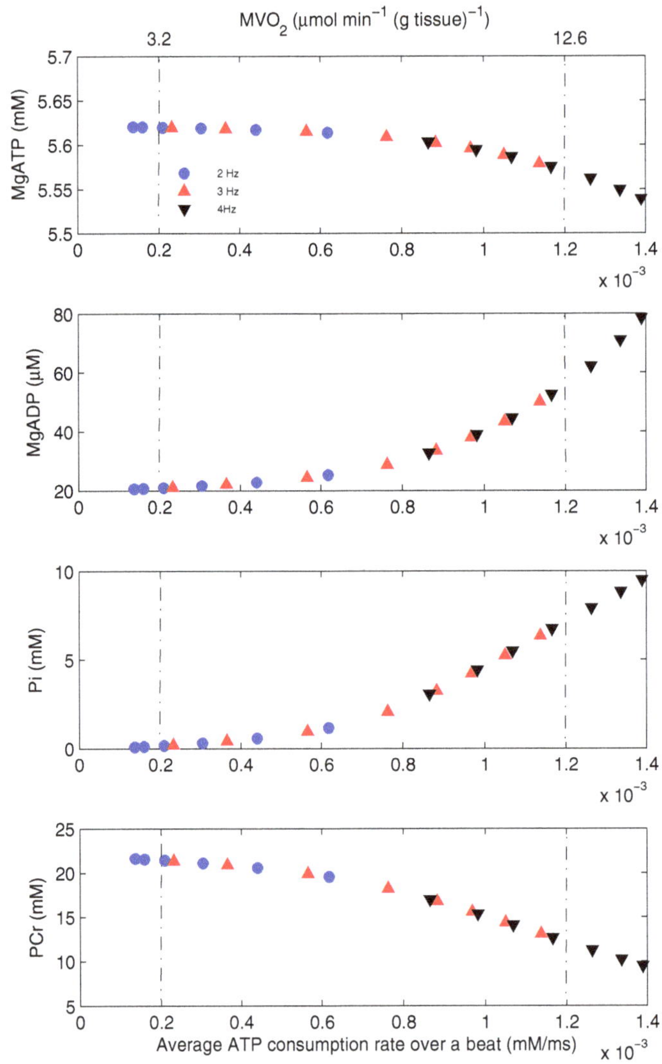

Figure 6.10. Metabolite concentrations predicted by the coupled model as a function of ATP consumption rate. The MVO_2 values shown above the top panel are directly calculated from the flux of the complex IV reaction.

the complex IV reaction flux in the model. Zhang and colleagues (Bache *et al* 1999, Gong *et al* 1999, 2003, Ochiai 2001, Zhang *et al* 2003, 2005) measured a maximum MVO_2 of 10.7 μmol min^{-1}(g tissue)$^{-1}$ in isolated perfused whole canine hearts, while Cooper *et al* (2001) reported 12 μmol min^{-1}(g tissue)$^{-1}$ and 6 μmol min^{-1}(g tissue)$^{-1}$ in isolated perfused rat and guinea pig hearts, respectively. In the case of the guinea pig heart, the maximum MVO_2 was obtained by putting it into KCl arrest and then subjecting it to very low extracellular Na^+ concentrations, causing the Na^+/Ca^+ exchanger to reverse and the cytosol to flood with Ca^+, fully activating the cross-bridges and SERCA pump.

A further experimental measure of cardiomyocyte energetic status is the inorganic phosphate (Pi) to creatine phosphate (CrP) ratio ΔPi/CrP. Experimental and model-predicted values are presented in figure 6.11. The canine data from Zhang and colleagues indicate that the metabolite concentrations (ΔPi/CrP) remain relatively constant until MVO_2 is greater than approximately 6 μmol min^{-1}(g tissue)$^{-1}$ before rising rapidly (figure 6.11(A)). Figure 6.11(B) shows very similar behaviour from the model. ΔPi/CrP was calculated from the model-predicted Pi and CrP values shown in figure 6.10, plotted against MVO_2.

6.4.6 Measuring and modelling mechano-energetics

The mechanical performance of the heart has been extensively studied. The relationship between mechanical performance—pressure–volume loop for the whole organ, and force-length loop for isolated cardiac tissue or individual cardiomyocytes—and energy metabolism is central to understanding the function of the heart. These relationships aim to characterise the work done by the heart muscle in relation to metabolic energy expended. An extensive recent review of theoretical and experimental aspects of cardiac mechano-energetics can be found in Han *et al* (2019), with a particular focus on determining the nature of, and the fundamental constraints on, the thermodynamic efficiency of the working heart, as examined through measurements of heat production and work rate in both whole organ and multicellular tissue preparations.

Notably, recent advances in instrumentation (Han *et al* 2009, Taberner *et al* 2011) have enabled measurement and control of force and length, and simultaneous measurement of heat output in isolated trabeculae (naturally occurring columnar tissue preparations containing aligned ventricular cardiomyocytes) mounted in a micro-calorimeter device. Measurements include isometric (constant length) contractions (Han *et al* 2010) and work loop experiments (isometric contraction followed by shortening against a preset load, or afterload, followed by isometric relaxation, and then lengthening at a set preload) representing physiologic heart function.

The Tran *et al* (2010) and Tran *et al* (2015) models have been used to help interpret experimental data from such micro-calorimeter experiments, to determine different components of measured heat, to explain the observed linear relationship between energy expenditure and work in the heart (Loiselle *et al* 2008, Han *et al* 2012), and to interpret observed heat production and thermodynamic efficiency of

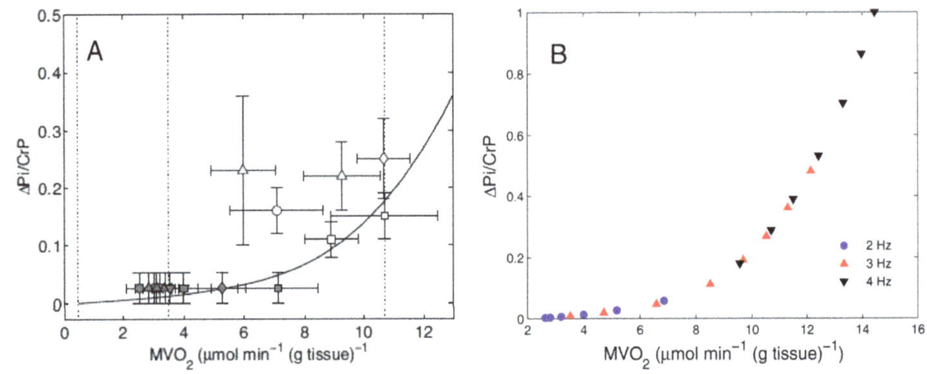

Figure 6.11. Panel A: Data from Zhang and colleagues (Bache *et al* 1999, Gong *et al* 1999, 2003, Ochiai 2001, Zhang *et al* 2003, 2005), modified from figure 1(B) of Wu *et al* (2009). The line is a prediction from the oxygen-transport coupled model of Wu *et al* (2008), copyright 2009 National Academy of Sciences, which also incorporates the Beard (2005) mitochondrial model of oxidative phosphorylation. Panel B: Tran *et al* (2015) cell model simulation of ΔPi/CrP as a function of MVO_2.

the heart, in normal (Tran *et al* 2017) and disease conditions (Han *et al* 2014, Tran *et al* 2016).

6.5 Current research, controversies, and future directions

The development of thermodynamically-constrained models of cardiac energetics has shed light on many aspects of cardiomyocyte function. Below we highlight some current research directions and new opportunities for energy-based modelling of cardiomyocytes.

6.5.1 How is mitochondrial energy generation regulated?

Predictions of metabolite concentration changes with cardiomyocyte workload (or oxygen consumption rate) from the Tran *et al* (2015) model are shown in figure 6.10. At low workloads the model predicts that steady-state metabolite concentrations change only slowly with increasing workload, however, it is nevertheless the change in metabolite concentration, specifically Pi, which signals to the mitochondria via negative feedback regulation of ATP production.

The Beard (2005) model of mitochondrial oxidative phophorylation utilises no other mechanisms of oxidative phosphorylation control than the feedback effects of the phosphate metabolites. Specifically, the dehydrogenase and complex III fluxes are dependent on Pi while magnesium-bound ATP, ADP and Pi govern the activity of the F_1F_0-ATP-synthase via mass action. Analysis by Tran *et al* (2015) showed that under physiological conditions the rate of mitochondrial ATP production can match varying demand through substrate availability alone; that ATP production rate is not limited by the supply of reducing equivalents from the TCA cycle, as a result of Ca^+ or Pi activation of the dehydrogenase; and that ATP production rate is sensitive to feedback activation of complex III by Pi. Furthermore, they showed that feedback regulation of mitochondrial complex III by Pi improves the coupling between energy demand and mitochondrial ATP production and stabilizes cytosolic ADP and Pi concentrations, which consequently stabilizes cytosolic ionic concentrations leading to reduced cellular energetic cost from other ATPases in maintaining cellular ionic homeostasis.

Investigating the regulation of energy supply by Pi

The computational models of Beard and colleagues predict that Pi is the major regulator of mitochondrial ATP generation, in direct contrast to the metabolic stability conclusions reached earlier by Balaban and co-workers (Balaban *et al* 1986, Katz *et al* 1989), who reported experimental observations that metabolite values remain constant over a wide range of workload, and hence concluded that mitochondrial ATP production cannot be regulated by metabolite levels. One explanation for the apparent discrepancy is, as Beard has argued, that the baseline Pi concentration in the myocardium is only marginally detectable using phosphate magnetic resonance spectroscopy (P-MRS) and that the constant baseline Pi measurements made by Katz *et al* (1989) are likely due to 2,3-diphosphoglycerate in the blood pool of the left ventricle cavity (Beard and Kushmerick 2009). More recent experimental data from Zhang and co-workers (Bache *et al* 1999, Gong *et al* 1999, 2003, Ochiai 2001, Zhang *et al* 2003, 2005) have demonstrated that inorganic phosphate does indeed increase during increased cardiac energy consumption.

Beard and colleagues have shown in a number of simulation studies (Beard 2006, Wu *et al* 2008, 2009) incorporating the Beard 2005 model, that Pi is the key regulator of oxidative phosphorylation in the heart. To test this hypothesis in the Tran *et al* 2015 model, the two Pi regulatory steps from the mitochondrial model were removed and the cell model was subjected to an increase in work rate. While the work rate was kept below the threshold hold value of 0.6×10^{-3} mM ms^{-1}, there was a significant difference between the simulations with and without the Pi regulatory steps. In the absence of the Pi regulatory steps, the Pi concentration increased by 40 fold and the MgADP concentration increased by almost 2 fold, whereas in the presence of Pi regulatory steps, the Pi and MgADP increases were 10 fold and 0.25 fold, respectively. The Pi-regulatory steps in the Beard 2005 model are thus shown to have a significant impact on keeping the metabolite concentrations within a narrow range in the Tran *et al* 2015 model. The 2 fold change in Pi is within the range which is low enough to avoid detection by P-MRS (Beard and Kushmerick 2009). Hence, over this range of work demand, small changes in the concentration of Pi are sufficient to stimulate the mitochondria to produce ATP and maintain homeostatic concentrations of metabolite, which also satisfies the metabolic hypothesis from Balaban and co-workers (Balaban *et al* 1986, Katz *et al* 1989) as such small metabolite changes (especially of MgADP) would be undetectable using P-MRS.

Other mitochondrial models

As well as those of the Beard group, modelling efforts by Saks and colleagues (Saks *et al* 2000, Aliev and Saks 1997) have also concluded that Pi plays a key role in regulating mitochondrial oxidative phosphorylation. At high work rates, their models estimate that Pi increases by approximately 8 fold which is similar to the predictions from the Tran *et al* (2015) study. The models from the Saks group, however, are constructed from three compartments, with spatially inhomogeneous reaction–diffusion equations governing the movement of phosphate metabolites between these compartments: compartmentalised energy transfer. These authors hypothesised that there is a coupling between mitochondrial creatine kinase and and adenine translocase in the mitochondrial intermembrane space which may play a role in maintaining metabolic concentrations within a narrow range by the diffusion of the smaller and much faster CrP molecule (Vendelin *et al* 2000), discussed further below.

Modelling studies from Korzeniewski *et al* (2005) and from Zhou (2006) suggested that activation of mitochondrial respiration via an increase in energy demand was not enough to maintain stable metabolite levels. They hypothesised that simultaneous parallel activation of processes up-stream and down-stream of the respiratory chain were necessary to stimulate the mitochondrial ATP generation. In simulations by Tran *et al* (2015), only metabolic product feedback from an increase in energy demand was necessary to maintain metabolite homeostasis.

Regulation by mitochondrial calcium
A number of other modelling studies have coupled models of the TCA cycle and mitochondrial oxidative phosphorylation into an electrophysiological framework and suggested Ca^+ to be the key regulator that stimulates substrate dehydrogenation as well as mitochondrial respiration (Nguyen *et al* 2004, Cortassa *et al* 2006). Although these models explicitly model the three main ATPases, they are, however, not constrained to be thermodynamically consistent.

The energy supply part of the Tran *et al* (2015) model consists solely of mitochondrial oxidative phosphorylation with Pi regulation. The model does not take into account any of the upstream TCA cycle processes or mitochondrial Ca^+ handling. Ca^+-dependent activation of the TCA and respiratory processes provides a feed forward mechanism as Ca^+ is intimately linked to cross-bridge cycling and is therefore ideally situated to act as a signalling molecule regulating energy demand and supply. However, experimental findings report that the degree of respiratory activation by Ca^+ is relatively small (Territo *et al* 2001). It is possible that Ca^+ regulation may become important as a secondary mechanism at higher work rates when cytosolic Ca^+ concentrations begin to rise.

6.5.2 Creatine kinase shuttle

The ATP generated by mitochondria through oxidative phosphorylation is transferred out of the mitochondrial matrix by the adenine nucleotide translocator (ANT). ANT exchanges one ATP molecule from the matrix for one ADP molecule present in the intermembrane space (IMS). Several parallel pathways have been proposed to facilitate the rest of the journey of ATP from the mitochondria to the site of hydrolysis in the myofibrils. These include direct diffusion of ATP or ADP (Vendelin *et al* 2000, Beard 2006); adenylate kinase (AK) mediated phospho-transfer through diffusion of AMP (Vendelin *et al* 2000, Dzeja *et al* 2011, Savabi 1994, Zeleznikar *et al* 1995); and creatine kinase (CK) mediated phospho-transfer through diffusion of creatine phosphate (CrP) and creatine (Cr) (Vendelin *et al* 2000, Meyer *et al* 1984, Guimarães-Ferreira 2014, Neubauer 1998).

The first two pathways can be broadly classified as diffusion of ANPs (adenine nucleotide phosphate groups). Due to higher concentration of ATP in the IMS, ATP can directly diffuse from the IMS to the myofibrils. On the other hand, ADP can diffuse in the reverse direction from the myofibrils to the IMS. Alongside the diffusion of ADP, diffusion of AMP can also facilitate the transfer of ADP from myofibrils to IMS. Cytosolic adenylate kinase, a phosphotransferase enzyme present in myofibrils, catalyses the following equilibrium reaction: $2ADP \rightleftharpoons ATP + AMP$. The generated AMP diffuses to the IMS, and it is converted back to ADP by the mitochondrial adenylate kinase enzyme which catalyzes the reaction $ATP + AMP \rightleftharpoons 2ADP$. This mechanism is often referred to as the adenylate kinase shuttle (Saks *et al* 1994).

The pathway involving the CK enzymes is also known as the creatine kinase shuttle, or the phosphocreatine shuttle (Vendelin *et al* 2000, Meyer *et al* 1984, Guimarães-Ferreira 2014, Neubauer 1998). In the last few decades, several research groups have suggested that the phosphocreatine shuttle is the dominant pathway for phospho-transfer between mitochondria and myofibrils (Dzeja *et al* 2011, Aliev *et al* 2011, Aliev and Saks 1997). The phosphocreatine shuttle involves four substrates: ATP, ADP, Cr and CrP. In the mitochondrial IMS (including the cristae space), a major proportion of the ATP released by the ANT carrier proteins is converted into CrP through a reaction mediated by mitochondrial creatine kinase enzymes (mt-CK). It has been proposed that restricted diffusivity of ATP in the IMS region leads to an effective coupling between ANT and mt-CK (Simson *et al* 2016, Jacobus and Saks 1982, Schlattner *et al* 2006). Consequently, most of the ATP transfer between IMS and myofibrils occurs through diffusion of CrP instead of direct diffusion of ATP. Following diffusion of CrP to myofibrils, myofibrillar creatine kinase (m-CK) converts CrP back to ATP from ADP formed by ATP hydrolysis in myofibrils. The Cr formed in the myofibrils diffuses to the IMS and is subsequently converted back to ADP by mt-CK, thus completing the phosphocreatine shuttle, see figure 6.12.

Previous studies have proposed several advantages of the phosphocreatine shuttle mechanism over direct diffusion of ATP or ADP. For instance, CrP and Cr molecules being smaller in size than ATP and ADP, encounter fewer diffusion restrictions. Beside this, the CrP shuttle also minimizes the concentration gradients of ATP and ADP. According to an experimental study by Dzeja *et al* (2011), the phosphocreatine shuttle is responsible for around 70%–80% of total phospho-transfer. This value is consistent with model predictions from Vendelin, Saks and co-workers (Aliev *et al* 2011, Aliev and Saks 1997). However, other modelling studies predict a much lower 10%–50% contribution of phosphocreatine shuttle to total ATP and ADP exchange (Hatano *et al* 2011, Hettling and van Beek 2011).

It is not yet clear whether phosphocreatine shuttle is necessary for normal functioning of cardiomyocytes. A computational study by Meyer *et al* 1984 suggested that the phosphocreatine shuttle is only a consequence of the spatial

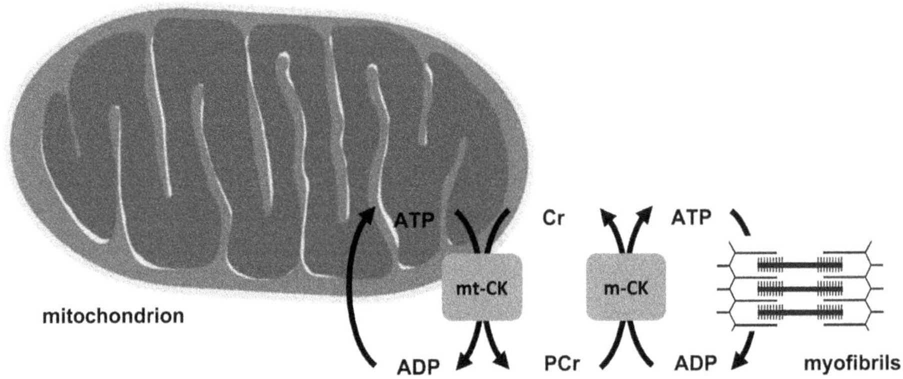

Figure 6.12. The phosphocreatine shuttle involves the activity of two different isoforms of the CK enzyme: myofibrillar m-CK and mitochondrial mt-CK. While mt-CK drives the reactions in favour of formation of PCr from mitochondrial ATP, m-CK acts in the opposite direction by regenerating ATP from PCr at the myofibrils.

buffering of ATP by CrP and not an essential mechanism. Indeed, several experimental studies on mice with genetic knockout of cytosolic and mitochondrial CK (Kaasik *et al* 2001, Saupe *et al* 1998), as well as Cr deficient mice (Branovets *et al* 2013), show that these mice can maintain normal cardiac function without a functional phosphocreatine shuttle system, suggesting that phosphocreatine-based energy transfer is a redundant process among several parallel pathways. The Kaasik *et al* (2001) study also reported ultra-structural changes to the cardiomyocytes in CK deficient mice that might compensate for the lack of a phosphocreatine shuttle, finding that the rows of myofibrils and mitochondria were narrower in CK deficient cardiomyocytes resulting in smaller intracellular diffusion distances. This possible relationship between cardiac energy transfer pathways and sub-cellular ultrastructure warrants further investigation.

6.5.3 Role of cellular architecture: spatial models of cardiomyocyte energetics

ATP demands at hydrolysis sites within myofibrils can vary five-fold between resting and active cardiac states (Bittl and Ingwall 1985, Katz *et al* 1989, Williamson *et al* 1976). Most of this ATP comes from longitudinal columns of cardiac mitochondria, which traverse in between columns of myofibrils. Previous studies have suggested that cardiac mitochondria are arranged in a 'crystal like regular pattern with very small deviation in the distances between neighbouring parallel strands of mitochondria (Vendelin *et al* 2005, Birkedal *et al* 2006). It was also predicted that this highly ordered mitochondrial arrangement further leads to the formation of functional units called intracellular energetic units (ICEU) (Birkedal *et al* 2006, Vendelin *et al* 2004, Seppet *et al* 2005). Within ICEUs, neighbouring mitochondria and myofibrils were proposed to interact via energy transfer mechanisms including phosphocreatine (CrP) shuttle-facilitated transport of ATP/ADP, direct diffusion of ATP/ADP as well as adenylate kinase reactions. It was also proposed that the presence of localized diffusion barriers within the ICEUs might lead to limited exchange of ATP/ADP

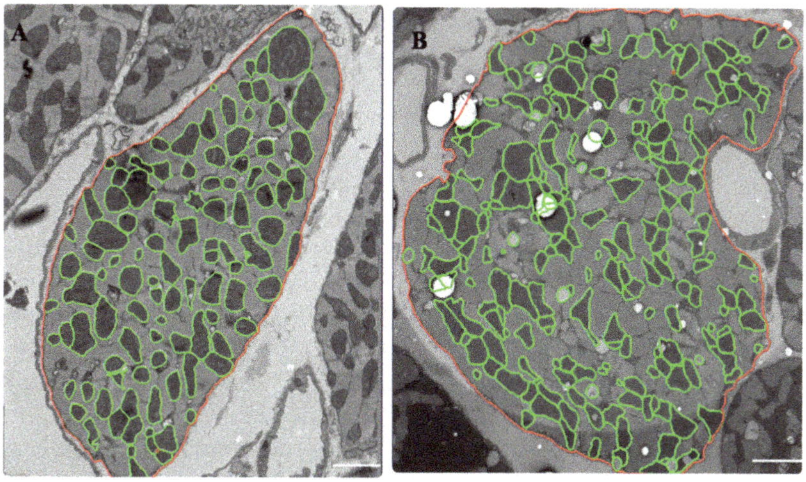

Figure 6.13. Transverse view of mitochondrial organization in left ventricular myocytes from control and STZ (induced diabetic) rat hearts. (A) transverse section from a control animal with outlines of the sarcolemma boundary (red) and mitochondrial outer membrane boundary (green). (B) transverse section from a STZ-induced type 1 diabetic rat heart cardiomyocyte. Scale bar is 2 microns. Reprinted with permission from Jarosz *et al* (2017), copyright by the American Physical Society.

with the larger cytosolic pool of high-energy phosphate compounds outside the ICEUs (Seppet *et al* 2005, 2006).

However, these earlier studies were based on confocal image data, which did not provide an accurate description of the mitochondrial arrangement. Recent analyses on 2D transmission electron microscopy (TEM) images and 3D serial block-face images have revealed that cardiomyocytes have a non-crystalline distribution of mitochondria across the cross sections (Jarosz *et al* 2017, Ghosh *et al* 2018, Lavorato *et al* 2017, Glancy *et al* 2017). Individual mitochondria vary in shape, size and exhibit variations in clustering between the myofibrillar bundles. Considering the previously proposed hypothesis of ICEUs, this observed mitochondrial heterogeneity could give rise to a non-uniform distribution of ATP/ADP which can have a significant impact on the contractile performance of the cell. Furthermore, recent studies have reported changes to the distribution of mitochondria as a consequence of changes to mitochondrial substrate selection and metabolism in disease conditions (Jarosz *et al* 2017, Glancy *et al* 2017, Tsushima *et al* 2018); see figure 6.13 which shows mitochondria and myofibrilar arrangement in normal and diabetic cardiomyocytes.

Thus, it is important to understand how the cell maintains a uniform distribution of ADP, ATP and acto-myosin force production with a heterogeneous mitochondrial distribution. Ghosh *et al* (2018) developed a spatially detailed computational model of mitochondrial energetics and metabolite diffusion that was derived from 2D electron microscopy images, like those shown in figure 6.13. The model shows that the heterogeneous distribution of mitochondria can lead to significant spatial gradients in inorganic phosphate, Cr and CrP at high workloads or hypoxic conditions (figure 6.14), but predicts that relatively uniform ATP and ADP distributions can be maintained. This suggests that the phosphocreatine shuttle mechanism can act as an effective distributor of energy from heterogeneously distributed mitochondria. However, the model also shows that this shuttle mechanism can be compromised in hypoxic conditions leading to altered cross-bridge kinetics.

Although Ghosh *et al* (2018) only simulated cardiomyocyte energetics in 2D cross-sections, they also measured the variation in mitochondrial area fraction and local mitochondrial density within each cross-section of 3D serial block-face microscopy data. The analysis revealed that these variables changed relatively little across the length of the cell, thus suggesting that the 2D simulation findings could be extended to the 3D cell.

The advent of high-throughput 3D electron microscopy techniques for imaging cardiac cells has also enabled visualisation and study of the mitochondrial networks (Hussain *et al* 2018, Pinali and Kitmitto 2014, Glancy *et al* 2017, Bleck *et al* 2018). Recent studies by Glancy *et al* (2015) and Bleck *et al* (2018) provide evidence for the functional role of the 3D connected mitochondrial network as a power distribution system. Eisner (2017) also measured increased mitochondrial fusion dynamics and demonstrated its dependence on contractile activity. Ghosh *et al* (2018) suggest that the phosphocreatine shuttle can act as the main mechanism for energy distribution, with the mitochondrial network mechanism providing a compensation mechanism to maintain uniform force production under hypoxic conditions. This remains, however, not completely resolved, and is an area that is ideally suited for further analysis through spatially detailed computational modelling based on high resolution imaging data.

6.5.4 Multiscale modelling of cardiac mechano-energetics: from cell to organ

Ultimately, understanding cardiomyocyte energetics, and how it relates to cellular electrical and mechanical performance, must to be linked to functioning of cardiac tissue and performance of the whole organ. Multiscale 'Physiome' modelling provides the computational framework within which this can be achieved (Crampin *et al* 2004, Smith *et al* 2007).

At the cellular level, spatial modelling is providing new insights into the relationships and interplay between structure and function, and in particular how changed structural properties may affect cell function, and how structure may alter in the attempt to compensate for reduced function in disease. Studies of cellular remodelling of this type, incorporating modelling of ECC, energetics and cellular mechanics, are now possible with super-resolution spatial detail (Rajagopal *et al* 2015, 2018).

On the other hand, increasingly comprehensive mathematical models of cellular metabolism and its regulation will enable further insights into cardiomyocyte bioenergetics, regulation of ATP production and consumption, under normal and pathological conditions (Fink *et al* 2011), and the bond graph framework is well suited for development of such whole cell computational modelling approaches.

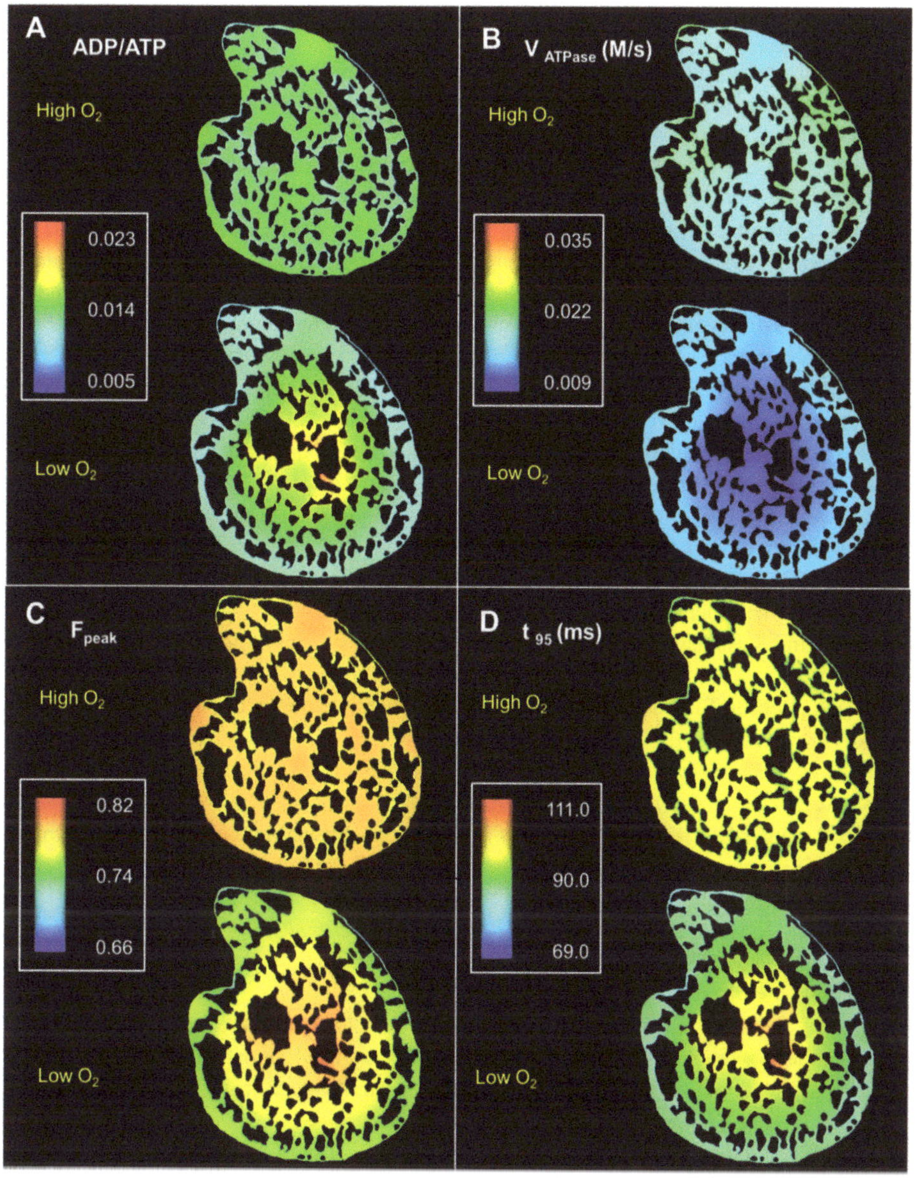

Figure 6.14. Spatial variation in cytosolic metabolite concentration and force dynamics under normoxic and hypoxic conditions. (A) steady state profiles of ADP/ATP ratio. ADP/ATP is significantly higher in the central anoxic core under low oxygen; (B) ATP hydrolysis rate is similarly lower in the cell core under hypoxic conditions; (C) peak twitch force (F_{peak}) is significantly higher in the cell core under hypoxia; (D) increased F_{peak} is accompanied by increased twitch duration in the cell core, which can lead to loss of energy in the form of shear strain. From Ghosh *et al* (2018) with permission from PLoS One.

References

Aliev M, Guzun R, Karu-Varikmaa M, Kaambre T, Wallimann T and Saks V 2011 Molecular system bioenergics of the heart: experimental studies of metabolic compartmentation and energy fluxes versus computer modeling *Int. J. Mol. Sci.* **12** 9296–331

Aliev M K and Saks V A 1997 Compartmentalized energy transfer in cardiomyocytes: use of mathematical modeling for analysis of *in vivo* regulation of respiration *Biophys. J.* **73** 18

Bache R J, Zhang J Y, Murakami Y, Zhang Y, Cho Y K, Merkle H, Gong G R, From A H L and Ugurbil K 1999 Myocardial oxygenation at high workstates in hearts with left ventricular hypertrophy *Cardiovasc. Res.* **42** 616–26

Balaban R S 1990 Regulation of oxidative-phosphorylation in the mammalian cell *Am. J. Physiol.* **258** C377–89

Balaban R S, Kantor H L, Katz L A and Briggs R W 1986 Relation between work and phosphate metabolite in the invivo paced mammalian heart *Science* **232** 1121–3

Bazil J N, Beard D A and Vinnakota K C 2016 Catalytic coupling of oxidative phosphorylation, ATP demand, and reactive oxygen species generation *Biophys. J.* **110** 962–71

Beard D A 2005 A biophysical model of the mitochondrial respiratory system and oxidative phosphorylation *PLoS Comput. Biol.* **1** e36

Beard D A 2006 Modeling of oxygen transport and cellular energetics explains observations on *in vivo* cardiac energy metabolism *PLoS Comput. Biol.* **2** e107

Beard D A and Kushmerick M J 2009 Strong inference for systems biology *PLoS Comput. Biol.* **5** e1000459

Beard D A, Babson E, Curtis E and Qian H 2004 Thermodynamic constraints for biochemical networks *J. Theor. Biol.* **228** 327–33

Bers D M 2001 *Excitation-contraction Coupling and Cardiac Contractile Force* 2nd edn (Dordrecht: Kluwer)

Birkedal R, Shiels H A and Vendelin M 2006 Three-dimensional mitochondrial arrangement in ventricular myocytes: from chaos to order *AJP: Cell Physiol.* **291** C1148–58

Bittl J A and Ingwall J S 1985 Reaction rates of creatine kinase and ATP synthesis in the isolated rat heart. A 31p NMR magnetization transfer study *J. Biol. Chem.* **260** 3512–7

Bleck C K E, Kim Y, Willingham T B and Glancy B 2018 Subcellular connectomic analyses of energy networks in striated muscle *Nat. Commun.* **9** 5111

Branovets J, Sepp M, Kotlyarova S, Jepihhina N, Sokolova N, Aksentijevic D, Lygate C A, Neubauer S, Vendelin M and Birkedal R 2013 Unchanged mitochondrial organization and compartmentation of high-energy phosphates in creatine-deficient GAMT-/- mouse hearts *Am. J. Physiol. Heart Circ. Physiol.* **305** H506–20

Cooling M T, Hunter P J and Crampin E J 2008 Modelling biological modularity with CellML *Syst. Biol. IET* **2** 73–9

Cooper P J, Ward M L, Hanley P J, Denyer G R and Loiselle D S 2001 Metabolic consequences of a species difference in Gibbs free energy of Na^+/Ca^{2+} exchange: rat versus guinea pig *Am. J. Physiol.-Regul. Integr. Comp. Physiol* **280** R1221–9

Cortassa S, Aon M A, O'Rourke B, Jacques R, Tseng H-J, Marban E and Winslow R L 2006 A computational model integrating electrophysiology, contraction, and mitochondrial bioenergetics in the ventricular myocyte *Biophys. J.* **91** 1564–89

Crampin E J and Smith N P 2006 A dynamic model of excitation-contraction coupling during acidosis in cardiac ventricular myocytes *Biophys. J.* **90** 3074–90

Crampin E J, Smith N P and Hunter P J 2004 Multi-scale modelling and the IUPS physiome project *J. Mol. Histol.* **35** 707–14

Crampin E J, Smith N P, Langham A, Clayton R and Orchard C H 2006 Acidosis in models of cardiac ventricular myocytes *Philos. Trans. R. Soc. A: Math. Phys. Eng. Sci.* **364** 1171–86

Curtin N A and Davies R E 1973 Chemical and mechanical changes during stretching of activated frog skeletal muscle *Cold Spring Harb. Symp. Quant. Biol.* **37** 619–26

Curtin N A, Drobnis D D, Larson R E and Davies R E 1969 Very low ATP usage with very high tension in activated muscles during a slow stretch *Fed. Proc.* **28** 711

Dzeja P P, Hoyer K, Tian R, Zhang S, Nemutlu E, Spindler M and Ingwall J S 2011 Rearrangement of energetic and substrate utilization networks compensate for chronic myocardial creatine kinase deficiency *J. Physiol.* **589** 5193–211

Ederer M and Gilles E D 2007 Thermodynamically feasible kinetic models of reaction networks *Biophys. J.* **92** 1846–57

Eisner V 2017 Mitochondrial fusion dynamics is robust in the heart and depends on calcium oscillations and contractile activity *Proc. Natl Acad. Sci. USA* **114** E859–68

Faber G M and Rudy Y 2000 Action potential and contractility changes in Na_i^+ overloaded cardiac myocytes: a simulation study *Biophys. J.* **78** 2392–404

Fink M *et al* 2011 Cardiac cell modelling: Observations from the heart of the cardiac physiome project *Prog. Biophys. Mol. Biol.* **104** 2–21

Garrahan P J and Glynn I M 1967 Incorporation of inorganic phosphate into adenosine triphosphate by reversal of sodium pump *J. Physiol. Lond.* **192** 237

Gawthrop P J 2017 Bond graph modeling of chemiosmotic biomolecular energy transduction *IEEE Trans. NanoBiosci.* **16** 177–88

Gawthrop P J 2018 Physically-plausible models for identification *Proc. Int. Conf. on Bond Graph Modelling and Simulation ICBGM* 1–6

Gawthrop P J and Crampin E J 2014 Energy-based analysis of biochemical cycles using bond graphs *Proc. R. Soc. A: Math. Phys. Eng. Sci.* **470** 20140459

Gawthrop P J and Crampin E J 2016 Modular bond-graph modelling and analysis of biomolecular systems *IET Syst. Biol.* **10** 1–15

Gawthrop P J and Smith L 1996 *Metamodelling: For Bond Graphs and Dynamic SystemsPrentice Hall International Series in Systems and Control Engineering* (London: Prentice Hall)

Gawthrop P J, Cursons J and Crampin E J 2015 Hierarchical bond graph modelling of biochemical networks *Proc. R. Soc. A: Math. Phys. Eng. Sci.* **471** 0642

Gawthrop P J, Siekmann I, Kameneva T, Saha S, Ibbotson M R and Crampin E J 2017 Bond graph modelling of chemoelectrical energy transduction *IET Syst. Biol.* **11** 127–38

Gawthrop P J, Cudmore P and Crampin E J 2019 Physically-plausible modelling of biomolecular systems: a simplified, energy-based model of the mitochondrial electron transport chain https://arxiv.org/abs/1905.129 58

Ghosh S, Tran K, Delbridge L M D, Hickey A J R, Hanssen E, Crampin E J and Rajagopal V 2018 Insights on the impact of mitochondrial organisation on bioenergetics in high-resolution computational models of cardiac cell architecture *PLoS Comput. Biol.* **14** e1006640

Glancy B, Hartnell L M, Malide D, Yu Z-X, Combs C A, Connelly P S, Subramaniam S and Balaban R S 2015 Mitochondrial reticulum for cellular energy distribution in muscle *Nature* **523** 617–20

Glancy B, Hartnell L M, Combs C A, Fenmou A, Sun J, Murphy E, Subramaniam S and Balaban R S 2017 Power grid protection of the muscle mitochondrial reticulum *Cell Rep.* **19** 487–96

Gong G, Ugurbil K and Zhang J Y 1999 Transmural metabolic heterogeneity at high cardiac work states *Am. J. Physiol.-Heart Circ. Physiol.* **277** H236–42

Gong G R *et al* 2003 Oxidative capacity in failing hearts *Am. J. Physiol.-Heart Circ. Physiol* **285** H541–8

Guimarães-Ferreira L 2014 Role of the phosphocreatine system on energetic homeostasis in skeletal and cardiac muscles *Einstein* **12** 126–31

Gupta R K, Benovic J L and Rose Z B 1978 Determination of free magnesium level in human red blood-cell by P-31-NMR *J. Biol. Chem.* **253** 6172–6

Guynn R W and Veech R L 1973 The equilibrium constants of the adenosine triphosphate hydrolysis and the adenosine triphosphate-citrate lyase reactions *J. Biol. Chem.* **248** 6966–72

Han J, Taberner A, Kirton R, Nielsen P M F, Smith N P and Loiselle D S 2009 A unique micromechanocalorimeter for simultaneous measurement of heat rate and force production of cardiac trabeculae carneae *J. Appl. Physiol.* **107** 946

Han J-C, Taberner A J, Nielsen P M F, Kirton R S, Ward M L and Loiselle D S 2010 Energetics of stress production in isolated cardiac trabeculae from the rat *Am. J. Physiol.-Heart Circ. Physiol.* **299** H1382–94

Han J C, Tran K, Taberner A J, Nickerson D P, Kirton R S, Nielsen P M F, Ward M L, Nash M P, Crampin E J and Loiselle D S 2012 Myocardial twitch duration and the dependence of oxygen consumption on pressure-volume area: experiments and modelling *J. Physiol.* **590** 4603–22

Han J-C, Tran K, Johnston C M, Nielsen P M F, Barrett C J, Taberner A J and Loiselle D S Reduced mechanical efficiency in left-ventricular trabeculae of the spontaneously hypertensive rat *Physiol. Rep.* **2** e12211

Han J-C, Tran K, Taberner A J, Chapman B and Loiselle D S 2019 Cardiac energetics *Muscle and Exercise Physiology* ed J A Zoladz (New York: Academic) 505–39

Haraldsdóttir H S and Fleming R M T 2019 Identification of conserved moieties in metabolic networks by graph theoretical analysis of atom transition networks *PLoS Comput. Biol.* **12** e1004999

Hatano A, Okada J-I, Washio T, Hisada T and Sugiura S 2011 A three-dimensional simulation model of cardiomyocyte integrating excitation-contraction coupling and metabolism *Biophys. J.* **101** 2601–10

Hettling H and van Beek J H G 2011 Analyzing the functional properties of the creatine kinase system with multiscale 'sloppy' modeling *PLoS Comput. Biol.* **7** e1002130

Hill T L 1989 *Free Energy Transduction and Biochemical Cycle Kinetics* (New York: Springer)

Hund T and Rudy Y 2004 Rate dependence and regulation of action potential and calcium transient in a canine cardiac ventricular cell model *Circulation* **110** 3168–74

Hund T, Kucera J, Otani N and Rudy Y 2001 Ionic charge conservation and long-term steady state in the Luo-Rudy dynamic cell model *Biophys. J.* **81** 3324–31

Hussain A, Ghosh S, Kalkhoran S B, Hausenloy D J, Hanssen E and Rajagopal V 2018 An automated workflow for segmenting single adult cardiac cells from large-volume serial block-face scanning electron microscopy data *J. Struct. Biol.* **202** 275–85

Huxley A F 1957 Muscle structure and theories of contraction *Prog. Biophys. Mol. Biol.* **7** 255

Infante A A, Klaupiks D and Davies R E 1964 Adenosine triphosphate: changes in muscles doing negative work *Science* **144** 1577–8

Ingwall J S 2002 *ATP and the Heart* (Dordrecht: Kluwer)

Jacobus W E and Saks V A 1982 Creatine kinase of heart mitochondria: Changes in its kinetic properties induced by coupling to oxidative phosphorylation *Arch. Biochem. Biophys.* **219** 167–78

Jansen M A, Shen H, Zhang L, Wolkowicz P E and Balschi J A 2003 Energy requirements for the Na^+ gradient in the oxygenated isolated heart: effect of changing the free energy of ATP hydrolysis *Am. J. Physiol. -Heart Circ. Physiol* **285** H2437–45

Jarosz J, Ghosh S, Delbridge L M D, Petzer A, Hickey A J R, Crampin E J, Hanssen E and Rajagopal V 2017 Changes in mitochondrial morphology and organization can enhance energy supply from mitochondrial oxidative phosphorylation in diabetic cardiomyopathy *Am. J. Physiol.-Cell Physiol* **312** C190–7

Kaasik A, Veksler V, Boehm E, Novotova M, Minajeva A and Ventura-Clapier R 2001 Energetic crosstalk between organelles: architectural integration of energy production and utilization *Circ. Res.* **89** 153–9

Katz A M 1977 *Physiology of the Heart* (New York: Raven)

Katz L A, Swain J A, Portman M A and Balaban R S 1989 Relation between phosphate metabolites and oxygen-consumption of heart *in vivo Am. J. Physiol.* **256** H265–74

Keener J P and Sneyd J 2009 *Mathematical Physiology: Springer Interdisciplinary Applied Mathematics* (Berlin: Springer)

Klipp E, Liebermeister W, Wierling C, Kowald A, Lehrach H and Herwig R 2009 *Systems Biology: A Textbook* (Weinheim: Wiley)

Korzeniewski B, Noma A and Matsuoka S 2005 Regulation of oxidative phosphorylation in intact mammalian heart *in vivo Biophys. Chem.* **116** 145–57

Lavorato M *et al* 2017 Increased mitochondrial nanotunneling activity, induced by calcium imbalance, affects intermitochondrial matrix exchanges *Proc. Natl Acad. Sci. USA* **114** E849–58

Lew V L, Glynn I M and Ellory J C 1970 Net synthesis of ATP by reversal of the sodium pump *Nature* **225** 865–6

Liebermeister W, Uhlendorf J and Klipp E 2010 Modular rate laws for enzymatic reactions: thermodynamics, elasticities and implementation *Bioinformatics* **26** 1528–34

Linari M, Woledge R C and Curtin N A 2003 Energy storage during stretch of active single fibres from frog skeletal muscle *J. Physiol. Lond.* **548** 461–74

Livshitz L and Rudy Y 2009 Uniqueness and stability of action potential models during rest, pacing, and conduction using problem-solving environment *Biophys. J.* **97** 1265–76

Loiselle D S, Crampin E J, Niederer S A, Smith N P and Barclay C J 2008 Energetic consequences of mechanical loads *Prog. Biophys. Mol. Biol.* **97** 348–66

Loiselle D S, Tran K, Crampin E J and Curtin N A 2010 Why has reversal of the actin-myosin cross-bridge cycle not been observed experimentally? *J. Appl. Physiol.* **108** 1465–71

Luo C H and Rudy Y 1994 A dynamic model of the cardiac ventricular action potential. I. Simulations of ionic currents and concentration changes *Circ. Res.* **74** 1071–96

Makinose M and Hasselbach W 1971 ATP synthesis by reverse of sarcoplasmic calcium pump *FEBS Lett.* **12** 271

Meyer R A, Sweeney H L and Kushmerick M J 1984 A simple analysis of the 'phosphocreatine shuttle *Am. J. Physiol.* **246** C365–77

Mitchell P 1961 Coupling of phosphorylation to electron and hydrogen transfer by a chemi-osmotic type of mechanism *Nature* **191** 144

Neubauer S 1998 Influence of left ventricular pressures and heart rate on myocardial high-energy phosphate metabolism *Basic Res. Cardiol.* **93** 102–7

Neubauer S 2007 The failing heart—an engine out of fuel *New Engl. J. Med* **356** 1140–51

Nguyen M, Ramay H, Dudycha S and Jafri M 2004 Modeling the cellular basis of cardiac excitation-contraction coupling and energy metabolism *Recent Res. Dev. Biophys* **3** 125–45

Nicholls D G and Ferguson S J 2002 *Bioenergetics 3* (New York: Academic)

Ochiai K 2001 Effects of augmented delivery of pyruvate on myocardial high-energy phosphate metabolism at high workstate *Am. J. Physiol.-Heart Circ. Physiol* **281** H1823–32

Oster G F, Perelson A S and Katchalsky A 1971 Network Thermodynamics *Nature* **234** 393–9

Oster G F, Perelson A S and Katchalsky A 1973 Network thermodynamics: dynamic modelling of biophysical systems *Q. Rev. Biophys.* **6** 1–134

Palsson B 2006 *Systems Biology: Properties of Reconstructed Networks* (Cambridge: Cambridge University Press)

Pan M, Gawthrop P J, Tran K, Cursons J and Crampin E J 2018a Bond graph modelling of the cardiac action potential: implications for drift and non-unique steady states *Proc. R. Soc. A: Math. Phys. Eng. Sci.* **474** 20180106–27

Pan M, Gawthrop P J, Tran K, Cursons J and Crampin E J 2018b A thermodynamic framework for modelling membrane transporters *J. Theor. Biol.* **481** 10–23

Pate E and Cooke R 1989 A model of crossbridge action–the effects of ATP, ADP and Pi *J. Muscle Res. Cell Motil.* **10** 181–96

Pinali C and Kitmitto A 2014 Serial block face scanning electron microscopy for the study of cardiac muscle ultrastructure at nanoscale resolutions *J. Mol. Cell. Cardiol.* **76** 1–11

Polettini M and Esposito M 2014 Irreversible thermodynamics of open chemical networks. I. Emergent cycles and broken conservation laws *J. Chem. Phys.* **141** 024117

Rajagopal V *et al* 2015 Examination of the effects of heterogeneous organization of RYR clusters, myofibrils and mitochondria on Ca^{2+} release patterns in cardiomyocytes *PLoS Comput. Biol.* **11** e1004417–31

Rajagopal V, Bass G, Ghosh S, Hunt H, Walker C, Hanssen E, Crampin E J and Soeller C 2018 Creating a structurally realistic finite element geometric model of a cardiomyocyte to study the role of cellular architecture in cardiomyocyte systems biology *J. Vis. Exp.* **134** e56817

Razumova M, Bukatina A E and Campbell K B 1999 Stiffness-distortion sarcomere model for muscle simulation *J. Appl. Physiol.* **87** 1861–76

Rice J J, Wang F, Bers D M and de Tombe P 2008 Approximate model of cooperative activation and crossbridge cycling in cardiac muscle using ordinary differential equations *Biophys. J.* **95** 2368–90

Roberts B N and Christini D J 2011 NHE inhibition does not improve Na^+ or Ca^{2+}. Overload during reperfusion: using modeling to illuminate the mechanisms underlying a therapeutic failure *PLoS Comput. Biol.* **7** e1002241

Saks V, Favier R, Guzun R, Schlattner U and Wallimann T 2006 Molecular system bioenergetics: regulation of substrate supply in response to heart energy demands *J. Physiol.* **577** 769–77

Saks V A, Khuchua Z A, Vasilyeva E V, Belikova O Y and Kuznetsov A V 1994 Metabolic compartmentation and substrate channelling in muscle cells *Cellular Bioenergetics: Role of Coupled Creatine Kinases, Developments in Molecular and Cellular Biochemistry* ed V A Saks and R Ventura-Clapier (Boston, MA: Springer) 155–92

Saks V A, Kongas O, Vendelin M and Kay L 2000 Role of the creatine/phosphocreatine system in the regulation of mitochondrial respiration *Acta Physiol. Scand.* **168** 635–41

Saupe K W, Spindler M, Tian R and Ingwall J S 1998 Impaired cardiac energetics in mice lacking muscle-specific isoenzymes of creatine kinase *Circ. Res.* **82** 898–907

Savabi F 1994 Interaction of creatine kinase and adenylate kinase systems in muscle cells *Mol. Cell. Biochem.* **133-134** 145–52

Schlattner U, Tokarska-Schlattner M and Wallimann T 2006 Mitochondrial creatine kinase in human health and disease *Biochim. Biophys. Acta* **1762** 164–80

Schramm M, Klieber H G and Daut J 1994 The energy expenditure of actomyosin-ATPase, Ca^{2+}-ATPase and Na^{+}/K^{+}-ATPase in guinea-pig cardiac ventricular muscle *J. Physiol.* **481** 647–62

Schuster S and Hilgetag C 1995 What information about the conserved-moiety structure of chemical reaction systems can be derived from their stoichiometry? *J. Phys. Chem.* **99** 8017–23

Seppet E K *et al* 2005 Intracellular energetic units in healthy and diseased hearts *Exp. Clin. Cardiol.* **10** 173–83

Seppet E K *et al* 2006 Structure-function relationships in the regulation of energy transfer between mitochondria and ATPases in cardiac cells *Exp. Clin. Cardiol.* **11** 189–94

Siekmann I, Sneyd J and Crampin E J 2012 MCMC can detect nonidentifiable models *Biophys. J.* **103** 2275–86

Simson P, Jepihhina N, Laasmaa M, Peterson P, Birkedal R and Vendelin M 2016 Restricted ADP movement in cardiomyocytes: Cytosolic diffusion obstacles are complemented with a small number of open mitochondrial voltage-dependent anion channels *J. Mol. Cell. Cardiol.* **97** 197–203

Smith D A 1998 A strain-dependent ratchet model for [phosphate]- and [ATP]-dependent muscle contraction *J. Muscle Res. Cell Motil.* **19** 189–211

Smith N P and Crampin E J 2004 Development of models of active ion transport for whole-cell modelling: cardiac sodium-potassium pump as a case study *Prog. Biophys. Mol. Biol.* **85** 387–405

Smith N P, Crampin E J, Niederer S A, Bassingthwaighte J B and Beard D A 2007 Computational biology of cardiac myocytes: proposed standards for the physiome *J. Exp. Biol.* **210** 1576–83

Taberner A J, Han J C, Loiselle D S and Nielsen P M F 2011 An innovative work-loop calorimeter for *in vitro* measurement of the mechanics and energetics of working cardiac trabeculae *J. Appl. Physiol.* **111** 1798–803

Terkildsen J R, Crampin E J and Smith N P 2007 The balance between inactivation and activation of the Na^{+}-K^{+} pump underlies the triphasic accumulation of extracellular K^{+} during myocardial ischemia *Am. J. Physiol.-Heart Circ. Physiol* **293** H3036–45

Terkildsen J R, Niederer, Crampin E J, Hunter P J and Smith N P 2008 Using Physiome standards to couple cellular functions for rat cardiac excitation-contraction *Exp. Physiol.* **93** 919–29

Territo P R, French S A, Dunleavy M C, Evans F J and Balaban R S 2001 Calcium activation of heart mitochondrial oxidative phosphorylation Rapid kinetics of $mV.O_2$, NADH, and light scattering *J. Biol. Chem.* **276** 2586–99

Tewari S G, Bugenhagen S M, Palmer B M and Beard D A 2015 Computational model of cross-bridge cycling and force generation to explain the effect of metabolites on cardiac muscle mechanics *Biophys. J* **108** 444a

Tewari S G, Bugenhagen S M, Palmer B M and Beard D A 2016 Dynamics of cross-bridge cycling, ATP hydrolysis, force generation, and deformation in cardiac muscle *J. Mol. Cell. Cardiol.* **96** 11–25

Tran K 2011 Mathematical modelling of energy demand and supply in the cardiac myocyte *PhD Thesis* University of Auckland

Tran K, Smith N P, Loiselle D S and Crampin E J 2009 A thermodynamic model of the cardiac sarcoplasmic/endoplasmic Ca^{2+} (SERCA) pump *Biophys. J.* **96** 2029–42

Tran K, Smith N P, Loiselle D S and Crampin E J 2010 A metabolite-sensitive, thermodynamically constrained model of cardiac cross-bridge cycling: implications for force development during ischemia *Biophys. J.* **98** 267–76

Tran K, Loiselle D S and Crampin E J 2015 Regulation of cardiac cellular bioenergetics: mechanisms and consequences *Physiol. Rep.* **3** e12464

Tran K, Han J-C, Taberner A J, Barrett C J, Crampin E J and Loiselle D S 2016 Myocardial energetics is not compromised during compensated hypertrophy in the Dahl salt-sensitive rat model of hypertension *Am. J. Physiol.-Heart Circ. Physiol* **311** H563–71

Tran K, Han J-C, Crampin E J, Taberner A J and Loiselle D S 2017 Experimental and modelling evidence of shortening heat in cardiac muscle *J. Physiol.* **595** 6313–26

Tsushima K *et al* 2018 Mitochondrial reactive oxygen species in lipotoxic hearts induce post-translational modifications of AKAP121, DRP1, and OPA1 that promote mitochondrial fission *Circ. Res.* **122** 58–73

Varghese A and Sell G R 1997 A conservation principle and its effect on the formulation of Na–Ca exchanger current in cardiac cells *J. Theor. Biol.* **189** 33–40

Vendelin M, Kongas O and Saks V 2000 Regulation of mitochondrial respiration in heart cells analyzed by reaction-diffusion model of energy transfer *Am. J. Physiol.-Cell Physiol.* **278** C747–64

Vendelin M, Eimre M, Seppet E, Peet N, Andrienko T, Lemba M, Engelbrecht J, Seppet E K and Saks V A 2004 Intracellular diffusion of adenosine phosphates is locally restricted in cardiac muscle *Mol. Cell. Biochem.* **256-257** 229–41

Vendelin M M, Béraud N N, Guerrero K K, Andrienko T T, Kuznetsov A V A, Olivares J J, Kay L L and Saks V A V 2005 Mitochondrial regular arrangement in muscle cells: a crystal-like pattern *Am. J. Physiol.-Cell Physiol* **288** C757–67

Williamson J R, Ford C, Illingworth J and Safer B 1976 Coordination of citric acid cycle activity with electron transport flux *Circ. Res.* **38** I39–51

Wu F, Jeneson J A L and Beard D A 2007 Oxidative ATP synthesis in skeletal muscle is controlled by substrate feedback *Am. J. Physiol.-Cell Physiol* **292** C115–24

Wu F, Zhang E, Zhang J, Bache R and Beard D A 2008 Phosphate metabolite concentrations and ATP hydrolysis potential in normal and ischaemic hearts *J. Physiol.* **586** 4193

Wu F, Zhang J and Beard D A 2009 Experimentally observed phenomena on cardiac energetics in heart failure emerge from simulations of cardiac metabolism *Proc. Natl Acad. Sci. USA* **106** 7143

Zeleznikar R J, Dzeja P P and Goldberg N D 1995 Adenylate kinase-catalyzed phosphoryl transfer couples ATP utilization with its generation by glycolysis in intact muscle *J. Biol. Chem.* **270** 7311–9

Zhang J Y, From A H L, Ugurbil K and Bache R J 2003 Myocardial oxygenation and high-energy phosphate levels during K-ATP channel blockade *Am. J. Physiol.-Heart Circ. Physiol.* **285** H1420–7

Zhang J Y *et al* 2005 Nitric oxide regulation of myocardial O_2 consumption and HEP metabolism *Am. J. Physiol.-Heart Circ. Physiol.* **288** H310–6

Zhou L 2006 Regulation of myocardial substrate metabolism during increased energy expenditure: insights from computational studies *Am. J. Physiol.-Heart Circul. Physiol.* **291** H1036–46

Chapter 7

Tissue and organ scale modeling: coupled cells, wave propagation, simulated arrhythmia

Edward J Vigmond

Cardiac electrical behavior is nonlinear, from the single cell to the tissue level. This chapter discusses how to model this behavior, starting from the channel level and connecting cells to form tissue, and the whole organ. Mathematical cell models are joined together using the bidomain equations, which describe electric field interaction. An action potential occurring in a cell propagates to neighboring cells by electrical current flow resulting in a wave of activation. It is also demonstrated how to induce abnormal propagation leading to reentry by specific pacing protocols.

7.1 Overview

Tissue behavior can be quite different from observed single cell behavior due to the effect of electrotonic coupling between cells. Even though differences in individual intrinsic properties may be significant, coupling tends to average and reduce differences measured. For example, automaticity seen in a single cell or group of cells may be suppressed when placed in the context of tissue [52]. That is not to say that tissue level behavior is less complex than that observed in single cells. Given the nonlinear nature of cells, connecting a large number of them together produces emergent behavior, activity which is not seen in single cell but nonetheless arises as a consequence of individual cell properties. Phenomena which only exist at the tissue level include reentrant arrhythmias, and cathode-break and anode-make excitation [8].

7.2 Tissue conductivity

Cardiac myocytes can be considered roughly cylindrical in shape, being on the order of 100 μm in length and 7 μm in diameter. They are mostly mononucleate but some regions express a higher proportion of binucleate cells. The myocytes have a brick-like arrangement, with flat ends and stepped edges to better abut neighboring cells,

doi:10.1088/978-0-7503-2064-1ch7

and are branched to connect to multiple cells, 9.1 on average [22]. In figure 7.1, this arrangement is shown in a thin two-dimensional slice. In three dimensions, additional branching would also occur out of plane. Adjacent cell membranes are separated by approximately 10 nm. In addition to gap junctions that link the two

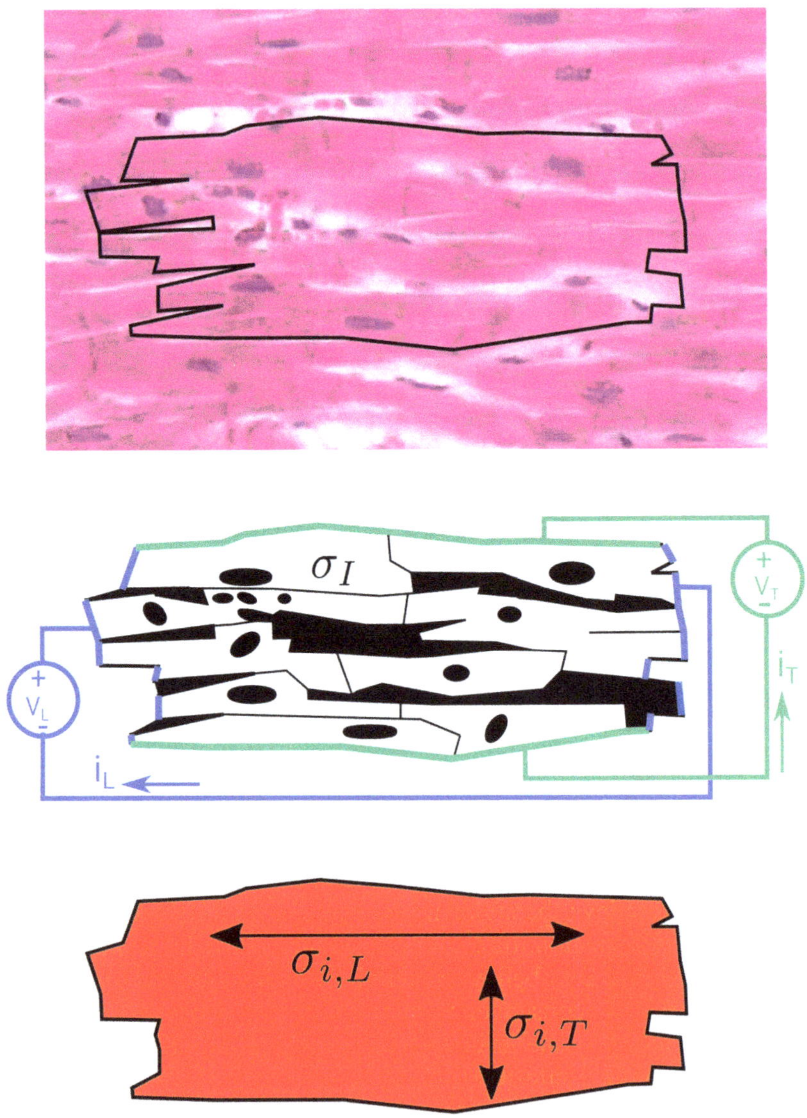

Figure 7.1. Homogenization. *Top*: tissue with typical region to be homogenized shown. *Middle*: skeletonized region indicting individual cells and nonconductive regions. A voltage is applied across the longitudinal direction, V_L, and the current, i_L, computed. Similarly, a voltage, V_T, is applied across the transverse direction and that current, i_T, computed. The conductivity of the myoplasm is a scalar, σ_I. *Bottom*: homogenized region. The region is now considered as uniform with a conductivity tensor having longitudinal and transverse components σ_L and σ_T. Reproduced with permission from David G King.

cells electrically, anchoring proteins extend across the gap to mechanically link the cells and provide a stable interface [32]. In the ventricles, thin collagen sheets separate thin layers of myocytes, about four cells thick. These sheets are roughly parallel to the cardiac surfaces, and are limited in size to allow myocytes from different layers to connect [31].

Gap junctions are the structures linking cells [10]. They can be electrically represented by resistors. The basic unit is a protein called a connexin, six of which assemble themselves to form a tubular structure with a central pore, called a connexon as shown in figure 7.2. Two connexons, one from each cell, align to form an intercellular pore, allowing the passage of ions and small weight molecules. The conductance of the pore is modulated by factors such as the voltage across the gap junction and pH, although the kinetics of opening and closing are on the order of

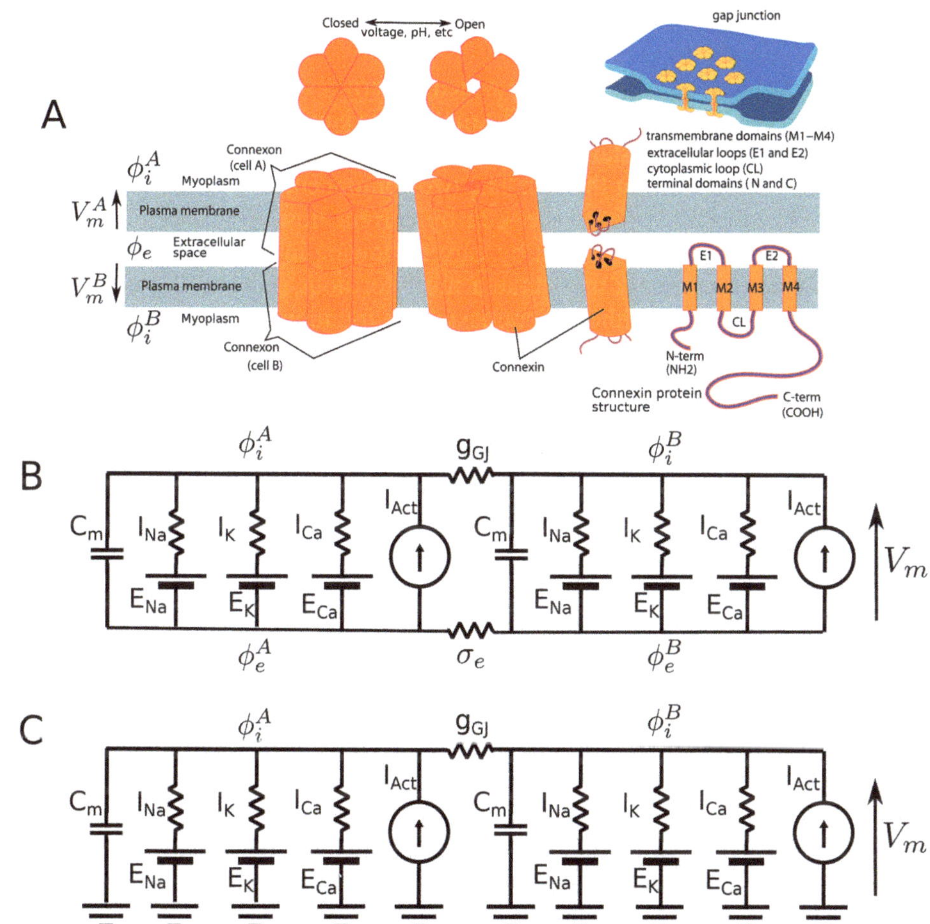

Figure 7.2. Structure of gap junction channel. (A) Structure of connexin protein, assembly into connexons, and formation off gap junctions. (B) Representation of two coupled cells taking into account extracellular current flow. Currents are generic for each ion type and I_{Act} encompasses all pumps and exchangers. (C) Monodomain representation of two coupled cells. Author: Mariana Ruiz LadyofHats.

hundreds of milliseconds and the degree of modulation is less than gated, membrane, ionic channels. The interface between two cardiac cells is an intercalated disk. The membrane within the intercalated disk is highly undulated to increased surface area, allowing more channels.

Whether in the atria [21] or ventricles [43], myocytes are arranged in layers in which the myocytes are parallel as defined by their long axes. This direction in each layer is called the fibre direction since from a certain scale, the cells appear to form one-dimensional chains of cells, i.e. fibres. In figure 7.1, the fibre direction is indicated as the longitudinal direction. However, microscopic examination reveals that cells branch and make connections off of the main axis. This has repercussions in terms of the directional dependence of the electrical conductivity seen by cardiac cells. We define a longitudinal direction or fibre direction, a transverse direction which is normal to both the longitudinal direction and the surface, and, finally, a sheet normal direction which is perpendicular to the intracellular collagen sheets and the other two directions [31].

There are two distinct components which govern resistance to electrical current flow in the tissue, one arising from resistance of the gap junctions between the cells, and the other from the resistance of the cytoplasm which carries the charge within the restricted confines within the cell. The two components are of the same order of magnitude. The electrical resistance seen by current flowing intracellularly in the fibre direction is less than that seen by current flowing in the other directions [40]. Current flowing transversely must take a more tortuous route and cross more gap junctions per unit length. Although connexins are found in the membrane outside of the intercalated disk, it remains an open question whether they form functional gap junctions which contribute to transverse propagation, or whether the branching morphology of cardiac myocytes is sufficient to account for observed propagation [5, 22]. Extracellularly, there is an anisotropy in current flow as well, though not as extreme. The tight packing of cells with their long axis makes current flow along the fibre direction easier than current flowing transversely or in the sheet normal direction. The collageneous sheets, which disrupt intercellular connections, also provide a barrier for extracellular current flow. Conductivity is greatest in the longitudinal direction and least in the sheet normal direction.

7.3 Continuum representation

A literal description of cardiac tissue would involve representing each cell as a distinct unit and coupling them resistively. Given the small dimensions of cardiac cells, their explicit representation in tissue or organ scale simulations leads to a prohibitive number of calculations [27]. Mathematically, it is more difficult to deal with a discontinuous domain when trying to develop formulae to describe phenomena which extend beyond a single cell. Thus, cardiac tissue is usually represented as a continuous domain, one in which individual cells are no longer considered, and discretization is larger than a single cell.

7.3.1 Homogenization

The process of finding equivalent properties that best replicate the discontinuous domain behavior in a continuous domain is called homogenization. It is outlined in figure 7.1. The discontinuous domain is assumed to be more or less spatially periodic, and the minimum continuous volume considered must be at least as large as the repeating unit. This way it is ensured that each equivalent continuous volume has the same properties. Continuum properties are adjusted so that behavior across the continuous volume best matches that across the discontinuous volume. For cardiac cells, this implies that the basic volume encompasses at least a cell and the gap junctions connecting it to the neighboring cells. Equivalent material properties can be determined for the continuous domain [18], and the continuous domain solved at a resolution finer than the basic unit. However, the interpretation of the results at a fine scale violates the homogenization assumption and may not properly capture the microscopic behavior [34]. Also, some microscopic details may be ignored which may have impact, like sawtooth potentials at cell junctions [14, 46].

In a continuum formulation, electrical conductivity, with units of conductance per unit length (S m^{-1}), is a homogenized quantity combining the myoplasmic conductivity and cellular structure with the gap junction resistance to describe the ease with which current flows intracellularly. This is easily computed for the case of connected parallelpipeds with length L and cross-sectional area A and myoplasmic conductivity $\sigma_{\rm I}$ and a gap junction resistance between cells of R_{gj}. We find the homogenized conductivity σ_i^* from considering a few realistically shaped cells and equating the resistance between the idealized block and the actually geometry using the following equation:

$$\frac{L}{A^2\sigma_i^*} = \int_L \frac{dx}{\sigma_{\rm I} A_M(x)} + R_{gj}, \tag{7.1}$$

where $A_M(x)$ is the myoplasmic cross-sectional area at position x.

A further homogenization is required for mathematical approaches which treat intracellular and extracellular spaces to be continuous and everywhere within the cardiac tissue. These approaches consider each point to be in both spaces, which is physically impossible. In effect, both domains are rescaled to fill the entire domain. Continuing the example of myocytes as blocks packed in a volume, the effective conductivity in the longitudinal direction becomes [19]

$$\sigma_i = \eta\sigma_i^*, \tag{7.2}$$

where η is the fraction of the volume occupied by the cells.

Likewise, for the extracellular space, the conductivity must be homogenized. While there are no organelles or gap junctions, there is fibrosis and fat, and the shape of the cells and fibrosis makes current flow easier along the cell axis. Finally, the volume fraction must also be considered, so

$$\sigma_e = (1 - \eta)\sigma_e^*, \tag{7.3}$$

Parameter values used in simulations may be quite different from experimentally measured specific material properties due to homogenization. In addition, measuring conductivities cannot be performed directly in cardiac tissue due to the spatially varying anisotropy [5, 26], so generally, conductivities are simply tuned to reproduce desired propagation velocities.

7.4 Governing equations

Due to how well electrically coupled cardiac tissue is, it may be considered a syncytium and continuum representation used with homogenized values. Current introduced into one cell will be split into two components, one flowing through the membrane and the other flowing through gap junctions into neighboring cells to affect their behavior. This interaction allows action potentials to propagate throughout the tissue.

We consider the electric potential fields inside the cell, ϕ_i, and outside, ϕ_e. There are, thus, two domains, the intracellular and the extracellular spaces, which are separated by the cellular membrane. From basic electromagnetic theory, in each domain

$$-\boldsymbol{\sigma}_{\rm i}\nabla\phi_{\rm i} = I_{\rm i} \tag{7.4}$$

$$-\boldsymbol{\sigma}_{\rm e}\nabla\phi_{\rm e} = I_{\rm e}, \tag{7.5}$$

where $I_{\rm i}$ and $I_{\rm e}$ are the intra- and extra-cellular source current densities (μA cm^{-3}). Note that the conductivities, $\boldsymbol{\sigma}_{\rm i/e}$, are tensors since they are direction dependent even though at the microscopic scale, there are scalars. The anisotropy arises due to the cellular structure and connectivity as explained previously. If there is a change in the current flowing inside the cell at a point, it must result from current crossing the membrane to flow outside of the cell, since charge is conserved. Mathematically, this is expressed as

$$\nabla \cdot I_{\rm i} = -\nabla \cdot I_{\rm e} = -\beta I_m, \tag{7.6}$$

where I_m is the transmembrane current density (μA cm^{-2}) and the sign corresponds to the convention that current flowing outward from the cell is positive. The current surface density is converted to a source volume current density by β, the amount of membrane surface area per unit of volume.

7.4.1 Bidomain equations

Two domains are preserved in the formulation, the extracellular or interstitial space comprising the volume outside of and between the cells, and the intracellular space within myocytes [45]. The membrane separates the intracellular from interstitial space. Due to the high degree of coupling, the intracellular space is considered a functional syncytium. The bidomain equations result from assuming that intracellular and extracellular domains exist everywhere, and that current flows between them by crossing the separating membrane, resulting in the following equations:

$$\nabla \cdot \boldsymbol{\sigma}_i \nabla \phi_i = \beta I_m \tag{7.7}$$

$$\nabla \cdot \boldsymbol{\sigma}_e \nabla \phi_e = -\beta I_m + I_e, \tag{7.8}$$

The conductivity tensors describe the governing orthotropic eigendirections of the tissue at a given location, **x**, and are constructed as

$$\boldsymbol{\sigma}_X = \sigma_{f,X} \mathbf{f} \otimes \mathbf{f} + \sigma_{s,X} \mathbf{s} \otimes \mathbf{s} + \sigma_{n,X} \mathbf{n} \otimes \mathbf{n}, \tag{7.9}$$

where $\sigma_{f,X}$, $\sigma_{s,X}$ and $\sigma_{n,X}$ are conductivities of the domain X along the local eigenaxes, **f** = 'fiber direction', **s** = 'transverse direction', and **n** is the 'sheet normal direction.'

$$V_m = \phi_i - \phi_e, \tag{7.10}$$

is the transmembrane voltage which is important for determining channel behavior; and I_e is an extracellular current density as delivered by a stimulating electrode, specified on a per volume basis.

At tissue boundaries, no-flux boundary conditions are imposed on ϕ_i and ϕ_e, i.e.

$$\int_{\partial\Omega} \boldsymbol{\sigma}_i \nabla \phi_i \, d\Gamma = \int_{\partial\Omega} \boldsymbol{\sigma}_e \nabla \phi_e \, d\Gamma = 0, \tag{7.11}$$

holds. In those cases where cardiac tissue is surrounded by a conductive medium, such as blood in the ventricular cavities or a perfusing saline solution, an additional Poisson equation has to be solved

$$\nabla \cdot \sigma_b \nabla \Phi_e = -I_e, \tag{7.12}$$

where σ_b is the conductivity of the conductive isotropic medium, blood or tissue bath, and I_e is the stimulation current injected into the conductive medium. In this case, no flux boundary conditions are assumed at the boundaries of the conductive medium, whereas continuity of the normal component of the extracellular current and continuity of ϕ_e are enforced at the tissue–bath interface. That is,

$$\int_{\partial\Omega} \boldsymbol{\sigma}_e \nabla \phi_e \, d\Gamma = \int_{\partial\Omega} \sigma_b \nabla \phi_e \, d\Gamma. \tag{7.13}$$

The no-flux boundary condition for ϕ_i along the tissue–bath interface is enforced as the membrane acts as a barrier there, stopping current. Equations (7.7)–(7.8) are expressed in the parabolic–parabolic form of the bidomain equations; however, most solution schemes use the elliptic–parabolic form obtained by adding equations (7.7) and (7.8) and substituting $V_m + \phi_e$ for ϕ_i. This yields

$$\begin{bmatrix} -\nabla \cdot (\boldsymbol{\sigma}_i + \boldsymbol{\sigma}_e) \nabla \phi_e \\ -\nabla \cdot \sigma_b \nabla \phi_e \end{bmatrix} = \begin{bmatrix} \nabla \cdot \boldsymbol{\sigma}_i \nabla V_m \\ I_e \end{bmatrix}, \tag{7.14}$$

$$-\left(\nabla \cdot \boldsymbol{\sigma}_i \nabla V_m + \nabla \cdot \boldsymbol{\sigma}_i \nabla \phi_e\right) = -\beta C_m \frac{\partial V_m}{\partial t} - \beta (I_{ion} - I_{tr}), \tag{7.15}$$

where V_m and ϕ_e, the quantities of primary concern, are retained as the independent variables. Note that equation (7.14) treats the bath as an extension of the interstitial space.

Monodomain equation

A computationally less costly monodomain model [38] can be derived from the bidomain model by assuming that intracellular and interstitial conductivity tensors can be related by $\boldsymbol{\sigma}_\mathrm{i} = \alpha\boldsymbol{\sigma}_\mathrm{e}$, a condition called equivalent conductivity ratios. In this scenario, equation (7.8) can be neglected and the intracellular conductivity tensor in equation (7.15) is replaced by the monodomain tensor, derived as the harmonic mean tensor:

$$\boldsymbol{\sigma}_\mathrm{m} = \boldsymbol{\sigma}_\mathrm{i}\boldsymbol{\sigma}_\mathrm{e}(\boldsymbol{\sigma}_\mathrm{i} + \boldsymbol{\sigma}_\mathrm{e})^{-1}.$$

That is, the monodomain equation is given by

$$-\nabla \cdot \boldsymbol{\sigma}_\mathrm{m}\nabla V_\mathrm{m} = -\beta I_\mathrm{m} + \beta I_\mathrm{tr}, \tag{7.16}$$

where I_tr is a transmembrane stimulus current, a current which is introduced in the intracellular space and immediately removed from the extracellular space. In this case, monodomain and bidomain formulations are equivalent in the sense that both models yield the same conduction velocities along the principal axes which, in turn, results in very similar activation patterns. Minor deviations are expected when wavefronts are propagating in any other directions [33], which results in slightly different activation patterns. Extracellular stimulation is also not possible with a monodomain formulation.

This formulation has the advantage that it is computationally an order of magnitude faster to solve than the bidomain formulation. Though the equal anisotropy assumption is known to not hold, the approximation is reasonably accurate in situations where extracellular fields are not imposed, which precludes its use for defibrillation studies. Nevertheless, extracellular potentials ϕ_e can be recovered from the monodomain solution through the following [37]:

$$\phi_\mathrm{e}(\mathbf{x}) = \int_\Omega \frac{I_\mathrm{m}(\mathbf{x}')}{4\pi\sigma_e|\mathbf{x} - \mathbf{x}'|}d\Omega' = \int_\Omega \frac{\nabla V_\mathrm{m}(\mathbf{x}') \cdot (\mathbf{x} - \mathbf{x}')}{4\pi\sigma_e|\mathbf{x} - \mathbf{x}'|^3}d\Omega'. \tag{7.17}$$

Eikonal equation

In some cardiac problems, it is only desired to model the propagation of a single wave of electrical activity through the cardiac domain and determine the activation time of each point of the tissue. Under these conditions, a single time independent equation can be solved, called the eikonal curvature equation: [7, 28]

$$c_\mathrm{f}\sqrt{\nabla\tau^T\boldsymbol{\sigma}_\mathrm{m}\nabla\tau} - k_0\nabla \cdot (\boldsymbol{\sigma}_\mathrm{m}\nabla\tau) = 1 \text{ on } \Omega \tag{7.18}$$

$$\boldsymbol{\sigma}_\mathrm{m}\nabla\tau\boldsymbol{n} = 0 \text{ on } \partial\Omega, \tag{7.19}$$

where τ is the activation time for which we would like to solve, c_f is the velocity along the fibre direction, and k_0 describes the effect of wavefront curvature on velocity since an expanding wavefront moves slower than a planar one. The solution is limited in that it only considers depolarization and only for a single activation. The form as stated above has been solved using the finite element method but it is not straightforward [25]. If curvature effects are ignored and the requirement that the wavefront be orthogonal to the boundary relaxed, the eikonal equation can be solved by a fast marching algorithm which drastically reduces computational demands [33], offering an attractive method to determine activation sequences with reasonable accuracy at trivial computational expense.

Space constant

For subthreshold phenomena, the membrane may be approximated as a passive circuit comprising a capacitor and a resistor. The current flow across it is then

$$I_m = \frac{V_m}{R_m} + C_m \frac{\partial V_m}{\partial t}. \tag{7.20}$$

Therefore,

$$\beta I_m = \nabla \cdot \sigma_i \nabla V_m, \tag{7.21}$$

where β is the amount of cellular membrane per unit volume, This factor converts the current surface density to a volume density. We are not considering the nanometer range, but a larger scale where representative volumes are sufficiently similar and contain an adequate number of ionic channels. If we align the local coordinates so that x is the direction along the fibres, we will consider only one direction and can rearrange the equation:

$$\lambda_x^2 \partial^2 V_m / \partial x^2 = \tau_m \frac{\partial V_m}{\partial t} + V_m, \tag{7.22}$$

which has the solution

$$V_m(x, t) = A e^{-x/\lambda_x} e^{-t/\tau} + B e^{x/\lambda_x} e^{-t/\tau}, \tag{7.23}$$

with

$$\tau_m = R_m C_m \tag{7.24}$$

$$\lambda_x = \sqrt{\frac{R_m \sigma_{i,x} \sigma_{e,x}}{\sigma_{i,x} + \sigma_{e,x}}}, \tag{7.25}$$

and a similar solution for Y.

Thus, if current is injected into a point, its effect will drop off exponentially with distance from the point of injection and respond temporally with a time constant of τ. The spatial rate of decay is characterized by λ, which is referred to as the space constant, and depends on the ratio of the membrane resistance to the resistance of flowing within the cell. Increasing the resistance of the cellular medium means that more current will flow through the membrane since current follows the path of least resistance. Thus, current injected will have a larger extent along the fibre direction than across it. Transmembrane potential can only be changed by current crossing the membrane which serves to charge or discharge the membrane capacitance. Due to cell shape and the directional dependence of the number of gap junctions encountered per unit length, λ will be different for each dimension.

While this analytical analysis offers insight into the workings of the tissue, actual cardiac tissue presents more difficulties which prevent the use of such solutions. The fibre directions of actual tissue vary with space, and the nonlinearity of the membrane cause deviations from the simplified behaviour here.

7.5 Action potential propagation

Once the membrane voltage has been raised high enough, an action potential is triggered which begins to propagate in a self perpetuating fashion. Sodium channels open to allow an inward flow of Na ions. These ions serve to charge the transmembrane capacitance to increase the transmembrane voltage of the cell. A portion of the current entering the cell flows to adjacent cells, via gap junctions, to increase their transmembrane voltage as well, above the threshold of the sodium channels.

Current must flow in loops. Since, a decrease in intracellular current becomes an increase in extracellular current and vice versa (equation (7.6)), they change at exactly the opposite rate. If one part of the cell is at rest and no current flows, it follows that in the absence of a stimulus, the intra- and extra-cellular currents must cancel each other out, and the net current is zero.

$$I_{\mathrm{i}} \approx -\sigma_{\mathrm{i}} \nabla V_m, \tag{7.26}$$

and

$$I_m \approx -\nabla \cdot \sigma_{\mathrm{i}} \nabla V_m. \tag{7.27}$$

This situation is shown in figure 7.3. At the steepest point of action potential upstroke, the axial intra- and extra-cellular currents are maximal. At this point, the transmembrane current is zero and represents the separation of the source region from the sink region. The sodium channels have rapidly activated and counteract the current loss through the membrane and to the downstream cells. The hyper-polarizing current loops (blue), thus, represent the source region and the red loops, the sink region. Current densities are strongest near this transition point.

7.5.1 Safety factor

Electrical propagation in cardiac tissue is based on the passing of charge from excited myocytes to nonexcited myocytes through gap junctions. A minimum amount of charge is required to elicit an action potential. Charge delivered above this threshold may be considered extra, a safety factor ensuring propagation even if conditions alter, such as changes in ionic channel expression, or local loading. The most robust definition of safety factor (SF) [3] has been defined as

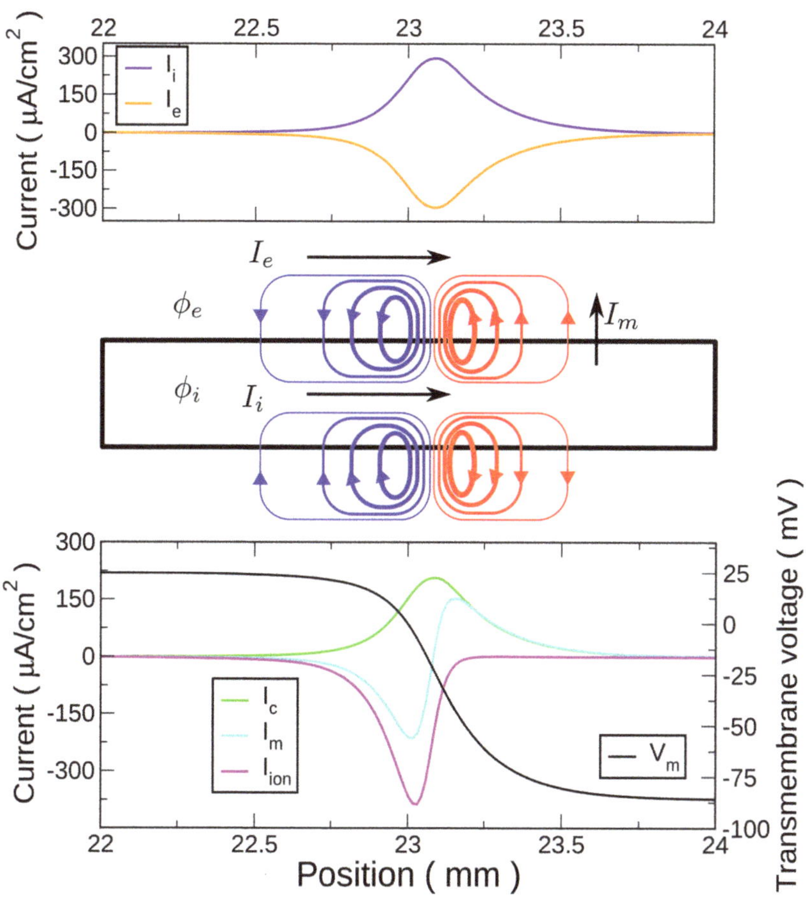

Figure 7.3. Spatial distribution of currents for plane wave propagating rightward. *Top*: intra- and extracellular axial currents as a function of position. *Middle*: schematic showing depolarizing (red) and hyperpolarizing (blue) current loops. The membrane is shown in black. Line width corresponds to current density. *Bottom*: the transmembrane voltage (V_m) is shown, along with the transmembrane current (I_m) broken into its constitutive components: capacitive current (I_c) and ionic current (I_{ion}). The ten Tusscher ionic model [44] was used on a grid with 25 μm discretization and 30 μs temporal output to properly capture the detail of the waveform. All graphs are spatially aligned.

$$SF = \frac{C_m \Delta V_m + Q_{\text{ion}} - Q_s}{Q_{\text{thresh}}(t_A)}, \tag{7.28}$$

where t_A is the interval from when the transmembrane current becomes positive until it passes through zero. ΔV_m is the change in transmembrane voltage over t_A, Q_{ion} is the net ionic charge delivered, Q_s is any external stimulus charge delivered to the cell, and $Q_{\text{thresh}}(t_A)$ is the threshold charge to trigger an action potential delivered to a single cell over a period t_A. A minimum SF of unity is required for propagation while a large SF indicates robust propagation. Global changes in tissue conductivities do not affect SF except at extremely low values where propagation fails. SF is, however, affected by boundaries between regions of different conductivities. It is a function of wavefront curvature, inexcitable boundaries, wavefront collision, and ionic channel expression, primarily Nav1.5.

7.5.2 Liminal length

Related to SF is the concept that for an action potential to propagate in a tissue, a minimum volume of tissue must be simultaneously excited so that there is sufficient current generated to excite all of the connected cells [15, 35]. These downstream cells must all be sufficiently excited or they will electrically drain the source cell without resulting in action potential propagation. The space constant determines the spatial extent of the excitatory load. In cardiac tissue, λ is on the order of a millimeter, meaning current spreads significantly up to this distance. Thus, the electrical load of all the surrounding cells is too great for the source current produced by a single cell. The minimum spatial extent of the excited cells necessary to perpetuate propagation is called the liminal length. Liminal length is important when initiating a wavefront. Stimulating a single node in a model will not result in propagation despite that the cell may be raised to extraordinarily large transmembrane voltages.

7.5.3 Propagation speed

The speed of wavefront propagation is determined by three factors: the sodium channel availability; gap junction coupling; and wavefront curvature [29]. Sodium channels have a refractory period, so if excited at a fast rate, they cannot totally remove inactivation and less current is available to drive the upstroke of the action potential. Sodium channel density is also intrinsic to cell type and may be affected by several factors. The more gap junctions, the better coupled the system is and the faster propagation will be. The relationship between speed and conductivity is $\theta \propto \sigma^2$. Wavefront curvature can be positive or negative, depending on whether the propagation wavefront is convex or concave.

If one considers a layer of activating cells, it is the current from this layer driving the activation of the immediately downstream cells. If the number of cells downstream is greater than the activated layer, there will be less stimulating current per cell, and propagation will be slower. This is a convex wavefront. Conversely, for a concave front, there are fewer downstream cells and speed accelerates. The relationship between speed and curvature is actually quite simple [53]

$$\theta = c - D \times K, \tag{7.29}$$

where c is the speed of a plane wave, K is the curvature, and D is the diffusion constant. D can be expressed as $\sigma_i / C_m \beta$ where C_m is the membrane capacitance per unit area.

7.6 Regional differences

When modelling at the organ level, the particularities of the different types of tissues must be taken into account. Considerable differences in action potential morphology, autorhythmicity, degree of coupling, and tissue structure are present, with important ramifications for cardiac function. Several highly specialized structures have developed to ensure optimal functioning of the organ. The major regions are discussed below.

7.6.1 Ventricles

The ventricles are the work horses of the heart, responsible for generating large pressures to pump blood throughout the body. They comprise a thicker, more than 1 cm, circular left ventricle, and a much thinner right ventricle appended to the side. The fibre architecture is complex and spatially varies [43]. Since current prefers to travel in the longitudinal than the transverse direction, a point stimulus will quickly become an ellipsoidal activation pattern and then the ellipse will rotate as deeper layers activate with a rotated fibre orientation. Activity propagates with a velocity of 40–70 cm s^{-1}.

Electrophysiological properties display heterogeneity. There is a transmural gradient with epicardial action potentials being shorter than endocardial ones. Epicardial action potentials also contain a deeper notch after the upstroke which allows for more rapid influx of Ca to reduce the latency between electrical activation and mechanical contraction, and better synchronize the whole organ's mechanical response. Midmyocardial cells with the longest action potential duration have been reported in canine wedge preparations [41], their presence on other preparations and species has not been confirmed [17]. There is a base to apex gradient with apical action potentials being shorter, as well as gradients in the left–right direction. No one direction dominates so that gradients in repolarization are difficult to categorize. However, the general pattern is preserved across individuals so that the T-wave polarity is common [36].

7.6.2 Atria

The atria show much more variability in shape than the ventricles, with a less well defined fibre orientation. The atria lack a specialized conduction system, yet conduction velocity within the atria is quite rapid, being 1–2 m s^{-1}. The atrial septum has an insulating layer within, preventing propagation of electrical activity across it. However, myocytes usually grow across the rim of the fossa ovalis to provide an interatrial pathway. Bachmann's bundle is most often the primary interatrial pathway, although its cells are not specialized but aligned in one direction. The right atrium has a thin (1–2 mm) epicardial layer of cell oriented roughly in the latitudinal direction (relative to the valve) and a set of coarse pectinate muscle on the endocardial side [21]. The sinoatrial node is situated within the wall but electrically isolated excepted for 4–5 exit sites [13]. The left atrium has two layers, with the endocardial being more trabeculated than the epicardial layer [20]. A third interatrial pathway is the muscular sheath of the coronary sinus which crosses from the right to left atrium.

7.6.3 Sinoatrial node

The SAN is a distinctive structure within the right atrium. It is composed of autorhythmic cells under strong neural influence. Gap junction expression differs within the SAN, comprising Cx43 and Cx40, instead of Cx45 as in regular atrial myocardium. The cell shape of SAN cells also differs, with the SAN cells being more spindle shaped and smaller than regular myocytes, without the stepped edges. In small mammals, the SAN makes contact with atrial tissue all along its periphery, except for a septal line of block, while in larger animals, the SAN is a disk-like, obliquely oriented structure isolated within the right atrial wall except for a small number of discrete connections. The septal line of block is still present.

Gradients in resting level, intrinsic beating frequency and neural response are found, especially along the cranial–caudal axis [2]. Increased sympathetic tone causes the leading pacemaker site, the site of earliest activation, to shift cranially, while increased vagal tone leads to a caudal shift. Conduction within the SAN is extremely slow, being on the order of 5 cm s^{-1}, indicating that coupling is extremely poor, or that the SAN acts as a population of coupled oscillators and propagation is only apparent. Automaticity within the cells is determined by the interaction of two clocks, one controlled by the balance of sarcolemmal currents with the funny current playing a major role in diastolic depolarization, and a second clock controlled by the uptake and release of calcium from the sarcoplasmic reticulum which causes a large membrane depolarization through the sodium–calcium exchanger [30].

Accurate modelling of the SAN thus, requires incorporating these many phenomena and carefully adjusting gradients in model parameters to reproduce known phenomena.

7.6.4 Atrioventricular node

In a normal heart, the atrioventricular node (AVN), is the only point of electrical contact between the upper and lower chambers of the heart. Conduction through the node is very slow and is responsible for a delay on the order of 80 ms, to allow optimal mechanical coordination of the atria and ventricles. There are actually two pathways within the AVN, one fast, and one slow, with different entrance points in the right atrium. Reentry within the AVN is seen clinically. Activity may initially descend down only one branch before ascending the other branch, leading to a reentrant loop within the AVN. Treatment consists of ablating the slow pathway. The propagation delay through the AVN is modulated by nervous input, as well as being a strong function of the time from the previous excitation. The AVN has an absolute refractory period, so that it acts as a filter for ventricular excitation when excited at rapid rates due to atrial tachycardias. It also exhibits autorhythmicity but at a slower rate than the SAN, so it can act as a backup pacemaker when the SAN fails to excite the atria. Propagation through the AVN is bidirectional, meaning that activity initiated in the ventricles may pass into the atria.

Two approaches have been used to model the AVN, either as a phenomenological network node with the current propagation delay based on the excitation history [6], or as a tissue with two pathways and representations of cellular activity [24].

Although histological examination has revealed the AVN to be a complex 3D structure, functionally it can be reduced to a fast conducting cable to which a slower conducting cable loop is attached.

7.6.5 His–Purkinje system

The His–Purkinje system rapidly distributes the electrical activity originating form the atria over the ventricles. Electrical propagation along this network is up to eight times faster than through regular myocardium. This is due to less lateral loading, a higher sodium current density, and higher conductance gap junctions formed by Cx40 and Cx45. Purkinje cells do not form intercalated disks and stepped edges as with regular myocytes, but they do form desmosomes for mechanical stability. These cells also display autorhythmicity although with a period much longer than that of the SAN or AVN.

The arrangement of the Purkinje system does not appear as a tree-like structure, but more fractal, since fibres rejoin after bifurcating, giving a mesh-like appearance that covers the endocardium. The point of first activation is about 2/3 of the way down the septum on the endocardium [11]. Given the large number of endocardial activation points, propagation is largely in the transmural direction in the lower 3/4 of the heart. Approaches to model the geometry of the PS include manually defining the major fascicles, and then using a fractal process [23] to compute the remainder of the PS [47].

The PS only connects with the LV myocardium at discrete points called Purkinje–Myocyte junctions. Propagation is bidirectional across the PMJ but asymmetric, with an anterograde delay (3–10 ms) longer than the retrograde delay (1–3 ms) [49]. This is due to the differing load of a one-dimensional network exciting a three-dimensional network as opposed to the converse. Modelling the junction requires distinct cell types for the Purkinje cells and the myocardium [9].

7.6.6 Coupling tissues

A schematic of the different tissues is shown in figure 7.4. The discrete nature of coupling between the tissues is illustrated. In many cases, there is effectively a change in dimensionality between the tissues. For example, the PS is a branching one-dimensional structure that must activate a three-dimensional load. When coupling two tissues of differing dimension, some provision must be made for the lower dimensional tissue if propagation is to travel into the higher dimensional system. The inherently larger electrotonic load will normally block propagation. Either a multiplication factor needs to be added and multiple nodes excited that exceed the liminal length [46], or structure must be finely represented to avoid local source–sink mismatch.

7.7 Arrhythmia

An arrhythmia is an abnormal electrical activation of the heart, any deviation from the normal sinus rhythm. A slower than normal heart rate is called a bradycardia, while a faster than normal heart beat is a tachyarrythmia. Tachyarrhythmia can be

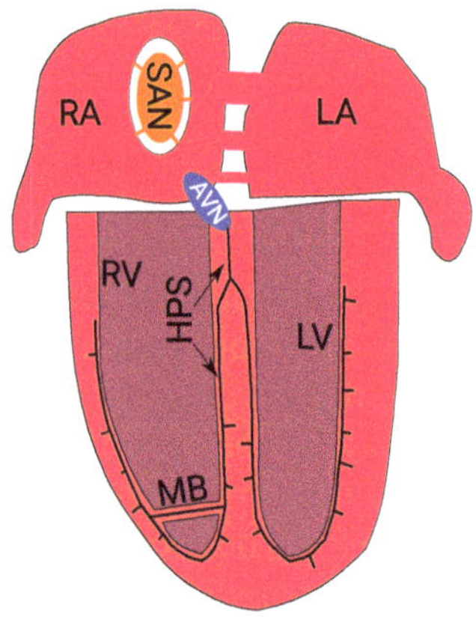

Figure 7.4. Schematic of four-chamber heart showing the different tissues and their coupling. Right atrium (RA), sinoatrial node (SAN), left atrium (LA), His–Purkinje system (HPS), atrioventricular node (AVN), left ventricle (LV), right ventricle (RV), moderator band (MB).

subdivided according to the mechanism responsible. If a region of the heart other than the sinoatrial node spontaneously fires an action potential, this is called an ectopic beat. These usually results from abnormal ionic channel abnormalities. The other mechanism responsible is called reentry, which refers to self-exciting circus activation. In other words, an activation front makes it back to a region of the tissue that it has excited before and finds enough time has passed such that the tissue can be re-excited, leading to a recurring circuit of activation.

7.7.1 Initiating reentry

All reentries start with a one way conduction block. This means, at some point, an activation wavefront reaches a portion of cardiac tissue where it was blocked from propagating towards tissue in one direction, but successfully propagated in another direction. The block may be temporary and a single instance can be sufficient to produce reentry. This block sets the stage for the activation front to travel and then come back around and excite the unexcited tissue which was blocked at another time. Reasons for the one way block include geometrical structures which have a source-sink mismatch only allowing propagation in one direction, a slower recovery time in a particular region of tissue, or after depolarizations which fail to propagate yet reduce sodium channel availability.

It is also important that a reentry circuit be of sufficient length for reentry. The path length must be greater than the wavelength, the product of the speed of propagation and action potential duration, which is the spatial extent of an action potential [42]. This distance is important since a cell is refractory during the action potential so it gives the spatial extent of refractory tissue. Wavelength can be reduced by altering either propagation speed or action potential duration. Some arrhythmias require rapid pacing beforehand to reduce the action potential duration to reduce the wavelength. When pacing a cell to reduce action potential duration, care must be taken to gradually reduce the pacing rate since a sudden decrease in pacing cycle length may lead to a 2:1 response, i.e. not every beat is captured, or a 2:2 response, alternating waveforms, instead of the desired 1:1 response [48] See figure 7.5.

A cross shock protocol uses two stimuli, denoted S1 and S2. A fixed first stimulus is applied, the S1, and after a certain amount of time, called the *coupling interval*, a second stimulus, the S2, is applied with a certain strength. The combinations of timings and strengths which produce reentry is called the window of vulnerability (figure 7.6). This is often plotted as a table with S2 timing and strength as the axes. A larger window of vulnerability implies tissue which is more susceptible to reentry induction.

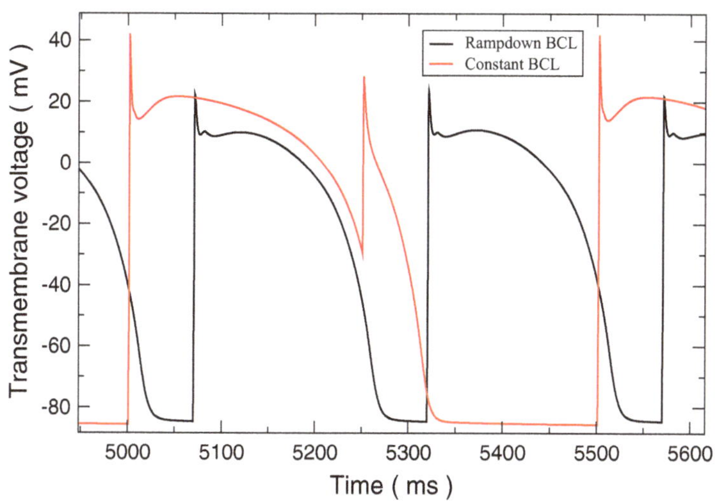

Figure 7.5. Ventricular ionic model paced (1) using a rampdown protocol starting at 350 ms and decreasing the stimulus interval by 10 ms until 250 was reached and a 1:1 response (black), or (2) at a constant pacing cycle length of 250 ms (red) with a 2:2 response.

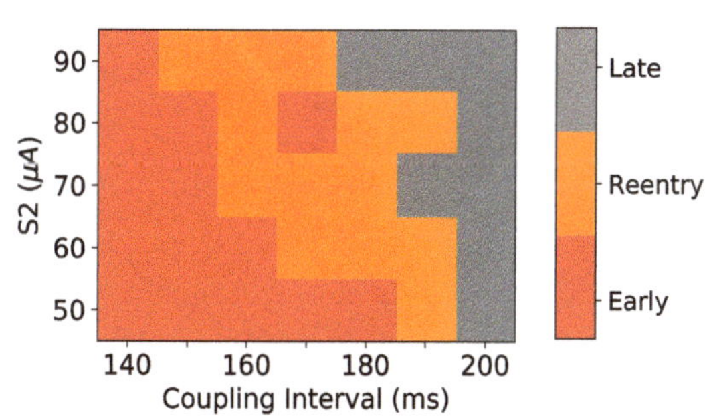

Figure 7.6. Window of vulnerability. The combination of S2 strength and timing needed to start reentry are shown in orange. *Late* regions mean that the S2 was applied after the repolarization wavefront had passed, resulting in a second wave following the first with no reentry. *Early* signifies an S2 that stimulated refractory tissue and did not initiate any new wavefronts.

7.7.2 Macroreentry

When the reentry circulates around a fixed inexcitable obstacle, it is referred to as a macroreentry. This is easily visualized as an annulus of tissue and shown in figure 7.7. One point on the ring is stimulated and a transient block permits propagation in one direction. By the time the activity travels to the blocked region, the annulus has regained excitability and the propagation can continue.

7.7.3 Rotors

Even without a central inexcitable zone, reentry can still be realized in the form of a rotor. Rotors have a functional core, a region of tissue that is excitable yet remains inexcitable due to a constant flow of current across it [50]. Activity resembles a spiral wave of activity. Since the spiral structure rotates with constant angular velocity, the propagation velocity along the wavefront must vary, and decrease towards the center. This occurs since the wavefront curvature increases as one approaches the core, resulting in slower propagation. Rotors are a two-dimensional phenomenon and are more suitable to describe atrial reentry. The three-dimensional analog of the rotor is the scroll wave. Running through the middle of the scroll wave is a line of phase singularity, called the filament, about which the scroll wave rotates [51]. Since these filaments, like the rotor

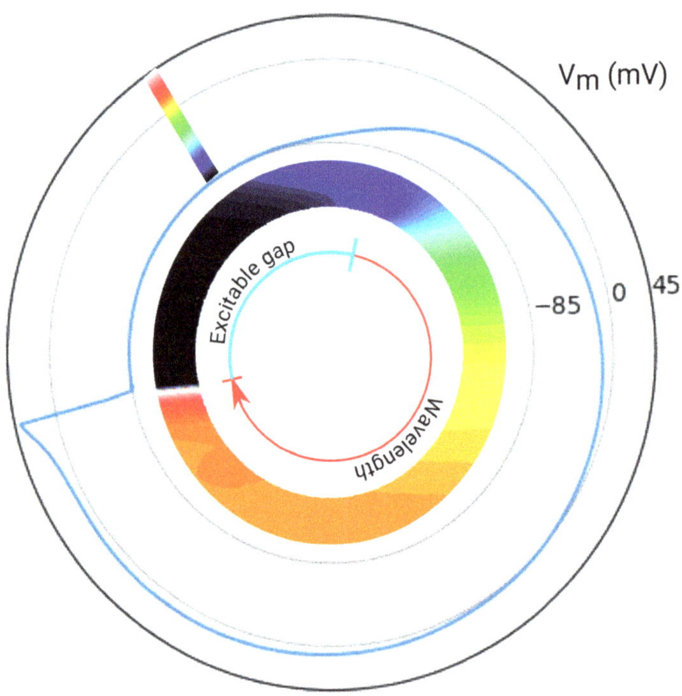

Figure 7.7. Macroanatomical reentry in a ring. Transmembrane voltage is displayed on the ring geometry. Outside of the ring, a polar plot shows transmembrane voltage around the ring. The tissue wavelength (red) and excitable gap (blue) are indicated on the interior of the ring. Propagation proceeds in the direction of the red arrow. The color bar is calibrated to the polar graph scale.

cores, are the organizing centers of the reentrant activity, tracking the filaments and their dynamics yields a simplified way to view the complicated electrical activity.

Rotors and scroll waves are initiated by a cross shock protocol [16]. A stimulating influence must be applied over a certain spatial extent across a recovery front. Either the recovery varies spatially or there is a gradient in stimulus. This results in a critical point being formed which separates the excited tissue from the nonexcited tissue. The nonexcited tissue will be driven deeper into refractoriness by the failed attempt at excitation. The new activation will progress and, since there is a gradient in recovery, portions will start to recover allowing activation to propagate into the unexcited region. Effectively, activity will begin to rotate around the critical point and a spiral wave will result. This process is illustrated in figure 7.8.

7.7.4 Fibrillation

Reentry may be stable, having a constant period and activity following a fixed path. Such activity corresponds to a simple tachyarrhythmia and would be classified as atrial flutter, or monomorphic ventricular tachycardia. Activity can, however, become disorganized, with ECGs showing a chaotic, unpredictable pattern. The amplitude of the ECG decreases, the rate is extremely fast (400 bpm) and timings of the waves on the ECG are unpredictable. Such a state is called defibrillation, and results from several

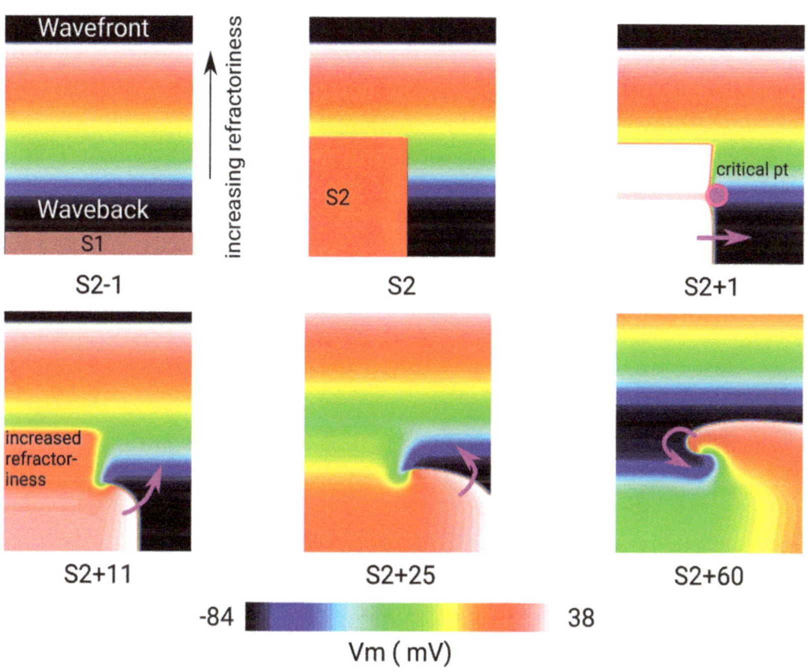

Figure 7.8. Rotor induction by cross shock protocol. Various instances are given as time relative to the application of S2. A plane wave is launched by S1 at S2−137. When S2 is applied, a critical point forms at the limit of where the S2 can induce a new action potential. When S2 is applied, a new planar wavefront initially propagates left-to-right but then turns to follow the waveback as tissue recovers excitability.

wavefronts moving throughout the myocardium, resulting in constant fractionation and annihilation of wavefronts.

Setting up such simulations requires a substrate capable of supporting such behaviour. One sufficient, but not necessary, condition is the slope of the APD restitution curve having a slope greater than one [4]. This means that if one plots APDs based on the interval between APDs, the slope of this curve will be greater than one. The slope is greatest at short cycle lengths. If the slope of this curve is greater, an instability occurs and the APD becomes unstable. Other features of this graph, like bistability, may also lead to fibrillation. Conduction velocity restitution may also contribute to spiral wave breakdown.

Spatial heterogeneity is another mechanism by which activity may become fibrillatory [12]. Regions of differing APD lead to situations in which activity sometimes propagates between regions and at other times, does not. Given the sensitive nonlinear nature of the excitability, propagation patterns can drastically change and lead to chaotic excitation.

Setting up a fibrillation simulation, thus, requires first determining the mechanism by which the spiral wave will break down. For tissue with a steep restitution curve, activity will spontaneously break down once a spiral wave is initiated. Interactions between the wavefront and waveback will lead to wavefront fractionation, an increase in excitation complexity, and fibrillation. For some spatial heterogeneities, spiral wave activity will also spontaneously revert to fibrillation. For other situations, a complex excitation pattern, as obtained through application of an external cross shock, will hasten fibrillation. An example of fibrillation initiated by field shock is shown in figure 7.9. Spatial heterogeneity maintains the fibrillatory nature of the system. Burst pacing from a single location, constantly reducing the interval between beats, may also be successful in initiating fibrillation [39].

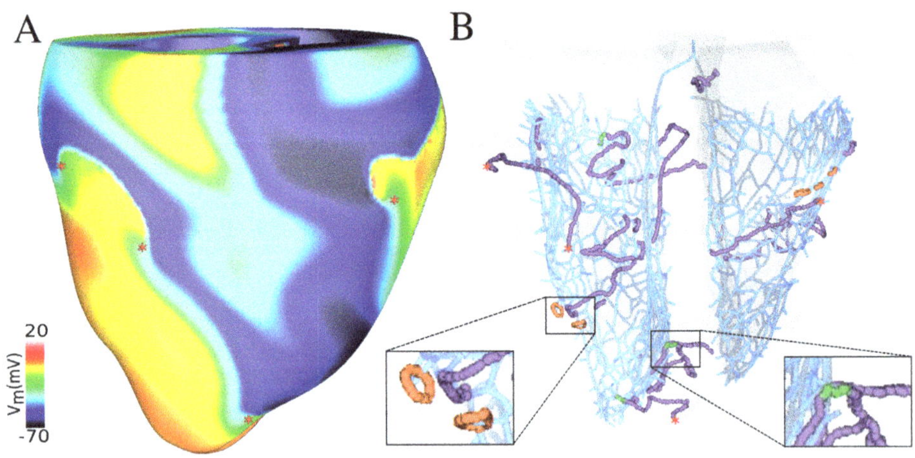

Figure 7.9. Purkinje system contribution during reentry. (A) Rotors on the epicardium surface along with phase singularities indicated by red asterisks. (B) Filaments in ventricles colocalized with Purkinje system endpoints. I-shaped filaments which terminate close to Purkinje system endpoints are shown in green, and O-shaped filaments which formed close to Purkinje system endpoints are indicated in orange. Surface phase singularities are as in A. Reprinted from [1] with permission from PLoS One.

7.8 Conclusion

Modelling of cardiac tissue and arrhythmia is a mature and well developed area of research. Current imaging methodologies allow us to easily construct geometries but assignment of material and ionic model properties remains more challenging. Matching the models with the type of tissue under consideration is important, as well as the mechanism implemented to initiate and maintain the arrhythmia.

References

[1] Behradfar E, Nygren A and Vigmond E 2014 The role of Purkinje-myocardial coupling during ventricular arrhythmia: a modeling study *PLoS One* **9** e88000

[2] Boyett M R, Dobrzynski H, Lancaster M K, Jones S A, Honjo H and Kodama I 2003 Sophisticated architecture is required for the sinoatrial node to perform its normal pacemaker function *J. Cardiovasc. Electrophysiol.* **14** 104–6

[3] Boyle P M, Franceschi W H, Constantin M, Hawks C, Desplantez T, Trayanova N A and Vigmond E J 2019 New insights on the cardiac safety factor: unraveling the relationship between conduction velocity and robustness of propagation *J. Mol. Cell. Cardiol.* **128** 117–28

[4] Cao J M, Qu Z, Kim Y H, Wu T J, Garfinkel A, Weiss J N, Karagueuzian H S and Chen P S 1999 Spatiotemporal heterogeneity in the induction of ventricular fibrillation by rapid pacing: importance of cardiac restitution properties *Circ. Res.* **84** 1318–31

[5] Clerc L 1976 Directional differences of impulse spread in trabecular muscle from mammalian heart *J. Physiol.* **255** 335–46

[6] Cohen R J, Berger R D and Dushane T E 1983 A quantitative model for the ventricular response during atrial fibrillation *IEEE Trans. Biomed. Eng.* **30** 769–81

[7] Colli Franzone P, Guerri L and Rovida S 1990 Wavefront propagation in an activation model of the anisotropic cardiac tissue: asymptotic analysis and numerical simulations *J. Math. Biol.* **28** 121–76

[8] Dekker E 1970 Direct current make and break thresholds for pacemaker electrodes on the canine ventricle *Circ. Res.* **27** 811–23

[9] Deo M, Boyle P M, Kim A M and Vigmond E J 2010 Arrhythmogenesis by single ectopic beats originating in the purkinje system *Am. J. Physiol. Heart Circ. Physiol* **299** H1002–11

[10] Desplantez T 2017 Cardiac Cx43, Cx40 and Cx45 co-assembling: involvement of connexins epitopes in formation of hemichannels and gap junction channels *BMC Cell Biol* **18** 3

[11] Durrer D, van Dam R T, Freud G E, Janse M J, Meijler F L and Arzbaecher R C 1970 Total excitation of the isolated human heart *Circulation* **41** 899–912

[12] Fareh S, Villemaire C and Nattel S 1998 Importance of refractoriness heterogeneity in the enhanced vulnerability to atrial fibrillation induction caused by tachycardia-induced atrial electrical remodeling *Circulation* **98** 2202–9

[13] Fedorov V V, Schuessler R B, Hemphill M, Ambrosi C M, Chang R, Voloshina A S, Brown K, Hucker W J and Efimov I R 2009 Structural and functional evidence for discrete exit pathways that connect the canine sinoatrial node and atria *Circ. Res.* **104** 915–23

[14] Fishler M G, Sobie E A, Tung L and Thakor N V 1996 Modeling the interaction between propagating cardiac waves and monophasic and biphasic field stimuli: the importance of the induced spatial excitatory response *J. Cardiovasc. Electrophysiol.* **7** 1183–96

[15] Fozzard H A and Schoenberg M 1972 Strength-duration curves in cardiac purkinje fibres: effects of liminal length and charge distribution *J. Physiol.* **226** 593–618

[16] Frazier D W, Wolf P D, Wharton J M, Tang A S, Smith W M and Ideker R E 1989 Stimulus-induced critical point. Mechanism for electrical initiation of reentry in normal canine myocardium *J. Clin. Invest* **83** 1039–52

[17] Glukhov A V, Fedorov V V, Lou Q, Ravikumar V K, Kalish P W, Schuessler R B, Moazami N and Efimov I R 2010 Transmural dispersion of repolarization in failing and nonfailing human ventricle *Circ. Res.* **106** 981–91

[18] Hand P E, Griffith B E and Peskin C S 2009 Deriving macroscopic myocardial conductivities by homogenization of microscopic models *Bull. Math. Biol.* **71** 1707–26

[19] Henriquez C S 1993 Simulating the electrical behavior of cardiac tissue using the bidomain model *Crit. Rev. Biomed. Eng.* **21** 1–77

[20] Ho S Y and Sánchez-Quintana D 2009 The importance of atrial structure and fibers *Clin. Anat.* **22** 52–63

[21] Ho S Y, Anderson R H and Sánchez-Quintana D 2002 Gross structure of the atriums: more than an anatomic curiosity? *Pacing Clin. Electrophysiol.* **25** 342–50

[22] Hoyt R H, Cohen M L and Saffitz J E 1989 Distribution and three-dimensional structure of intercellular junctions in canine myocardium *Circ. Res.* **64** 563–74

[23] Ijiri T, Ashihara T, Yamaguchi T, Takayama K, Igarashi T, Shimada T, Namba T, Haraguchi R and Nakazawa K 2008 A procedural method for modeling the Purkinje fibers of the heart *J. Physiol. Sci.* **58** 481–6

[24] Li J, Inada S, Schneider J E, Zhang H, Dobrzynsk H and Boyett M R 2014 Three-dimensional computer model of the right atrium including the sinoatrial and atrioventricular nodes predicts classical nodal behaviours *PLoS One* **7** e112547

[25] Jacquemet V 2010 An eikonal approach for the initiation of reentrant cardiac propagation in reaction-diffusion models *IEEE Trans. Biomed. Eng.* **57** 2090–8

[26] Johnston B M and Johnston P R 2013 A multi-electrode array and inversion technique for retrieving six conductivities from heart potential measurements *Med. Biol. Eng. Comput.* **51** 1295–303

[27] Jæger K H, Edwards A G, McCulloch A and Tveito A 2019 Properties of cardiac conduction in a cell-based computational model *PLoS Comput. Biol.* **15** e1007042

[28] Keener J P 1991 An eikonal-curvature equation for action potential propagation in myocardium *J. Math. Biol.* **29** 629–51

[29] Kléber A G and Rudy Y 2004 Basic mechanisms of cardiac impulse propagation and associated arrhythmias *Physiol. Rev.* **84** 431–88

[30] Lakatta E G, Maltsev V A and Vinogradova T M 2010 A coupled system of intracellular Ca^{2+} clocks and surface membrane voltage clocks controls the timekeeping mechanism of the heart's pacemaker *Circ. Res.* **106** 659–73

[31] Legrice I J, Hunter P J and Smaill B H 1997 Laminar structure of the heart: a mathematical model *Am. J. Physiol.* **272** H2466–76

[32] Leo-Macias A, Agullo-Pascual E and Delmar M 2016 The cardiac connexome: Non-canonical functions of connexin43 and their role in cardiac arrhythmias *Semin. Cell Dev. Biol.* **50** 13–21

[33] Neic A, Campos F O, Prassl A J, Niederer S A, Bishop M J, Vigmond E J and Plank G 2017 Efficient computation of electrograms and ecgs in human whole heart simulations using a reaction-eikonal model *J. Comput. Phys.* **346** 191–211

[34] Neu J C and Krassowska W 1993 Homogenization of syncytial tissues *Crit. Rev. Biomed. Eng.* **21** 137–99

[35] Noble D 1972 The relation of Rushton's 'liminal length' for excitation to the resting and active conductances of excitable cells *J. Physiol.* **226** 573–91
[36] Opthof T, Janse M J, Meijborg V M F, Cinca J, Rosen M R and Coronel R 2016 Dispersion in ventricular repolarization in the human, canine and porcine heart *Prog. Biophys. Mol. Biol.* **120** 222–35
[37] Plonsey R and Barr R 2007 *Bioelectricity: A Quantitative Approach* (Berlin: Springer)
[38] Pollard A E, Hooke N and Henriquez C S 1992 Cardiac propagation simulation *Crit. Rev. Biomed. Eng.* **20** 171–210
[39] Roney C H, Bayer J D, Cochet H, Meo M, Dubois R, Jaïs P and Vigmond E J 2018 Variability in pulmonary vein electrophysiology and fibrosis determines arrhythmia susceptibility and dynamics *PLoS Comput. Biol.* **14** e1006166
[40] Roth B J 1997 Electrical conductivity values used with the bidomain model of cardiac tissue *IEEE Trans. Biomed. Eng.* **44** 326–8
[41] Sicouri S, Fish J and Antzelevitch C 1994 Distribution of M cells in the canine ventricle *J. Cardiovasc. Electrophysiol.* **5** 824–37
[42] Smeets J L, Allessie M A, Lammers W J, Bonke F I and Hollen J 1986 The wavelength of the cardiac impulse and reentrant arrhythmias in isolated rabbit atrium. the role of heart rate, autonomic transmitters, temperature, and potassium *Circ. Res.* **58** 96–108
[43] Streeter D D, Spotnitz H M, Patel D P, Ross J and Sonnenblick E H 1969 Fiber orientation in the canine left ventricle during diastole and systole *Circ. Res.* **24** 339–47
[44] ten Tusscher K H W and Panfilov A V J 2006 Cell model for efficient simulation of wave propagation in human ventricular tissue under normal and pathological conditions *Phys. Med. Biol.* **51** 6141–56
[45] Tung L 1978 A bi-domain model for describing ischemic myocardial d-c potentials *PhD Thesis* MIT, Cambridge, MA
[46] Vigmond E J and Clements C 2007 Construction of a computer model to investigate sawtooth effects in the Purkinje system *IEEE Trans. Biomed. Eng.* **54** 389–99
[47] Vigmond E J and Stuyvers B D 2016 Modeling our understanding of the His-Purkinje system *Prog. Biophys. Mol. Biol.* **120** 179–88
[48] Vinet A and Roberge F A 1994 Analysis of an iterative difference equation model of the cardiac cell membrane *J. Theor. Biol.* **170** 201–14
[49] Wiedmann R T, Tan R C and Joyner R W 1996 Discontinuous conduction at Purkinje-ventricular muscle junction *Am. J. Physiol.* **271** H1507–16
[50] Winfree A T 1989 Electrical instability in cardiac muscle: phase singularities and rotors *J. Theor. Biol.* **138** 353–405
[51] Winfree A T 1994 Electrical turbulence in three-dimensional heart muscle *Science* **266** 1003–6
[52] Xie Y, Sato D, Garfinkel A, Qu Z and Weiss J N 2010 So little source, so much sink: requirements for afterdepolarizations to propagate in tissue *Biophys. J.* **99** 1408–15
[53] Zykov V S 1980 Analytic evaluation of the relationship between the speed of a wave of excitation in a two-dimensional excitable medium and the curvature of its front *Biofizika* **25** 888–92

Chapter 8

Clinical translation of patient-specific organ level cardiac models

Caroline Mendonca Costa, Caroline Roney, Marina Strocchi Orod Razeghi and Steven Niederer

Patient-specific cardiac models are increasingly being applied to clinical problems. However, building these models is still a major challenge. Several methods have been developed to create patient-specific geometries and scar/fibrosis morphology from clinical images, to include myofibre architecture, as well as to assign model parameters based on clinical electrophysiology data. However, translating these models into the clinical setting will require high quality clinical data acquisition and scaling up modelling pipelines to build a large number of models in a fast and standardized fashion. The latter will require substantial effort into the development methods and software, especially for automatic image segmentation.

8.1 Introduction

In the past decade, substantial effort has been invested in building patient-specific models of the heart based on clinical data and translating these into clinical knowledge and practice [1–3]. These models have been applied to simulate cardiac resynchronization therapy (CRT) [4], ventricular [5, 6] and atrial [7–9] arrhythmias and to predict arrhythmia risk in infarct patients [10]. Early modelling studies show that patient-specific models can be used prospectively in the clinic to predict targets for ventricular tachycardia ablation [11] and to predict the activation sequence of patients undergoing CRT [12]. In this chapter, we provide an overview of current modelling and parameterization workflows to build patient-specific models and some of the challenges and future perspectives for translating these models into clinical practice.

doi:10.1088/978-0-7503-2064-1ch8

8.2 Current anatomy modelling workflows

Typical anatomy modelling workflows start with segmenting the heart from clinical images, such as CT and MRI, generating meshes, and assigning fibre orientations. In addition, scars and fibrosis may also be segmented from contrast enhanced images and included in the models. In this section, we present and discuss examples of recently developed workflows to build models with patient-specific anatomies.

8.2.1 Geometry

Image segmentation consists of labelling regions of an image, where segmenting the heart typically consists of labelling the heart's chambers, valves, and vessels. This is a non-trivial and often tedious and time-consuming task that requires knowledge of cardiac anatomy and image processing techniques and software. The image resolution, artefacts and noise mean that in many cases there is no gold standard and a heuristic or manual approach is required. Several segmentation methods have been developed, from manual [13] through to semi-automatic [14, 15] and fully automatic four chamber [16] segmentation.

Segmentation of the (left) ventricular wall is often employed in the clinical setting to segment 2D image stacks and used for diagnosis of cardiovascular diseases. While ventricular segmentation of 2D image stacks can be automated in specific cases [17], manual segmentation is often used to segment 2D image stacks for small to medium patient numbers. These image stacks typically consist of only 5–10 slices and manually segmenting them requires only basic image processing knowledge. However, manual segmentation is subject to user interpretation and often leads to substantial inter-operator variability [18].

A manual segmentation method implemented in the scientific software Eidolon [13] was used in a recent study by our group [19] to segment the left ventricular wall and infarct scar from late-gadolinium-enhancement (LGE) MRI. The segmentation process consists of manually drawing endocardial and epicardial contours on each 2D image slice and manual selection of the apex in different image views, such as two, three, and four chambers long-axis views. The contours are then interpolated in 3D and combined with selected apex points to generate a left ventricular anatomical mesh and a binary image of the left ventricular segmentation in the 2D image stack.

Semi-automatic segmentation methods have also been developed and applied [10, 20] to cardiac modelling. For example, a segmentation method based on cubic splines and variational functions [15] was developed, where the myocardial boundaries of the left and right ventricles are manually contoured with cubic splines fit through landmark points on 2D slices of an LGE-MRI stack. A patient-specific 3D ventricular wall geometry is then reconstructed from the landmark points using variational implicit functions interpolation. This method offers the advantage of allowing segmentation of both right and left ventricular walls. In addition, the manual contouring can be performed on a few slices and automatically interpolated to the remaining slices, thus, speeding up the segmentation step and reducing user errors. A similar method has been implemented in the segmentation and visualization software CardioViz3D [14], an interactive tool that allows the user to model

3D surfaces using control points, and applied to segment the ventricular myocardium from cine-MRI [21]. Another semi-automatic method [22] implemented within the MUSIC software (IHU Lyric Bordeaux, Inria Sophia Antipolis, France) consists of a region-growing algorithm to segment the endocardium and manually drawing epicardial contours on a few image slices that are then interpolated to all slices. Alternatively, the Seg3D [23] software (Sci Institute, Salt Lake City, Utah, USA) offers general purpose semi-automatic segmentation methods, however, it requires advanced image processing expertise.

More advanced segmentation methods allow automatic segmentation of the four chambers of the heart. A model-based segmentation [16] method has been used by our group to build patient-specific models for electromechanics simulations [24, 25]. This automatic segmentation contains the four chambers' blood pools and the left ventricular wall. Such methods provide high quality and detailed automatic segmentation, but require high resolution and quality images, which are often not widely available in the clinical setting, as they often require long acquisition times. As such, these images are often obtained within clinical research studies for a small number of patients. Another drawback of such automatic methods is the need to add the right ventricular and atrial walls. The walls are thin and thus are usually not visible on MRI due to low image resolution, typically 1.8×1.8 mm in-plane resolution. Thus, the walls are often added by dilating their corresponding blood pools to a pre-defined thickness and then subtracting the original blood pool to obtain a wall. Moreover, the method does not provide a segmentation of the pulmonary veins (PVs) and the mitral valve (MV), which are of clinical interest.

Automatic segmentation of the left atrium is particularly challenging, as the left atrial appendage (LAA), the PVs, and MV have highly varied anatomy and size. Our group has recently developed a multi-label convolutional neural network (CNN) for segmenting the left atrium (LA) blood pool, pulmonary veins (PVs), and the mitral valve (MV). The network is a specialisation of U-Net [26]. The input of our convolutional network is an MRI, where the labels are hot-encoded with three channels for blood pool, PVs and MV. Our method was validated using segmentations manually produced by five expert clinicians and achieved a 91% dice score. Figure 8.1 shows an example of manual segmentation and the corresponding automatic segmentation generated with the CNN.

In summary, image segmentation is a highly active field offering a large variety of segmentation methods. When choosing a segmentation method for building patient-specific models of anatomy, the type of image available (MRI, CT), image resolution, and the level of anatomical detailed required should be considered.

8.2.2 Scar and fibrosis

Recent simulation studies have used patient-specific models that include scar [10, 20] and fibrosis [7, 8] to investigate their role on arrhythmogenesis. Identifying scar and fibrosis requires the use of contrast enhanced MRI, typically LGE-MRI, where the scar and fibrotic tissue appear as high-intensity (white) regions, as illustrated in figure 8.2.

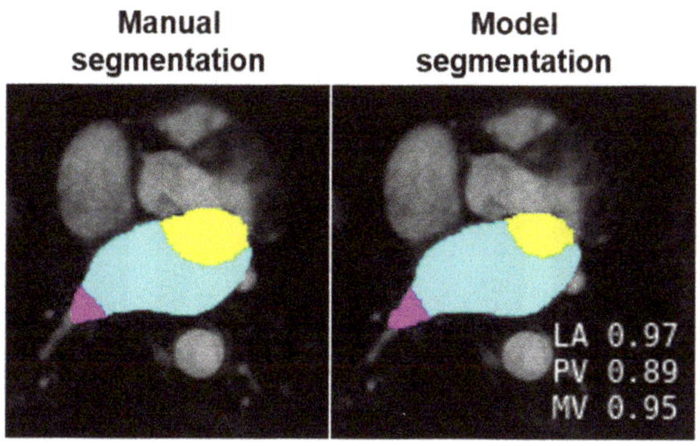

Figure 8.1. Automatic segmentation of the left atrium (LA), pulmonary veins (PV), and mitral valve (MV). Left: example of manual segmentation performed by an expert clinician. Right: corresponding automatic segmentation generated by the convolutional neural network developed by our group and the respective dice scores for each label. The numbers indicate dice scores for each label.

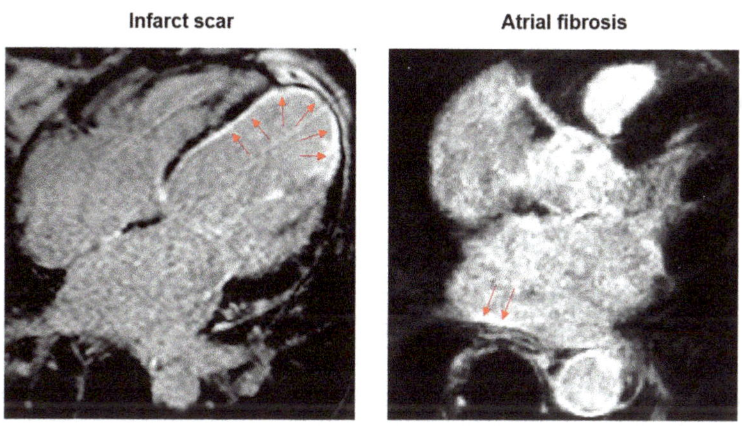

Figure 8.2. Examples of late gadolinium enhancement MRI showing a large infarct scar (left panel) and atrial fibrosis (right panel) as regions of high intensity (white) areas, as indicated by the red arrows.

Most infarct scar segmentation methods are based on defining image intensity thresholds and labelling the regions above these as scar. A typical technique consists of manually selecting a region of healthy myocardium and computing the mean and standard deviation of the voxel intensities within that region. The dense scar core and the scar border zone are then labelled as the regions with voxel intensity above the mean intensity plus 3 and 2 standard deviations, respectively [27]. The full-width-half-max (FWHM) method has also been widely used [28–30] and it consists of identifying the maximum voxel intensity within the myocardium and labelling the scar as the region above 50% of the maximum intensity. It has been demonstrated to be the most reproducible method [31], as it is fully automatic and does not require defining a region of healthy myocardium, which is subject to user interpretation and experience. However, scar and border zone identified via the standard deviation method were stronger predictors of the risk of implantable cardioverter defibrillator

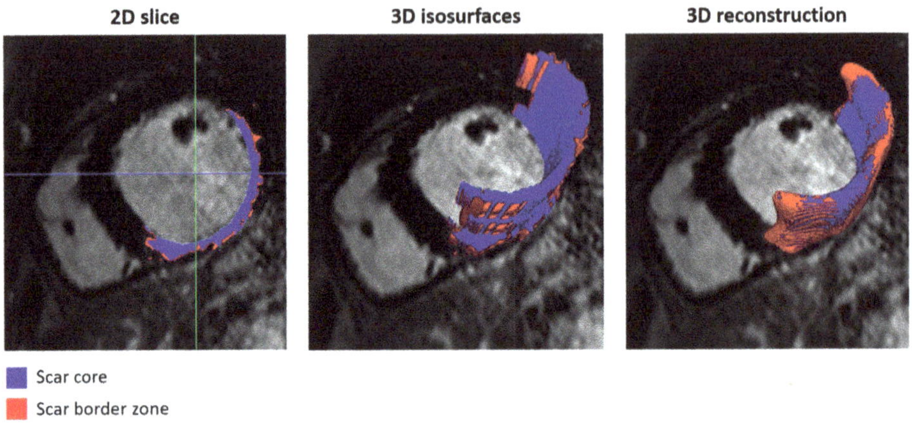

Figure 8.3. Scar segmentation using the standard deviation method and 3D reconstruction using the LogOdds method. The left panel shows the scar core (blue) and border zone (red) segmentation on a LGE-MRI slice. The middle panel shows the 3D isosurfaces of the 2D slice segmentations. The right panel shows the 3D reconsctruction of the 2D segmentation using the LogOdds [32] method.

(ICD) therapy in patients with ischemic and non-ischemic cardiomyopathy than the FWHM method [27].

LGE-MRI is typically acquired as a series of 2D slices with slice thickness varying from 6 to 20 mm. Thus, the scars segmented from the 2D slices need to be reconstructed in 3D before they can be combined with the heart's anatomy segmentation. Scar reconstruction is a challenging problem and different methods lead to different results with varying accuracy [32]. In a modelling pipeline recently developed by our group [33], we employed a statistical shape reconstruction method based on the LogOdds function, as it was shown to yield the most accurate 3D scar reconstruction and ventricular arrhythmia morphology in electrophysiology simulations [32]. An example of 3D scar reconstruction using this methodology is shown in figure 8.3.

When segmenting atrial fibrosis, it is important to restrict the assessment to the body of the LA, and exclude the PVs and MV, which have different tissue characteristics and unreliable LGE intensities. Our group has recently applied the CNN LA automatic segmentation to automate the quantification of fibrosis by analysing maximum transmural values from LGE-MRI scans. The methodology was validated against ground truth manual segmentations performed by expert clinicians. An example of atrial meshes from a test set with its associated LGE-CMR intensities projected on the surface are shown in figure 8.4. LGE intensity values are normalized by blood pool signal intensity within a region of interest for each voxel within the wall. Results are presented for both manual and automatic segmentations.

8.2.3 Mesh generation

After the anatomy and scar/fibrosis morphology of the heart have been segmented, the next step in the modelling workflow is to create a mesh or grid representation of the anatomy. If the finite element method (FEM) is used, a finite element mesh can be built from the segmented images using a myriad of meshing tools available, such

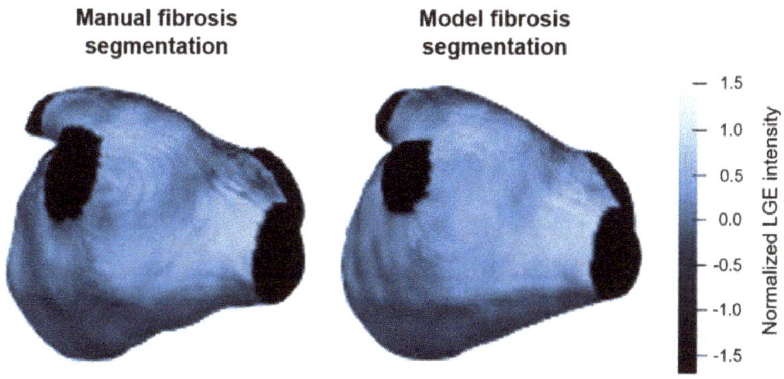

Figure 8.4. Automatic segmentation of left atrial (LA) fibrosis. Manual (left) and automatic (right) segmentations projected onto the LA suface. Colors represent the normalized LGE intensity.

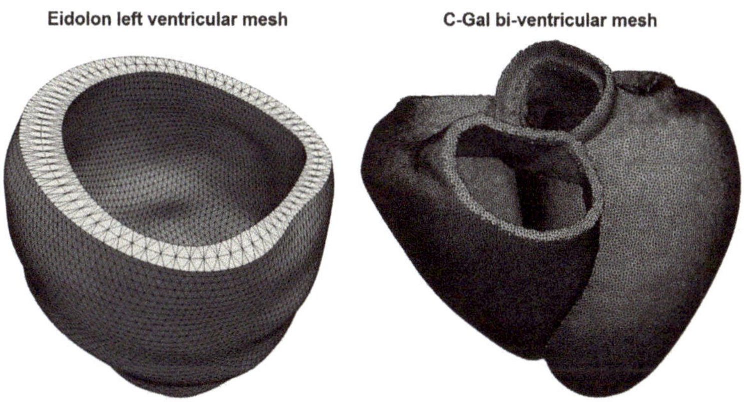

Figure 8.5. Example of meshes generated from clinical cardiac images. Left panel: left ventricular mesh generated with Eidolon [13]. Right panel: Bi-venricular mesh generated with C-Gal [34].

as the C-Gal library [34], Tarantula [35], Cleaver [36], Eidolon [13], Continuity [37], and TetGen [38]. Using FEM has the advantage of allowing the use of meshes with elements of irregular shape to obtain a smooth surface, even at thin trabeculae [39]. However, the finite volume [40, 41] and finite difference [42, 43] methods are still used by the modelling community. These methods offer good numerical accuracy and can easily be generated from medical images, but the use of regular element grids leads to jagged surfaces, which can cause spurious polarizations when simulating extracellular shocks [44]. Examples of meshes generated with Eidolon and C-Gal are shown in figure 8.5.

8.2.4 Fibres

In addition to anatomical detail, it is important to include fibre orientations in computational models, because structural anisotropy affects electrical propagation patterns, including paced activation sequences and conduction velocities [45–47], morphology of re-entrant circuits [48], and anchor locations of atrial drivers [49].

While substantial progress has been made in building models of patient-specific anatomy and scar/fibrosis, acquiring patient-specific fibres orientation is still a major challenge. Diffusion-tensor MRI allows estimation of detailed fibre orientations, however, poor signal-to-noise ratio, long acquisition times, and motion artefacts mean that imaging the whole heart and incorporating patient-specific fibre orientations into computer models of the heart is currently unfeasible. Therefore, most modelling studies [10, 33] have used rule-based [50–52] or atlas-based [53] methods to incorporate ventricular and atrial fibre orientations into anatomical models.

8.2.4.1 Ventricular fibres

Fibre orientation is relatively regular in the ventricles, particularly in the left ventricle. This allows the definition of rule sets to assign ventricular fibres. Rule-based methods have been developed for the ventricles and often combine a series of Laplace solves that are combined using a pre-defined set of rules to derive fibre orientations that conform to the ventricular wall. Bayer *et al* [50] developed a Laplace–Dirichlet rule-based method to assign ventricular fibres based on DT-MRI and histological data. Their method consists of (1) solving the Laplace equation within the ventricles from the apex to the base and from the endocardial to the epicardial surface and (2) a set of six rules defining how the fibres rotate from the endocardial to the epicardial surface and from the apex to the base. The Laplace solutions and fibre rotation angles are combined to create a fibre field within the ventricular walls. An example of rule-based fibres generated with this method is shown in figure 8.6 (left panel).

Fibre orientation is more complex in the right than in the left ventricle [51]. The longitudinal fibre orientation in the right ventricular (RV) endocardium is aligned from the apex towards the pulmonary and tricuspid valves. In addition, fibres in the RV epicardial outflow tract have a circular fibre structure. To address these specific structural differences, Doste *et al* [51] extended Bayer's [50] rule-based method to also include a set of rules specific for the RV outflow tract. Validation of both rule-based methods show good agreement with DT-MRI [50, 51] and histology [51] data.

8.2.4.2 Atrial fibres

Atrial fibre structure is not as organised as ventricular structure and demonstrates greater inter-subject variability [54]. For example, atrial fibre direction within the pulmonary veins (PV) and around the junction of the PV with the left atrial body are particularly variable between subjects, with reported fibre arrangements including spiralling, circular, chaotic and longitudinal fibres, and either similar or different fibres on the endocardial and epicardial surfaces [55–57].

Computational modelling groups use different methods for including atrial fibres in either 2D surface or 3D volumetric meshes. These include rule-based techniques to capture histological descriptions of atrial anatomy and fibre structure. For example, Fastl *et al* developed a modelling pipeline for constructing fibre fields on both the endocardial and epicardial left atrial surfaces based on morphological data by using local solutions to Laplace's equation, and these fields were then interpolated transmurally [52]. An example of fibres generated using this method is shown figure 8.6 (right panel). Similarly, Labarthe *et al* used semi-automatic techniques for including atrial fibres in a biatrial bilayer model based on histological data [58]. In addition, Krueger *et al* presented a semi-automated method in which 22 distinct seed-points are selected from which auxiliary lines are constructed to define a biatrial fibre field [59].

An alternative approach for incorporating fibre direction in computational models is by registering a fibre atlas to a patient specific geometry. For example, McDowell *et al* use an image-based registration method to transform an atlas geometry to a target geometry [53]. The calculated optimal deformation field was then applied to assign a personalised fibre field to a biatrial model [60]. Our group recently published an atrial coordinate system, which was used to map the fibre field from an atrial bilayer model to individual patient geometries [61]. Our group is also investigating an alternative approach by constructing a functional fibre atlas using electrical anisotropy maps estimated from multiple activation maps [49].

8.3 Current parameterization techniques for electrophysiology models

Patient-specific anatomical models may be combined with mathematical models to simulate patient-specific EP. This requires parameterizing the mathematical models based on clinical EP data of a patient. In this section, we describe and discuss examples of recent parametrization techniques used to build computational models of patient-specific EP.

8.3.1 Electrophysiology models and data

Most simulation studies employ the Monodomain model, which is a simplified and computationally less expensive version of the Bidomain model. Alternatively, a reaction-eikonal [62] model may be used, which offers a trade-off between the eikonal and the monodomain models and allows the use of coarse meshes, thus, substantially decreasing computational cost. These models are coupled to a cell model describing action potential and ionic current dynamics, which span from simple phenomenological models to detailed biophysical action potential models. Cellular automaton [63] models may also be used, although these are less commonly employed. The most appropriate model choice depends on the available clinical data, which may characterise conduction properties at a single pacing rate, across

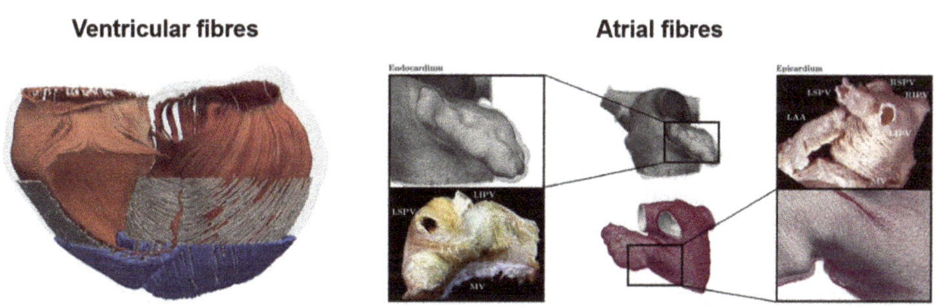

Figure 8.6. Rule-based fibres. Left: ventricular fibres generated with Bayer's method. Right: atrial fibres generated with Fastl *et al*'s [52] method (adapted from [52], copyright 2018, with permission from Elsevier).

pacing rates to include restitution properties or during fibrillation, measured at either single or multiple locations within the heart.

Personalisation of EP models is important for understanding arrhythmia mechanisms and for making patient specific predictions. To accurately simulate a patient's EP, model parameters must be fitted to electrical measurements, such as electrocardiography (ECG), electro-anatomical mapping (EAM), and body surface potential map (BSPM) data, and validated. Parameterizing these models is a challenging problem, which requires tissue conductivity values and ion current parameters that approximate conduction velocity and action potential duration. In addition, simulating accurate activation/repolarization sequences within the ventricles also requires including Purkinje network [62, 64] models or a layer of fast endocardial conduction [12, 65] and estimating early activation sites [66]. Atrial electrophysiology exhibits both regional and spatial heterogeneity, and varies between patients [67], which must be accounted for in patient-specific models.

Several methods are available for model parameterization depending on the type of EP data available. Substantial effort has been put into parameterization methods that rely solely on non-invasive clinical data, such as ECG and BSPM. This is a very challenging problem, which may only be solvable in specific cases, where the EP properties are well known.

8.3.2 Personalized ventricular models

Our group has recently developed and validated a rule-based method for predicting the activation sequence of CRT patients using non-invasive clinical EP data [12]. The method consists of running several eikonal simulations of electrical activation in the ventricles to fit the ventricular conduction velocity to the QRS duration of a patient's ECG. A simple model with a fast endocardial conduction layer is used and activation is started at the location of the right-ventricular CRT pacing lead. The method has been trained and validated using independent clinical data sets from two international centres.

The Sermesant group has recently developed a novel method for personalizing EP in patients with infarct scars [22]. The method uses the ventricular wall thickness obtained from CT images to parameterize the eikonal model and shows good agreement with EAM data. Parameterization techniques based on both invasive and non-invasive EP data have also been developed by the Sermesant group [21], where Ensite EP maps are used to estimate early activation times and the QRS duration is used to estimate a mean conduction velocity. In a second step, the local conduction velocities are adjusted to match the Ensite data and give a detailed approximation of the activation sequence and local conduction velocities.

Although EP personalization is possible, invasive or non-invasive EP data is seldom available in the clinical setting. Moreover, even when such data is available, the resolution is very coarse and does not allow accurate EP parameterization. Consequently, most simulation studies use experimentally measured EP parameters from the literature. However, very few measurements of tissue conductivities are available and these vary substantially between studies [68]. Measurements of conduction velocity are more common, as it is easier to measure in the experimental setting [69], and it is possible to compute the tissue conductivities by solving an inverse problem [68, 70]. Infarct models are even more challenging to parameterize, as experimentally measured conduction velocities and action potential characteristics vary between studies and depend on the time after infarct measurements were done [71]. Nevertheless, computational models of patient-specific anatomy and infarct morphology, where EP parameters obtained from the literature were used, have been shown to accurately predict the risk of arrhythmias in infarct patients [10].

8.3.3 Personalized atrial models

Our group developed and validated a work flow for calibrating patient specific atrial EP models to clinically recorded restitution data, representing the first validated cohort of models that include clinically measured EP heterogeneity [72]. Conduction velocity restitution and effective refractory period were measured at multiple locations on the endocardium of the left atrium using a pentaray catheter with pacing from the coronary sinus. The four parameters of the modified Mitchell–Schaeffer phenomenological model [73] and the tissue conductivity were tuned locally to these conduction velocity restitution and effective refractory period data. To validate the calibration process, simulated activation times from high right atrium pacing were compared to a second set of clinical data consisting of activation time maps from high right atrium pacing. The modified Mitchell–Schaeffer phenomenological model has the advantage that it has only four parameters, which represent different phases of the action potential. We are currently investigating the effects of uncertainty in both the spatial and temporal properties of electroanatomic mapping data on this calibration [74].

Lombardo *et al* compared a simplified phenomenological model to a biophysical model for capturing clinically recorded action potential duration restitution and conduction velocity restitution data for five patients with AF [75]. They found that both models accurately reproduced clinical data and produced similar spiral wave dynamics to each other, suggesting that phenomenological models may be sufficient. Importantly, these models varied significantly from the published model and also for each of the calibrated datasets, highlighting that personalisation of EP models is important.

Atrial EP properties affect predictions of arrhythmia properties. Loewe *et al* used an eikonal approach to identify potential atrial flutter pathways [76]. Both Deng *et al* [77] and Saha *et al* [78] showed that action potential duration and conduction velocity properties have a large enough effect on predictions of atrial driver locations during AF for atrial fibrosis models that models should personalise EP properties if such data are available, or account for this uncertainty in model prediction. In addition, challenges associated with interpreting atrial LGE imaging data mean that currently EP personalisation may be a more promising strategy for patient-specific prediction.

8.4 Universal coordinate systems

When building patient-specific models it is often necessary to map data across different geometries and to define specific regions within the heart. However, defining a consistent mapping method for several geometries can be challenging. Standardized mapping may be useful for mapping invasively measured EP data onto anatomical models [21] and Purkinje networks [79] from one geometry to another; for assigning regional EP heterogeneities [12, 19, 80] and constructing patient specific geometries with atrial structures and fibre directions [61]; and for 2D visualisation [81]. In this section, we describe two recently developed methods to define universal coordinate systems within the heart to facilitate these tasks.

8.4.1 Universal ventricular coordinates

Bayer *et al* [79] recently defined the universal ventricular coordinates (UVC) system to uniquely identify global positioning within the ventricles. This method extends the parameterization previously described by the same author to define ventricular fibre orientation [82] at the ventricular junction and the septum, giving a global positioning system within an arbitrary biventricular geometry. The UVC defines four coordinates (see figure 8.7): (1) apico-basal coordinate, varying between 0 and 1 from apex to base; (2) transmural coordinate, varying between 0 at the endocardium and 1 at the epicardium; (3) rotational coordinate, rotating around the left ventricle, ranging between $-\pi$ at the free wall, 0 at the septum and π back to the free wall; (4) an intra-ventricular coordinate, set at -1 at the left ventricle and $+1$ at the right ventricle. Similar to their previous method used to assign fibre orientations, the UVC system is obtained via a set of Laplace solves (one for each coordinate), whose solutions are combined at each node of the mesh to uniquely define its global position within the ventricles.

8.4.2 Universal atrial coordinates

Integration of spatial information about atrial physiology, either from multiple data modalities for a single subject, or across multiple subjects is particularly challenging in the atria because of both the complexity of the atrial anatomy and the great variability in atrial shape between patients [83]. Therefore, our group developed an

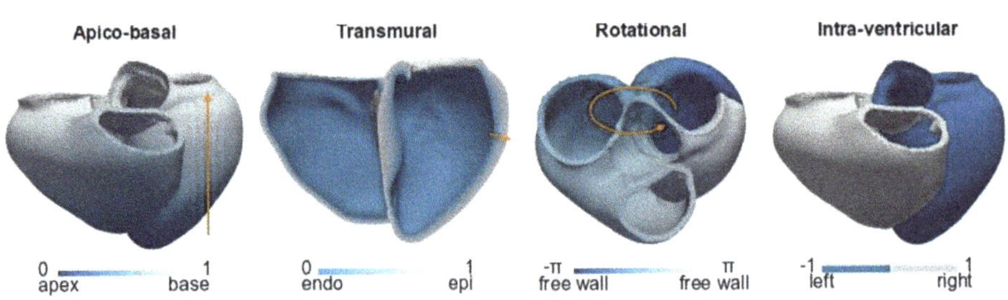

Figure 8.7. Universal ventricular coordinates. Showing the apico-basal, transmural, rotational, and intra-ventricular coordinates. The arrows indicate the direction of change.

analogous system to the UVC system [79] for the atria called the universal atrial coordinate system (UAC) [84].

Similar to UVC, the UAC are calculated as solutions to Laplace's equation on a left atrial or right atrial mesh with Dirichlet boundary conditions of zero and one applied along two sets of boundary nodes, to construct two coordinates per atria chosen to be as close to orthogonal as possible everywhere. For the left atrium, the coordinates are a septal to lateral coordinate, and a coordinate from the posterior mitral valve over the roof to the anterior mitral valve. In a similar way, for the right atrium, the coordinates are from the lateral tricuspid valve to the septal tricuspid valve, and from the inferior vena cava to the superior vena cava. Calculation of the UAC requires the manual selection of three landmark points for the left atrium, and two landmarks for the right atrium. An additional transmural coordinate can be included by solving Laplace's equation with 0 on the endocardial nodes of the mesh, and 1 on the epicardial nodes. This extension to three dimensions allows us to assign transmural tissue properties and fibres to volumetric meshes (figure 8.8).

8.5 Simulation costs

Recent advances in clinical imaging techniques and image processing and mesh generation methods now allow construction of highly detailed computational models of the heart. While detailed models are appealing for clinical use [1], a high level of detail substantially increases computational costs and often requires the use of specialised high-performance computing facilities. This can make the use of these models challenging in clinical timelines and limits the number of simulations that can be used for validation, sensitivity analysis and predictions. To address this problem, modellers have to compromise between the features included in the model and required simulation and post-processing times. When building patient-specific models, a trade-off between the level of personalisation/detail of the model and keeping simulations trackable has to be done at each step of modelling workflows.

Starting from image segmentation, modellers need to decide which cardiac chambers to include in the geometry, and that depends on the clinical problem

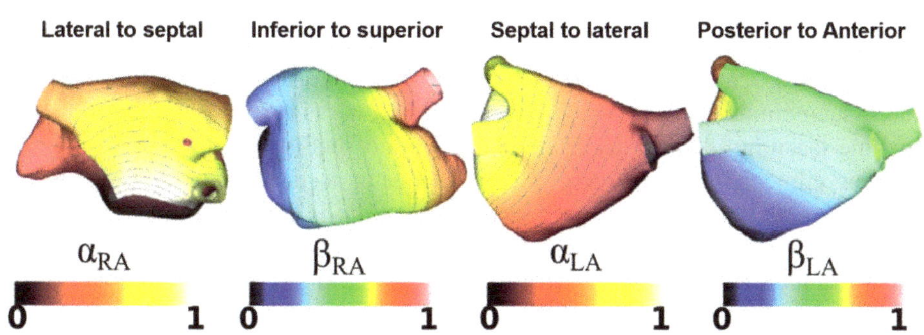

Figure 8.8. Universal atrial coordinates. Showing the two right atrium (RA) coordinates—from the lateral tricuspid valve to the septal tricuspid valve and from the inferior vena cava to the superior vena cava—and two left atrial (LA) coordinates—from septal to lateral walls and from the posterior mitral valve to the anterior mitral valve. Adapted from Roney *et al* [84], copyright 2019 with permission from Elsevier.

being addressed. The majority of computational models focus on the left ventricle [19, 85], both ventricles [10–12, 65] or the left atrium [52]. In the last decade, the modelling community has been moving towards four-chamber geometries [24], which leads to larger meshes and thus longer simulation times.

At the mesh generation stage, mesh resolution is the most important parameter to consider. A fine mesh (average edge length of 250 μm) is required when detailed structures are included in the geometry (ventricular papillary muscles and trabeculations, left atrial appendage), or if monodomain or bidomain formulations are used to model electrical activation [68, 86]. Coarse meshes (average edge length of 1 mm) can be used to solve reaction-eikonal models [62], which offer a compromise between the eikonal and mono- or bi-domain formations. However, such coarse meshes cannot account for complex small structures of the heart and are not appropriate for arrhythmia simulations.

Parameterizing computational models may require running a large number of simulations to identify parameter sets that best fit the data. To keep parameter sweeps trackable, eikonal simulations [12, 21, 66] are often employed to narrow down the search space before fitting mono- or bidomain conductivities.

8.6 Scaling-up modelling pipelines

In the previous sections, we described current modelling and parameterization workflows for building patient-specific models, including methods for image segmentation, mesh generation, fibre assignment, and parameter fitting, as well as the importance of universal coordinate systems and the impact of anatomical and EP detail on simulations costs.

Translating cardiac models to the clinical setting will require major investments into developing standard methodologies and software that allows fast model creation in a consistent fashion. For a binary outcome clinical trial, for example testing for response to CRT or not, where the response was expected to improve from 65% to 80% would require 276 patients [[1]], to have an 80% change of reaching a significance level of 0.05. In contrast, the largest virtual model cohorts to date applied to the clinic included 10–50 patients [10, 61, 87].

Based on recent modelling studies and experience within our group, the major challenges for scaling-up modelling pipelines are good quality (semi-) automatic image segmentation and model parameterization. The former is possible using machine learning [88] if a large number of good quality images can be acquired, however, clinical image acquisition is a challenge in itself. Conversely, the latter may not be solvable in the near future. While a few methods have demonstrated potential for EP model personalization based on clinical EP data, these are limited to a few specific scenarios where early activation sites [12] and tissue characteristics [22] are (relatively) well known.

[1] https://clincalc.com/stats/samplesize.aspx.

8.7 Future perspective

Developing workflows for creating patient-specific models at scale will be key to the use of patient-specific models in clinical trials, the use of models in clinical practice and the creation of virtual patient cohorts. Developing sufficient technical skills within a single group to achieve this goal will be challenging and will likely require a consortium of groups working together. This work may be facilitated by new hardware options, including cloud computing where complex research software can be set up by a single group and that machine setup can then be easily replicated as required. This also moves the computational cost of using software from the developer and onto the user. It could also facilitate linking multiple codes for specific elements of the model development and simulation workflow.

As the workflows for creating computational models become more complex, the barriers to entry to patient-specific cardiac models will increase. In the case of cardiac cell models, complexity has continued to increase but through a culture of publishing the code and the availability of public repositories [89] it has been possible for many groups to develop and use these models. In the case of whole organ heart models this has not always been the case. There have been a small number of specific species single anatomies published (dog [90], pig [91] and rabbit [92]). If groups publish and share their patient-specific models this will hopefully increase the use of models in research and foster the development of improved model development, simulation and analysis workflows.

8.8 Conclusion

Models and simulation have a clear role in the future of precision cardiology. To realise this potential will require a move towards scaling our technologies to meet the realities of clinical translation. There are very real challenges in moving from a single patient case study to replicating the inherent variability present in most patients' populations. There used to be a question about how you could simulate a single heart. Then we moved to asking can we model a specific patient. Now the challenge is how we model every patient, how we do this efficiently and how we communicate the results to clinicians.

References

[1] Niederer S A, Lumens J and Trayanova N A 2019 Computational models in cardiology *Nat. Rev. Cardiol.* **16** 100–11

[2] Niederer S A and Smith N P 2016 Using physiologically based models for clinical translation: predictive modelling, data interpretation or something in-between? *J. Physiol.* **594** 6849–63

[3] Niederer S, Mitchell L, Smith N and Plank G 2011 Simulating human cardiac electrophysiology on clinical time-scales *Front. Physiol.* **2** 14

[4] Lee A W C, Costa C M, Strocchi M, Rinaldi C A and Niederer S A 2018 Computational modeling for cardiac resynchronization therapy *J. Cardiovasc. Transl. Res.* **11** 92–108

[5] Arevalo H, Plank G, Helm P, Halperin H and Trayanova N 2013 Tachycardia in post-infarction hearts: insights from 3D image-based ventricular models *PLoS One* **8** e68872

[6] Relan J, Chinchapatnam P and Sermesant M *et al* 2011 Coupled personalization of cardiac electrophysiology models for prediction of ischaemic ventricular tachycardia *Interface Focus.* **1** 396–407
[7] Roney C H, Bayer J D and Zahid S *et al* 2016 Modelling methodology of atrial fibrosis affects rotor dynamics and electrograms *Europace* **18** iv146–55
[8] McDowell K S, Vadakkumpadan F and Blake R *et al* 2013 Mechanistic inquiry into the role of tissue remodeling in fibrotic lesions in human atrial fibrillation *Biophys. J.* **104** 2764–73
[9] McDowell K S, Arevalo H J, Maleckar M M and Trayanova N A 2011 Susceptibility to arrhythmia in the infarcted heart depends on myofibroblast density *Biophys. J.* **101** 1307–15
[10] Arevalo H J, Vadakkumpadan F and Guallar E *et al* 2016 Arrhythmia risk stratification of patients after myocardial infarction using personalized heart models *Nat. Commun.* **7** 11437
[11] Prakosa A, Arevalo, Hermenegild J, Deng D and Boyle P M *et al* 2018 Personalized virtual-heart technology for guiding the ablation of infarct-related ventricular tachycardia *Nat. Biomed. Eng.* **2** 732–40
[12] Lee A W C, Nguyen U C and Razeghi O *et al* 2019 A rule-based method for predicting the electrical activation of the heart with cardiac resynchronization therapy from non-invasive clinical data *Med. Image Anal.* **57** 197–213
[13] Kerfoot E, Fovargue L and Rivolo S *et al* 2016 Eidolon: visualization and computational framework for multi-modal biomedical data analysis *Medical Imaging and Augmented Reality. MIAR 2016 Lecture Notes in Computer Science* vol 9805 (Berlin: Springer) pp 425–37
[14] Toussaint N, Mansi T, Delingette H, Ayache N and Sermesant M 2008 An integrated platform for dynamic cardiac simulation and image processing: application to personalised tetralogy of fallot simulation *EG VCBM 2008 - Eurographics Workshop on Visual Computing for Biomedicine* 21–8
[15] Prakosa A, Malamas P and Zhang S *et al* 2014 Methodology for image-based reconstruction of ventricular geometry for patient-specific modeling of cardiac electrophysiology *Prog. Biophys. Mol. Biol.* **115** 226–34
[16] Peters J, Ecabert O and Meyer C *et al* 2007 Automatic whole heart segmentation in static magnetic resonance image *Medical Image Computing and Computer-Assisted Intervention—MICCAI 2007Lecture Notes in Computer Science* ed N Ayache, S Ourselin and A Maeder (Berlin: Springer)
[17] Ruijsink B, Puyol-Antón E and Oksuz I *et al* 2020 Fully automated, quality-controlled cardiac analysis from CMR *JACC Cardiovasc. Imaging* **13** 684–95
[18] Danilouchkine M G, Westenberg J J M, De Roos A, Reiber J H C and Lelieveldt B P F 2005 Operator induced variability in cardiovascular MR: left ventricular measurements and their reproducibility *J. Cardiovasc. Magn. Reson.* **7** 447–57
[19] Mendonca Costa C, Neic A and Kerfoot E *et al* 2019 Pacing in proximity to scar during cardiac resynchronization therapy increases local dispersion of repolarization and susceptibility to ventricular arrhythmogenesis *Heart Rhythm.* **16** 1475–83
[20] Deng D, Arevalo H J, Prakosa A, Callans D J and Trayanova N A 2016 A feasibility study of arrhythmia risk prediction in patients with myocardial infarction and preserved ejection fraction *Europace* **18** iv60–66
[21] Sermesant M, Chabiniok R and Chinchapatnam P *et al* 2012 Patient-specific electromechanical models of the heart for the prediction of pacing acute effects in CRT: a preliminary clinical validation *Med. Image Anal.* **16** 201–15

[22] Cedilnik N, Duchateau J and Dubois R *et al* 2018 Fast personalized electrophysiological models from CT images for ventricular tachycardia ablation planning *Europace* **20** iii94–101

[23] CIBC *Seg3D: Volumetric Image Segmentation and Visualization* Scientific Computing and Imaging Institute (SCI) http://seg3d.org

[24] Augustin C M, Neic A and Liebmann M *et al* 2016 Anatomically accurate high resolution modeling of human whole heart electromechanics: a strongly scalable algebraic multigrid solver method for nonlinear deformation *J. Comput. Phys.* **305** 622–46

[25] Crozier A, Augustin C M and Neic A *et al* 2016 Image-based personalization of cardiac anatomy for coupled electromechanical modeling *Ann. Biomed. Eng.* **44** 58–70

[26] Ronneberger O, Fischer P and Brox T 2015 U-Net: convolutional networks for biomedical image segmentation *Medical Image Computing and Computer-Assisted Intervention – MICCAI* ed N Navab, J Hornegger, W M Wells and A F Frangi (Cham: Springer) pp 234–41

[27] Chen M B B S Z, Sohal M B B S M and Voigt T *et al* 2015 Myocardial tissue characterization by cardiac magnetic resonance imaging using T1 mapping predicts ventricular arrhythmia in ischemic and non-ischemic cardiomyopathy patients with implantable cardio- verter defibrillators *Heart Rhythm.* **12** 792–801

[28] Chen Z, Relan J and Schulze W H *et al* 2013 Simultaneous non-contact mapping fused with CMR derived grey zone to explore the relationship with ventricular tachycardia substrate in ischaemic cardiomyopathy *J. Cardiovasc. Magn. Reson* **15** P64

[29] Flett A S, Hasleton J and Cook C *et al* 2011 Evaluation of techniques for the quantification of myocardial scar of differing etiology using cardiac magnetic resonance *JACC Cardiovasc. Imaging* **4** 150–6

[30] Amado L C, Gerber B L and Gupta S N *et al* 2004 Accurate and objective infarct sizing by contrast-enhanced magnetic resonance imaging in a canine myocardial infarction model *J. Am. Coll. Cardiol.* **44** 2383–9

[31] Karim R, Housden R J and Balasubramaniam M *et al* 2013 Evaluation of current algorithms for segmentation of scar tissue from late gadolinium enhancement cardiovascular magnetic resonance of the left atrium: an open-access grand challenge *J. Cardiovasc. Magn. Reson* **15** 105

[32] Ukwatta E, Rajchl M and White J *et al* 2015 Image-based reconstruction of 3D myocardial infarct geometry for patient specific applications *Proc SPIE* **73** 94132W

[33] Costa C M, Neic A and Kerfoot E *et al* 2019 Pacing in proximity to scar during cardiac resynchronization therapy increases local dispersion of repolarization and susceptibility to ventricular arrhythmogenesis *Heart Rhythm.* **16** 1475–83

[34] TC Project 2018 *{CGAL} User and Reference Manual* https://doc.cgal.org/4.12/Manual/packages.html

[35] Kickinger F, Kickinger F, Prassl A J and Prassl A J 2008 Automatically generated anatomically accurate mesh generation techniques for cardiac electrophysiology problems *IEEE Trans. Biomed. Eng.* **56** 1318–30

[36] CIBC 2016 *Cleaver: A MultiMaterial Tetrahedral Meshing Library and Application* Scientific Computing and Imaging Institute (SCI), Download from: http://sci.utah.edu/cibc/software.html

[37] Vincent K P, Gonzales M J and Gillette A K *et al* 2015 High-order finite element methods for cardiac monodomain simulations *Front. Physiol.* **6** 217

[38] Si H 2015 TetGen, a delaunay-based quality tetrahedral mesh generator *ACM Trans. Math. Softw.* **41** 1–36

[39] Bishop M J, Plank G and Burton R A B *et al* 2010 Development of an anatomically detailed MRI-derived rabbit ventricular model and assessment of its impact on simulations of electrophysiological function *AJP Heart Circ. Physiol.* **298** H699–718

[40] Trew M, Le Grice I, Smaill B and Pullan A 2005 A finite volume method for modeling discontinuous electrical activation in cardiac tissue *Ann. Biomed. Eng.* **33** 590–602

[41] Zozor S, Blanc O and Jacquemet V *et al* 2003 A numerical scheme for modeling wavefront propagation on a monolayer of arbitrary geometry *IEEE Trans. Biomed. Eng.* **50** 412–20

[42] Tveito A, Jæger K H, Kuchta M, Mardal K-A and Rognes M E 2017 A cell-based framework for numerical modeling of electrical conduction in cardiac tissue *Front. Phys.* **5** 1–18

[43] Trew M L, Smaill B H, Bullivant D P, Hunter P J and Pullan A J 2005 A generalized finite difference method for modeling cardiac electrical activation on arbitrary, irregular computational meshes *Math. Biosci.* **198** 169–89

[44] Trayanova N, Constantino J, Ashihara T and Plank G 2011 Modeling defibrillation of the heart: approaches and insights *IEEE Rev. Biomed. Eng.* **4** 89–102

[45] Krul S P J, Berger W R and Smit N W *et al* 2015 Atrial fibrosis and conduction slowing in the left atrial appendage of patients undergoing thoracoscopic surgical pulmonary vein isolation for atrial fibrillation *Circ. Arrhythm. Electrophysiol.* **8** 288–95

[46] Hooks D A, Trew M L, Caldwell B J, Sands G B, LeGrice I J and Smaill B H 2007 Laminar arrangement of ventricular myocytes influences electrical behavior of the heart *Circ. Res.* **101** e103–12

[47] Hooks D A, Tomlinson K A and Marsden S G *et al* 2002 Cardiac microstructure: implications for electrical propagation and defibrillation in the heart *Circ. Res.* **91** 331–8

[48] Smaill B H, Zhao J and Trew M L 2013 Three-dimensional impulse propagation in myocardium: arrhythmogenic mechanisms at the tissue level *Circ. Res.* **112** 834–48

[49] Roney C H, Whitaker J and Sim I *et al* 2019 A technique for measuring anisotropy in atrial conduction to estimate conduction velocity and atrial fibre direction *Comput. Biol. Med.* **104** 278–90

[50] Bayer J D, Blake R C, Plank G and Trayanova N A 2012 A novel rule-based algorithm for assigning myocardial fiber orientation to computational heart models *Ann. Biomed. Eng.* **40** 2243–54

[51] Doste R, Soto-Iglesias D and Bernardino G *et al* 2019 A rule-based method to model myocardial fiber orientation in cardiac biventricular geometries with outflow tracts *Int. J. Numer. Method Biomed. Eng.* **35** e3185

[52] Fastl T E, Tobon-Gomez C and Crozier A *et al* 2018 Personalized computational modeling of left atrial geometry and transmural myofiber architecture *Med. Image Anal.* **47** 180–90

[53] McDowell K S, Vadakkumpadan F and Blake R *et al* 2012 Methodology for patient-specific modeling of atrial fibrosis as a substrate for atrial fibrillation *J. Electrocardiol.* **45** 640–5

[54] Pashakhanloo F, Herzka D A and Ashikaga H *et al* 2016 Myofiber architecture of the human atria as revealed by submillimeter diffusion tensor imaging *Circ. Arrhythm. Electrophysiol.* **9** 1–9

[55] Hocini M, Ho S Y and Kawara T *et al* 2002 Electrical conduction in canine pulmonary veins: Electrophysiological and anatomic correlation *Circulation* **105** 2442–8

[56] Verheule S, Wilson E E, Arora R, Engle S K, Scott L R and Olgin J E 2002 Tissue structure and connexin expression of canine pulmonary veins *Cardiovasc. Res.* **55** 727–38

[57] Ho S Y 2001 Architecture of the pulmonary veins: relevance to radiofrequency ablation *Heart* **86** 265–70

[58] Labarthe S, Coudiere Y, Henry J and Cochet H 2012 A semi-automatic method to construct atrial fibre structures: a tool for atrial simulations *Comput. Cardiol.* **39** 881–4

[59] Krueger M W, Schmidt V and Tobón C *et al* 2011 Modeling atrial fiber orientation in patient-specific geometries: A semi-automatic rule-based approach *Functional Imaging and Modeling of the Heart. FIMH 2011 Lecture Notes in Computer Science* vol 6666 (Berlin: Springer) pp 223–32

[60] Zahid S, Cochet H and Boyle P M *et al* 2016 Patient-derived models link re-entrant driver localization in atrial fibrillation to fibrosis spatial pattern *Cardiovasc. Res.* **110** 443–54

[61] Roney C H, Williams S E and Cochet H *et al* 2018 Patient-specific simulations predict efficacy of ablation of interatrial connections for treatment of persistent atrial fibrillation *Europace* **20** III55–68

[62] Neic A, Campos F O and Prassl A J *et al* 2017 Efficient computation of electrograms and ECGs in human whole heart simulations using a reaction-eikonal model *J. Comput. Phys.* **346** 191–211

[63] Miri R, Graf I M and Dössel O 2009 Efficiency of timing delays and electrode positions in optimization of biventricular pacing: a simulation study *IEEE Trans. Biomed. Eng.* **56** 2573–82

[64] Pashaei A, Romero D, Sebastian R, Camara O and Frangi A F 2011 Fast multiscale modeling of cardiac electrophysiology including Purkinje system *IEEE Trans. Biomed. Eng.* **58** 2956–60

[65] Hyde E R, Behar J M and Claridge S *et al* 2015 Beneficial effect on cardiac resynchronization from left ventricular endocardial pacing is mediated by early access to high conduction velocity tissue: electrophysiological simulation study *Circ. Arrhythm. Electrophysiol.* **8** 1164–72

[66] Gillette K, Prassl A, Bayer J, Vigmond E J, Neic A and Plank G 2010 Patient-specific parameterization of a left-ventricular model of cardiac electrophysiology using electrocardiographic recordings *Comput. Cardiol.* **2017** 1–4

[67] Krummen D E, Bayer J D and Ho J *et al* 2012 Mechanisms of human atrial fibrillation initiation clinical and computational studies of repolarization restitution and activation latency *Circ. Arrhythm. Electrophysiol.* **5** 1149–59

[68] Costa C M, Hoetzl E, Rocha B M, Prassl A J and Plank G 2010 Automatic parameterization strategy for cardiac electrophysiology simulations *Comput. Cardiol.* **2013** 373–6

[69] Caldwell B J, Trew M L, Sands G B, Hooks D A, LeGrice I J and Smaill B H 2009 Three distinct directions of intramural activation reveal nonuniform side-to-side electrical coupling of ventricular myocytes *Circ. Arrhythm. Electrophysiol.* **2** 433–40

[70] Johnston B M 2016 Six conductivity values to use in the bidomain model of cardiac tissue *IEEE Trans. Biomed. Eng.* **63** 1525–31

[71] Mendonca Costa C, Plank G, Rinaldi C A, Niederer S A and Bishop M J 2018 Modeling the electrophysiological properties of the infarct border zone *Front. Physiol.* **9** 356

[72] Corrado C, Williams S, Karim R, Plank G, O'Neill M and Niederer S 2018 A work flow to build and validate patient specific left atrium electrophysiology models from catheter measurements *Med. Image Anal.* **47** 153–63

[73] Corrado C and Niederer S A 2016 A two-variable model robust to pacemaker behaviour for the dynamics of the cardiac action potential *Math. Biosci.* **281** 46–54

[74] Coveney S, Corrado C and Roney C *et al* 2019 Probabilistic interpolation of uncertain local activation times on human atrial manifolds *IEEE Trans. Biomed. Eng.* **67** 99–109

[75] Lombardo D M, Fenton F H, Narayan S M and Rappel W J 2016 Comparison of detailed and simplified models of human atrial myocytes to recapitulate patient specific properties *PLoS Comput. Biol.* **12** e1005060
[76] Loewe A, Poremba E and Oesterlein T *et al* 2019 Patient-specific identification of atrial flutter vulnerability—a computational approach to reveal latent reentry pathways *Front. Physiol.* **10**
[77] Deng D, Murphy M J and Hakim J B *et al* 2017 Sensitivity of reentrant driver localization to electrophysiological parameter variability in image-based computational models of persistent atrial fibrillation sustained by a fibrotic substrate *Chaos* **27** 093932
[78] Saha M, Roney C H and Bayer J D *et al* 2018 Wavelength and fibrosis affect phase singularity locations during atrial fibrillation *Front. Physiol.* **9** 1–12
[79] Bayer J, Prassl A J and Pashaei A *et al* 2018 Universal ventricular coordinates: a generic framework for describing position within the heart and transferring data *Med. Image Anal.* **45** 83–93
[80] Gillette K, Prassl A, Bayer J, Vigmond E, Neic A and Plank G 2018 Automatic generation of bi-ventricular models of cardiac electrophysiology for patient specific personalization using non-invasive recordings *2018 Computing in Cardiology Conf.* vol 45 pp 1–4
[81] Roney C H, Bayer J D and Cochet H *et al* 2018 Variability in pulmonary vein electrophysiology and fibrosis determines arrhythmia susceptibility and dynamics *PLoS Comput. Biol.* **14** e1006166
[82] Bayer J D, Blake R C, Plank G and Trayanova N A 2012 A novel rule-based algorithm for assigning myocardial fiber orientation to computational heart models *Ann. Biomed. Eng.* **40** 2243–54
[83] Cates J, Bieging E and Morris A *et al* 2015 Computational shape models characterize shape change of the left atrium in atrial fibrillation *Clin. Med. Insights Cardiol.* **8** 99–109
[84] Roney C H, Pashaei A and Meo M *et al* 2019 Universal atrial coordinates applied to visualisation, registration and construction of patient specific meshes *Med. Image Anal.* **55** 65–75
[85] Gsell M A F, Augustin C M and Prassl A J *et al* 2018 Assessment of wall stresses and mechanical heart power in the left ventricle: Finite element modeling versus Laplace analysis *Int. J. Numer. Method Biomed. Eng.* **34** 34
[86] Krishnamoorthi S, Sarkar M and Klug W S 2013 Numerical quadrature and operator splitting in finite element methods for cardiac electrophysiology *Int. J. Numer. Method Biomed. Eng* **29** 1243–66
[87] Kayvanpour E, Mansi T and Sedaghat-Hamedani F *et al* 2015 Towards personalized cardiology: Multi-scale modeling of the failing heart *PLoS One* **10** e0134869
[88] Duan J, Bello G and Schlemper J *et al* 2019 Automatic 3D bi-ventricular segmentation of cardiac images by a shape-refined multi-task deep learning approach *IEEE Trans. Med. Imaging* **38** 2151–64
[89] Cuellar A A, Lloyd C M, Nielsen P F, Bullivant D P, Nickerson D P and Hunter P J 2003 An overview of CellML 1.1, a biological model description language *Simulation* **79** 740–7
[90] Nielsen P M, Le Grice I J, Smaill B H and Hunter P J 1991 Mathematical model of geometry and fibrous structure of the heart *Am. J. Physiol.* **206** H1365–78
[91] Stevens C and Hunter P J 2003 Sarcomere length changes in a 3D mathematical model of the pig ventricles *Prog. Biophys. Mol. Biol.* **82** 229–41
[92] Vetter F J and McCulloch A D 1998 Three-dimensional analysis of regional cardiac function: a model of rabbit ventricular anatomy *Prog. Biophys. Mol. Biol.* **69** 157–83

Lightning Source UK Ltd.
Milton Keynes UK
UKHW051254231220
375661UK00002B/40

9 780750 320627